THE NIMBUS 7 USERS' GUIDE

Charles R. Madrid, et al

Goddard Space Flight Center
Greenbelt, Maryland

August 1978

THE NIMBUS 7 USERS' GUIDE

Prepared by

The Landsat/Nimbus Project
Goddard Space Flight Center
National Aeronautics and Space Administration

Edited by

Charles R. Madrid
Management and Technical
Services Company
Beltsville, Maryland

August 1978

THE NIMBUS 7 USERS' GUIDE

FOREWORD

This document provides users with background information on the Nimbus 7 spacecraft and instruments to help them understand, and obtain Nimbus 7 data.

The basic spacecraft system operation, mission, and scientific objectives of the Nimbus 7 flight are outlined, followed by a detailed discussion of each of the instruments. The formats, archiving, and access to the data are described in detail since data handling responsibilities denote a major departure from previous data processing methods. Additionally, each section contains a brief description of archived tapes and examples of image displays. The National Space Science Data Center and the National Oceanic and Atmospheric Administration will archive these products and issue catalogs containing information on Nimbus 7 data.

The principal authors for each section were selected by each Nimbus Experiment Team. All team members are listed in Appendix B of this document. The assembly and editing of this publication was accomplished by the Management and Technical Services Company (MATSCO), under contract No. NAS 5-23740 with the Goddard Space Flight Center, NASA, Greenbelt, Maryland.

Ronald K. Browning
Project Manager
Landsat/Nimbus Project
Goddard Space Flight Center

Table of Contents

	Page
FOREWORD	iii
SECTION 1. THE NIMBUS 7 SPACECRAFT SYSTEM	1
1.1 Introduction	1
1.2 Nimbus 7 Mission Objectives	1
1.3 Spacecraft Components	3
1.3.1 Experiments	3
1.3.2 Attitude Control Subsystem	4
1.3.3 Instrument Power	4
1.3.4 Communications and Data Handling Subsystem	5
1.3.5 Thermal Control Subsystem	8
1.4 Data Handling and Processing Complex	9
1.4.1 Nimbus Data Application System	12
1.4.2 Nimbus Experiment Teams	12
1.5 Data Archiving	15
SECTION 2. THE COASTAL ZONE COLOR SCANNER (CZCS)	19
2.1 Introduction	19
2.2 Scientific and Technical Objectives	20
2.2.1 Scientific Objectives	20
2.2.2 Technical Objectives	20
2.3 Instrument Description	21
2.3.1 Operation	21
2.3.2 Viewing Geometry	21
2.3.3 Channel Characteristics	21
2.4 Calibration	26
2.4.1 Prelaunch Calibration	26
2.4.2 In-Flight Calibration	26
2.5 Operational Modes	26

Table of Contents (Continued)

Page

2.6 Data Processing Formats and Availability 27

    2.6.1 Data Processing 27
    2.6.2 Tape Products 28
    2.6.3 Film Format 28
    2.6.4 Data Availability 30

2.7 Planned NET Experiment Investigations and Data Application 30

2.8 References and Bibliography 31

SECTION 3. THE EARTH RADIATION BUDGET (ERB) EXPERIMENT 33

3.1 Scientific Objective 33

3.2 Experiment Description 33

    3.2.1 Solar Channels 33
    3.2.2 Fixed Wide-Angle FOV Channels 37
    3.2.3 Narrow-Angle FOV Scanning Channels 37

3.3 Instrument Description 43

    3.3.1 Physical Layout 43
    3.3.2 Solar Channels 43
    3.3.3 Fixed Wide-Angle FOV Channels 46
    3.3.4 Thermopiles in Channels 1 through 14 47
    3.3.5 Interference Filters in Channels 1 through 14 47
    3.3.6 Narrow-Angle FOV Scanning Channels 47

3.4 Pre-launch Calibration 53

    3.4.1 Solar Channels 53
    3.4.2 Fixed Wide-Angle Channels 55
    3.4.3 Shortwave Scan Channels 56
    3.4.4 Longwave Scan Channels 56

3.5 In-Flight Calibration 56

3.6 Data Processing, Products, and Availability 57

    3.6.1 Data Processing 57
    3.6.2 Tape Products 57
    3.6.3 Display Products 58

Table of Contents (Continued)

	Page
3.6.4 Data Availability	58
SECTION 4. THE LIMB INFRARED MONITOR OF THE STRATOSPHERE (LIMS) EXPERIMENT	71
4.1 Introduction	71
4.2 Scientific and Technical Objectives	72
4.2.1 Scientific Objectives	72
4.2.2 Limb Scan Features	76
4.2.3 Technical Objectives	77
4.3 Instrument Description	77
4.3.1 LIMS Operation	77
4.3.2 LIMS Location and Limb Viewing Geometry	82
4.3.3 LIMS Channel Characteristics	82
4.4 Calibration	88
4.4.1 Preflight Calibration	88
4.4.2 In-flight Calibration	88
4.5 LIMS Operational Modes	89
4.6 LIMS Data Processing, Products, and Availability	89
4.6.1 Data Processing	89
4.6.2 Tape Products	89
4.6.3 Display Products	92
4.6.4 Data Availability	102
4.7 Planned Investigations and Data Applications	102
4.8 References	103
SECTION 5. THE STRATOSPHERIC AEROSOL MEASUREMENT (SAM II) EXPERIMENT	105
5.1 Introduction	105
5.2 Scientific Objectives	105

Table of Contents (Continued)

Page

5.3 Experiment Concept 107

5.4 SAM II Instrument 109

    5.4.1 Physical Description 109
    5.4.2 Functional Description 113
    5.4.3 Data System 116
    5.4.4 Viewing Geometry and Instantaneous Field of View 116
    5.4.5 Performance Characteristics 118
    5.4.6 Calibration 118

5.5 Data Validation 119

    5.5.1 Airborne Lidar 120
    5.5.2 Dustsondes 120
    5.5.3 Balloon-borne Sun Photometer 120
    5.5.4 Airborne Polar Nephelometer and Impactors 120
    5.5.5 Ground Truth Experiments 120

5.6 SAM II Data Processing, Formats, and Availability 121

    5.6.1 Data Processing 121
    5.6.2 Tape Products 123
    5.6.3 Display Products 123
    5.6.4 Data Availability 136

5.7 Planned NET Experiment Investigations and Data Applications 137

SECTION 6. THE STRATOSPHERIC AND MESOSPHERIC SOUNDER (SAMS) EXPERIMENT 139

6.1 Introduction and Objectives 139

    6.1.1 Introduction 139
    6.1.2 Scientific Objectives 139
    6.1.3 Technical Objectives 140

6.2 Principles of Operation 141

    6.2.1 Basic Theory 141
    6.2.2 Conventional Radiometry 143
    6.2.3 Pressure Modulation Radiometry 143
    6.2.4 Tuning of Pressure Modulated Channels 144
    6.2.5 Temperature and Concentration Measurements 145
    6.2.6 Zonal Wind Measurement 145

Table of Contents (Continued)

Page

    6.2.7 Reference Pressure Determination 148
    6.2.8 Elimination of Contaminating Signals 148
    6.2.9 Basic Measuring Procedure 150

6.3 The SAMS Radiometer 150

    6.3.1 Optics 150
    6.3.2 Thermal Design 152
    6.3.3 Detectors 152
    6.3.4 Scan Mirror Assembly 154
    6.3.5 Blackbody 159
    6.3.6 Black Chopper 159
    6.3.7 Pressure Modulators 159
    6.3.8 Cooler Door Release 159
    6.3.9 Signal Processing 161

6.4 Ground Calibration 161

    6.4.1 General 161
    6.4.2 Basic Radiance Calibration 165
    6.4.3 Scan Angle Dependence Checks 165
    6.4.4 Response of PMR Channels to Atmospheric Paths 165
    6.4.5 Field of View Measurement 165

6.5 Data Processing Formats and Availability 166

    6.5.1 Data Processing 166
    6.5.2 Tape Products 167
    6.5.3 Film Products 167
    6.5.4 Data Availability 174

6.6 References 174

SECTION 7. THE SOLAR BACKSCATTER ULTRAVIOLET (SBUV) AND TOTAL OZONE
MAPPING SPECTROMETER (TOMS) EXPERIMENT 175

7.1 Introduction 175

7.2 Scientific Objectives 175

7.3 Instrumentation 176

    7.3.1 SBUV Subsystem 176
    7.3.2 TOMS Subsystem 180

Table of Contents (Continued)

Page

	7.3.3 SBUV/TOMS Operating Modes	184
	7.3.4 Electronic System	187
7.4	Calibration	188
	7.4.1 Prelaunch Calibration	188
	7.4.2 In-flight Calibration	189
7.5	Data Processing, Formats, and Availability	190
	7.5.1 Data Processing	190
	7.5.2 Tape Products	194
	7.5.3 Display Products	195
	7.5.4 Data Availability	210
7.6	References	211

SECTION 8. THE SCANNING MULTICHANNEL MICROWAVE RADIOMETER (SMMR) EXPERIMENT ... 213

8.1	Introduction	213
8.2	Scientific and Technical Objectives	214
8.3	Instrument Description	214
8.4	Calibration	218
	8.4.1 Prelaunch Calibration	218
	8.4.2 Post-launch Calibration	223
8.5	Operational Modes	223
8.6	Data Processing Formats and Availability	223
	8.6.1 Data Processing	233
	8.6.2 Tape Products	223
	8.6.3 Display Products	223
	8.6.4 Data Availability	235
8.7	Planned NET Experiment Investigations and Data Applications	237
	8.7.1 Validation Investigations	237
	8.7.2 Application Investigations	243

Table of Contents (Continued)

Page

8.8 References and Bibliography 243

SECTION 9. THE TEMPERATURE HUMIDITY INFRARED RADIOMETER (THIR)
SUBSYSTEM 247

9.1 Introduction 247

9.2 Instrument Description 247

    9.2.1 THIR Operation 247
    9.2.2 Scan Sequence 251
    9.2.3 Scan Geometry 255
    9.2.4 THIR Data Flow 255

9.3 Calibration 259

    9.3.1 Laboratory Calibration 259
    9.3.2 Equipment Blackbody Temperature 259

9.4 Data Formats and Availability 259

    9.4.1 World Montage 259
    9.4.2 Tape Data 261
    9.4.3 Data Availability 262

APPENDIX A ABBREVIATIONS AND ACRONYMS A1

APPENDIX B NIMBUS EXPERIMENT TEAM (NET) MEMBERS B1

List of Figures

Figure		Page
1-1	Nimbus 7 Observatory	2
1-2	Nimbus 7 Spacecraft Data Handling System	6
1-3	NOPS Data Flow and Facility Plan	10
1-4	Relationship Between the Nimbus Observation Processing System (NOPS) and the Nimbus Data Applications System (NDAS)	11
1-5	Nimbus 7 Data Flow	13
1-5a	Nimbus 7 Data Flow	14
2-1	CZCS Viewing Geometry and Earth Scan Pattern	22
2-2	CZCS Optical Arrangement	23
2-3	CZCS Spectral Response for Channels 1 through 5	25
2-4	CZCS Image Display Format	29
3-1	Spectral Internvals Monitored by the ERB Solar Channels	36
3-2	ERB Scan Grid Earth Patterns	38
3-3	ERB Scan Models	39
3-4	ERB Scanning Channel Views of a Geographical Area Near the Subpoint	40
3-5	ERB Scanning Channel Views of a Geographical Area Away from the Subpoint	41
3-6	ERB Radiometer Unit	44
3-7	Transmittance of ERB Suprasil W and Schott Colored Glasses	45
3-8	Channel 6S ERB Flight Model	49
3-9	Channel 7S ERB Flight Model	50
3-10	Channel 8S ERB Flight Model	51
3-11	Channel 9S ERB Flight Model	52
3-12	ERB Microfilm Map Display Format	62
3-13	Mean Normal Solar Irradiance Display Format	63
3-14	Earth Albedo From WFOV Observations Display Format	64
3-15	Earth Albedo From NFOV Observations Display Format	65
3-16	Net Radiation From NFOV Observations Display Format	66
3-17	Net Radiation From NFOV Observations Display Format	67
3-18	Terrestrial Flux From WFOV Observations Display Format	68
3-19	Terrestrial Flux From NFOV Observations Display Format	69
4-1	LIMS Viewing Geometry	72
4-2	LRIR Temperature Retrieval (Antiqua, Dec. 17, 1975) Compared with Rocket Results	73
4-3	LRIR Ozone Retrieval Compared with Rocket Results	74
4-4	LRIR Temperature Map Results for 10 mb Level–Descending Node	75
4-5	LIMS Instrument Configuration	77
4-6	LIMS Proto-Flight Model Instrumentaion Installed on Mounting Yoke	78
4-7	LIMS Optical Schematic	79
4-8	LIMS Instantaneous Fields of View	80
4-9	LIMS Scan Mode Wave Form	81
4-10	LIMS System Block Diagram	83
4-11	Location of the LIMS Instrumentation on the Nimbus 7 Sensory Ring and Associated Viewing Geometry	84
4-12	LIMS Sub-Tangent Point Tracks for the Northern Hemisphere	85
4-13	LIMS Sub-Tangent Pont Tracks for the Southern Hemisphere	86

List of Figures (Continued)

Figure		Page
4-14	LIMS Data Processing Flow at NCAR	90
4-15	LIMS Microfilm Map Display Format for One Day Products	93
4-16	LIMS Microfilm Map Display Format for One Month and Three Months Products	94
4-17	LIMS Microfilm Cross Section Display Format for 60 Degree Longitude Zones	95
4-18	LIMS Microfilm Cross Section Display Format for One Day, One Month, and Three Month Products	96
4-19	LIMS Microfilm Profile Display Format of Radiances versus Scan Angle	97
4-20	LIMS Microfilm Display Format of Parameter and Radiance Profiles	98
4-21	LIMS Microfilm Display Format of Cloudtop Height Plots	99
5-1	SAM II Orbital and Viewing Geometry	106
5-2	Latitude Coverage of SAM II Tangent Points For Sun-Synchronous High-Noon Orbit	106
5-3	Sun Top and Bottom for Standard Atmosphere	108
5-4	Inversion Results from Synthetic Data	110
5-5	SAM II Instrument	111
5-6	SAM II Instrument Cross Section	112
5-7	SAM II System Block Diagram	114
5-8	Timing Diagram For T_{ON}, T_{OFF}, and ΔT	115
5-9	SAM II Field of View and Viewing Geometry	117
5-10	System Spectral Response	119
5-11	SAM II Data Processing	122
5-12	SAM II Microfilm Display Format of Solar Irradiance versus Time Profiles	125
5-13	SAM II Microfilm Display Format of Solar Irradiance versus Altitude Profiles	126
5-14	SAM II Microfilm Display Format of Aerosol Coefficient of Extinction versus Altitude Profiles	127
5-15	SAM II Microfilm Display Format of Total Extinction Ratio versus Altitude Profiles	128
5-16	SAM II Microfilm Display Format of Aerosol Coefficient versus Altitude Profiles	129
5-17	SAM II Microfilm Display Format of Longitude Cross Section of Aerosol Coefficient of Extinction versus Altitude	130
5-18	SAM II Microfilm Display Format of Latitude Cross Section of Aerosol Coefficient of Extinction versus Altitude	131
5-19	SAM II Microfilm Map Display Format	132
5-20	SAM II Microfilm Display Format of Optical Depth versus Time Plots	133
5-21	SAM II Microfilm Display Format of Altitude and Peak of Total Extinction Ratio versus Time Plots	134
6-1	Typical Line Modulation Spectra	142
6-2	Pressure Modulator Optical Arrangement	143
6-3	Effect of Doppler Shift on Line Spectra	145
6-4	Azimuth Scan Angles	146
6-5	Emissivity of CO versus Tangent Height in the Atmosphere	148
6-6	Variation of E_B/E_L with Height	149
6-7	Central Section through Sensor	151
6-8	Fields of View and Line-of-Sight Angles	153
6-9	Orbital Temperature Response	155
6-10	Block Schematic of SAMS Radiometer	157
6-11	Overall Timing Diagram	158
6-12	Pressure Modulator Cylinder	160

List of Figures (Continued)

Figure		Page
6-13	Block Schematic of Signal Electronics	162
6-14	Signal Channel Transfer Characteristics for Channel B1	163
6-15	SAMS Test and Calibration Facility	164
6-16	SAMS Microfilm Map Display	168
6-17	SAMS Microfilm Map Display Format for One Month and Three Month Products	169
6-18	SAMS Microfilm Cross Section Display Format for all Parameters except Nitric Oxide	170
6-19	SAMS Microfilm Cross Section Display Format for Nitric Oxide	171
7-1	Functional Block Diagram of SBUV/TOMS Modules	177
7-1a	Functional Block Diagram of SBUV/TOMS Modules	178
7-2	SBUV Optics Diagram	179
7-3	Polarization Sensitivities for SBUV/TOMS	181
7-4	Stray Light Rejection Capability of SBUV	182
7-5	TOMS Optics Diagram	183
7-6	TOMS Chopper Timing Sequence	184
7-7	IFOV of SBUV/TOMS in the Nadir	185
7-8	TOMS Projected Fields of View in a Cross-Scan	186
7-9a	SBUV/TOMS Process Flow Chart	193
7-9b	SBUV/TOMS Process Flow Chart	194
7-10	SBUV/TOMS Microfilm Map Format for One Month and Three Months Display	198
7-11	SBUV Microfilm Map Format for One Day Display	199
7-12	SBUV Microfilm Cross-Section Format of Ozone Mass Mixing Ratio	200
7-13	SBUV Microfilm Table Format of Solar Irradiance	201
7-14	SBUV Microfilm Table Format of Solar Irradiance	202
7-15	SBUV Microfilm Table Format of Total Ozone	203
7-16	SBUV Microfilm Table Format of High Level Ozone Mixing Ratio	204
7-17	TOMS Microfilm Tabel Format of Total Ozone	205
7-18	SBUV Microfilm Format of Solar Irradinace Plots	206
7-19	SBUV Microfilm Format of Terrestrial Radiance Plots	207
7-20	SBUV Microfilm Format of Terrestrial Albedo versus Wavelength Plots	208
7-21	TOMS Montage Format	209
8-1	Block Diagram of the SMMR Electronics	216
8-2	SMMR Instrument Configuration Showing Antenna, Feed Horn Drive Assembly, and Electronic Boxes	219
8-3	SMMR Instrument Showing Front View	220
8-4	SMMR Instrument Showing Rear View	221
8-5	SMMR Instrument Configuration Close-up Drive and Feed Details	222
8-6	SMMR Data Processing Flow Chart in the Meterological Operations Control Center	224
8-7	SMMR Data Processing Flow Chart	226
8-8	SMMR Data Processing Flow Chart	227
8-9	SMMR Data Processing Flow Chart	228
8-10	SMMR Data Processing Flow Chart	229
8-11	SMMR Data Processing Flow Chart	230
8-12	SMMR Data Processing Flow Chart	231
8-13	SMMR Output Product Flow Chart	232
8-14	Format for SMMR Short-term Polar Displays to 50°N and 50°S	234

List of Figures (Continued)

Figure		Page
8-15	Format for SMMR Short-term Polar Displays to 30°N and 30°S	235
8-16	Format for SMMR One-month Polar Displays to 50°N and 50°S	236
8-17	Format for SMMR One-month Polar Displays to 30°N and 30°S	237
8-18	Format for SMMR Mercator Displays	238
9-1	The Temperature Humidity Infrared Radiometer	248
9-2	THIR Optical Schematic	252
9-3	Relative Spectral Response (Transmission) of the 6.7 μm and 11.5 μm Channels	253
9-4	THIR Scan Angle Information	254
9-5	Relationship Between Ground Resolution and a Scan Sample with a 0.40° x 0.40° FOV for the THIR 11.5 μm Channel at 955 Km	256
9-6	Spacecraft/THIR Data Flow	257
9-7	Ground/THIR Data Flow	258
9-8	THIR Montage Film Display Format	260

List of Tables

Table		Page
1-1	Instrument Power Requirements for the Percentage of Operational Time Allotted Each Sensor	5
1-2	Tape Recorder Modes Summary	8
1-3	Film Data Products Available Through EDIS and NSSDC	16
1-4	Tape Types Available Through EDIS and NSSDC	17
2-1	CZCS Performance Parameters	24
3-1	Characteristics of ERB Solar Channels	34
3-2	Characteristics of ERB Fixed Wide-Angle FOV Channels	35
3-3	Characteristics of ERB Scanning Channels	35
3-4	ERB Scanning Channel Target Areas	42
3-5	Characteristics of the Fixed Channels	48
3-6	ERB Scan Head Positions During Scan	54
3-7	Titles and Specifications Numbers for ERB Microfilm Products	59
4-1	Optical Characteristics of LIMS Channels	87
4-2	ERB Microfilm Product Titles and Specification Numbers	100
5-1	SAM II Display Products	124
6-1	Molecular Species and Spectral Bands	140
6-2	Signal Channel Functions	154
6-3	Signal Channel Parameters	156
6-4	Titles and Film Specification Numbers for SAMS Microfilm Products	172
7-1	Actual Wavelengths of SBUV Wavelength Steps	191
7-2	Actual Wavelengths of TOMS	192
7-3	Titles and Film Specification Numbers for SBUV/TOMS Microfilm and Montage Products*	196
8-1	SMMR Performance Characteristics	215
8-2	SMMR Sensor Design Characteristics	217
8-3	SMMR Science Algorithms	225
8-4	SMMR Film Products	239
9-1	THIR Subsystem Specifications	249
9-2	Relative Spectral Response for the $6.7\mu m$ and $11.5\mu m$ Channels	252
9-3	Effective Radiance (N) Versus Equivalent Blackbody Temperature (T_B)	263

SECTION 1

THE NIMBUS 7 SPACECRAFT SYSTEM

by

Staff Members, Landsat/Nimbus Project
National Aeronautics and Space Administration
Goddard Space Flight Center

The purpose of this section is to present the Nimbus mission objectives, the scientific objectives of the Nimbus 7 experiments, outline the component subsystems and experiments of the spacecraft, and present information on data products and their availability.

1.1 Introduction

The Nimbus 7 spacecraft will be launched from the Western Test Range at Vandenberg Air Force Base, California, by a thrust-augmented Delta vehicle. The satellite will be placed in an 955 kilometer, sun-synchronous polar orbit, having local noon (ascending) and midnight (descending) equator crossings, with 26.1 degrees of longitude separation. The orbital period will be about 104.16 minutes.

In orbit, with its solar panels unfolded, the Nimbus 7 appears as shown in Figure 1.1. The earth-viewing sensors are mounted below the torus structure. The attitude control system and the solar array are supported above the torus by a truss.

1.2 Nimbus 7 Mission Objectives

The Nimbus 7 mission affords the opportunity to conduct a variety of experiments in the pollution, oceanographic and meteorological disciplines. It provides an opportunity to assess each instrument's operation in the space environment and to collect a sizable body of data with the global and seasonal coverage needed for support of each experiment. This mission also extends and refines the sounding and atmospheric structure measurement capabilities demonstrated by experiments on previous Nimbus observatories. The mission objectives of the Nimbus 7 are:

- To observe gases and particulates in the atmosphere for the purpose of determining the feasibility to map sources, sinks, and dispersion mechanisms of atmospheric pollutants (SBUV/TOMS – SAM II – SAMS-LIMS)

- To observe ocean color, temperature, and ice conditions, particularly in coastal zones, with sufficient spatial and spectral resolution to determine the feasibility of application such as:

    (a) detecting pollutants in the upper level of the oceans,

    (b) determining the nature of materials suspended in the water,

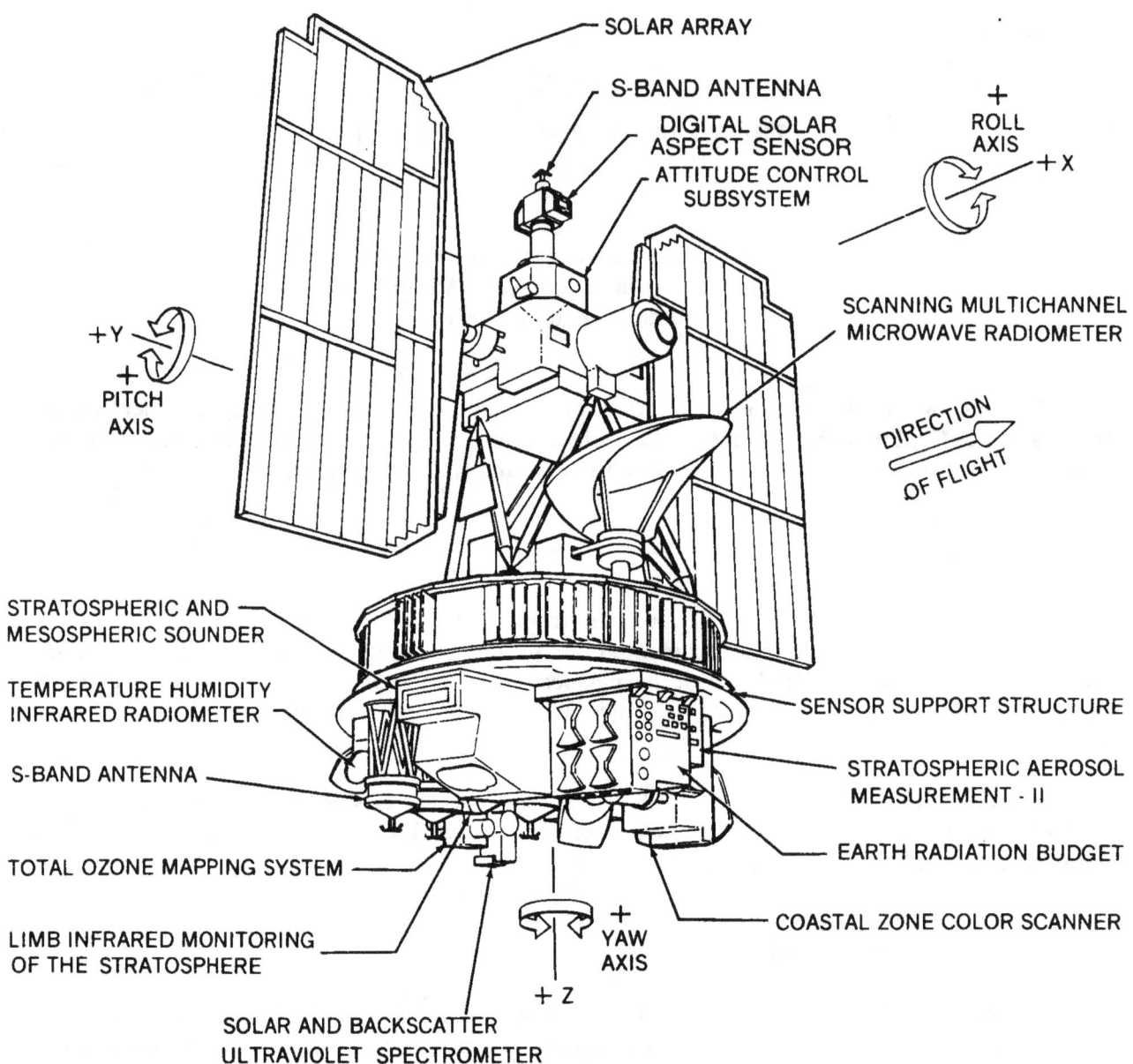

Figure 1-1. Nimbus 7 Observatory

(c) applying the observations to the mapping of sediments, biologically productive areas, and interactions between coastal effluents and open ocean waters (CZCS),

(d) demonstrating improvement in ship route forecasting (SMMR).

- To make quantitative measurements of air-surface boundary conditions (e.g., soil moisture, snow and ice cover, sea surface temperature and roughness, and albedo) or of precipitation, and to improve long-range weather forecasting in support of the Global Atmospheric Research Program (SMMR-ERB)

- To continue to make baseline measurements of variations of long wave radiation fluxes outside the atmosphere and of atmospheric constituents for the purpose of determining the effect of these variations on the earth's climate (ERB-SBUV/TOMS – LIMS).

1.3 Spacecraft Components

The Nimbus 7 spacecraft components, consisting of the integrated subsystems that provide the power, attitude control, and information flow required to support the payload for a period of one year in orbit, are contained within the three major structures of the spacecraft. These three structures consist of: a hollow torus-shaped sensor mount, the solar paddles, and a control housing unit that is connected to the sensor mount by a tripod truss structure.

The spacecraft weighs 965 kilograms and has a configuration similar to an ocean buoy. It is 3.04 meters tall, 1.52 meters in diameter at the base, and 3.96 meters wide with the solar paddles fully extended. The sensor mount that forms the satellite base houses the electronics equipment and battery modules. The lower surface of the torus provides mounting space for sensors and antennas. A box-beam structure mounted within the center of the torus provides support for the larger sensor experiments. The control housing unit is located on top of the spacecraft and above this unit are the sun sensors, horizon scanners, and a command antenna.

1.3.1 Experiments

Seven experiments and one subsystem (THIR) are on board the Nimbus 7 spacecraft. These experiments and a brief description of their scientific objectives are as follows:

- Coastal Zone Color Scanner (CZCS). The objective of the CZCS is to map chlorophyll concentration, sediment distribution, gelbstroffe concentrations as a salinity indicator, and temperature of coastal waters and the open ocean.

- Earth Radiation Budget (ERB). An instrument very similar to the ERB of Nimbus 6, is to determine over a period of one year the radiation budget of the earth on both synoptic and planetary scales by simultaneous measurement of incoming solar radiation and outgoing earth reflected (shortwave) and emitted (longwave) radiation. Both fixed wide angle sampling of terrestrial fluxes at the satellite altitude, and scanned narrow-angle sampling of the radiance components, dependent on angle are used to determine outgoing radiation.

- Limb Infrared Monitor of the Stratosphere (LIMS). The objective of LIMS is to obtain vertical profiles and maps of temperature and the concentration of ozone, water vapor, nitrogen dioxide, and nitric acid for the region of the stratosphere bounded by the upper troposphere and the lower mesosphere.

- Stratospheric Aerosol Measurement II (SAM II). The SAM II objective is to map the concentration and optical properties of stratospheric aerosols as a function of altitude, latitude, and longitude. When no clouds are present in the instrument's instantaneous field of view (IFOV), the tropospheric aerosols can also be mapped.

- Stratospheric and Mesospheric Sounder (SAMS). The SAMS objective is to observe the limb of the atmosphere through various pressure modulator radiometers in order to measure vertical concentrations of H_2O, CH_4, CO, and NO; observe resonant scattering of solar radiation in spectral bands H_2O, CO_2, CO and NO; measure the temperature of the stratosphere and mesosphere to ≈ 90 kilometers altitude; investigate source function and departure from the thermodynamic equilibrium between 80 and 130 kilometers associated with CO_2 emission bands, and measure the zonal wind velocity component along the line of sight.

- Solar Backscatter Ultraviolet/Total Ozone Mapping (SBUV/TOMS). The SBUV/TOMS objectives are to determine the vertical distribution of ozone, map the total ozone and 200-mb height fields, and monitor the incident solar ultraviolet irradiance and ultraviolet radiation backscattered from the earth.

- Scanning Multichannel Microwave Radiometer (SMMR). The primary purpose of the SMMR is to obtain and use ocean momentum and energy-transfer parameters on a nearly all-weather operational basis. Derived low altitude parameters are winds, water vapor, liquid water content, and mean cloud droplet size.

- Temperature Humidity Infrared Radiometer (THIR). The primary objective of the THIR is to measure the infrared radiation from the earth in two spectral bands during both day and night portions of the orbit; to provide pictures of the cloud cover, three-dimensional mappings of cloud cover, temperature mappings of clouds, land, and ocean surface, cirrus cloud content, atmospheric contamination and relative humidity.

1.3.2 Attitude Control Subsystem

The attitude control subsystem (ACS) provides stabilization about the spacecraft's roll, pitch, and yaw axis and control of the solar paddles orientation, maintaining them nearly perpendicular to the nominal sunline.

The ACS consists of four attitude control loops and associated switching logic, telemetry and test modes, electrical manifolding, and thermal environmental control. This system maintains spacecraft alignment with the local orbital reference axes to within 0.7 degree of the pitch axis and one degree of the roll and yaw axis. The system keeps the instantaneous angular rate changes about any axis to less than 0.01 degree per second.

The three-axis active ACS uses horizon scanners for roll and pitch attitude error sensing. The rate gyros sense yaw rate and, in a gyro compassing mode, sense yaw attitude. A torquing system uses a combination of reaction jets to provide spacecraft momentum control and large control torques when required; flywheels are utilized for fine control and residual momentum storage.

1.3.3 Instrument Power

The spacecraft power subsystem consists of solar arrays, nickel-cadmium batteries, charge and discharge regulators, and voltage regulators to operate all spacecraft support subsystems and to provide maximum power for the instrument payload.

It is anticipated that the orbit average regulated power provided by the observatory power subsystem will be approximately 300 watts, of which 123 watts are allocated to the spacecraft subsystems. If all the instruments were on full-time, the power requirements would exceed the available supply. Because of this power limitation, the subsystems will operate for approximately the percentage of time given in Table 1-1. Only THIR is scheduled to operate on a full-time basis. This schedule is in accordance with the specific objectives of the Nimbus Project.

Table 1-1
Instrument Power Requirements for the Percentage of Operational Time Allotted Each Sensor

Instrument	Power Requirements (Watts)	Operational Mode (%)
CZCS	11.4	30
ERB	36.3	80
LIMS	24.5	80
SAM II	0.8	8
SAMS	23.0	80
SBUV/TOMS	20.0	80
SMMR	61.6	50
THIR	8.5	100
Subsystem Total =	186.1	
Basic Spacecraft =	123.6	
Observatory Total =	309.7	

1.3.4 Communications and Data Handling Subsystem

The communications and data handling subsystem (CDHS) is composed of the S-band communications system and tape recorder subsystem and handles all spacecraft information flow. The S-band communication system includes the S-band command and telemetry system, the data processing system (DPS) and the command clock. The S-band command and telemetry system consists of two S-band transponders, a command and data interface unit (CDIU), four earth view antennas, a sky view antenna, and two S-band transmitters (2211 MHz). Commands are transmitted to the observatory by pulse code modulation (PCM), phase-shift keying (PSK)/frequency modulation (FM)/phase modulation (PM) of the assigned 2093.5 MHz S-band uplink carrier. Stored command capability provides for command execution at predetermined times. Figure 1-2, a diagram of the spacecraft data handling system, shows the routing of the sensor data to the versatile information processor (VIP), the digital information processor (DIP), and the Coastal Zone Color Scanner (CZCS) information processor (ZIP).

1.3.4.1 Telemetry and Ranging

Command, telemetry, and ranging signals are handled by a unified S-band (USB) transponder. Stored data are played back to the ground station using the S-band links (see Figure 1-2). Specifically, the telemetry system includes two transponders which are interlocked to prevent simultaneous transmission and also two wideband transmitters that are interlocked to prevent dual transmissions. Single or dual downlink transmissions may be commanded at the transponder downlink frequency of

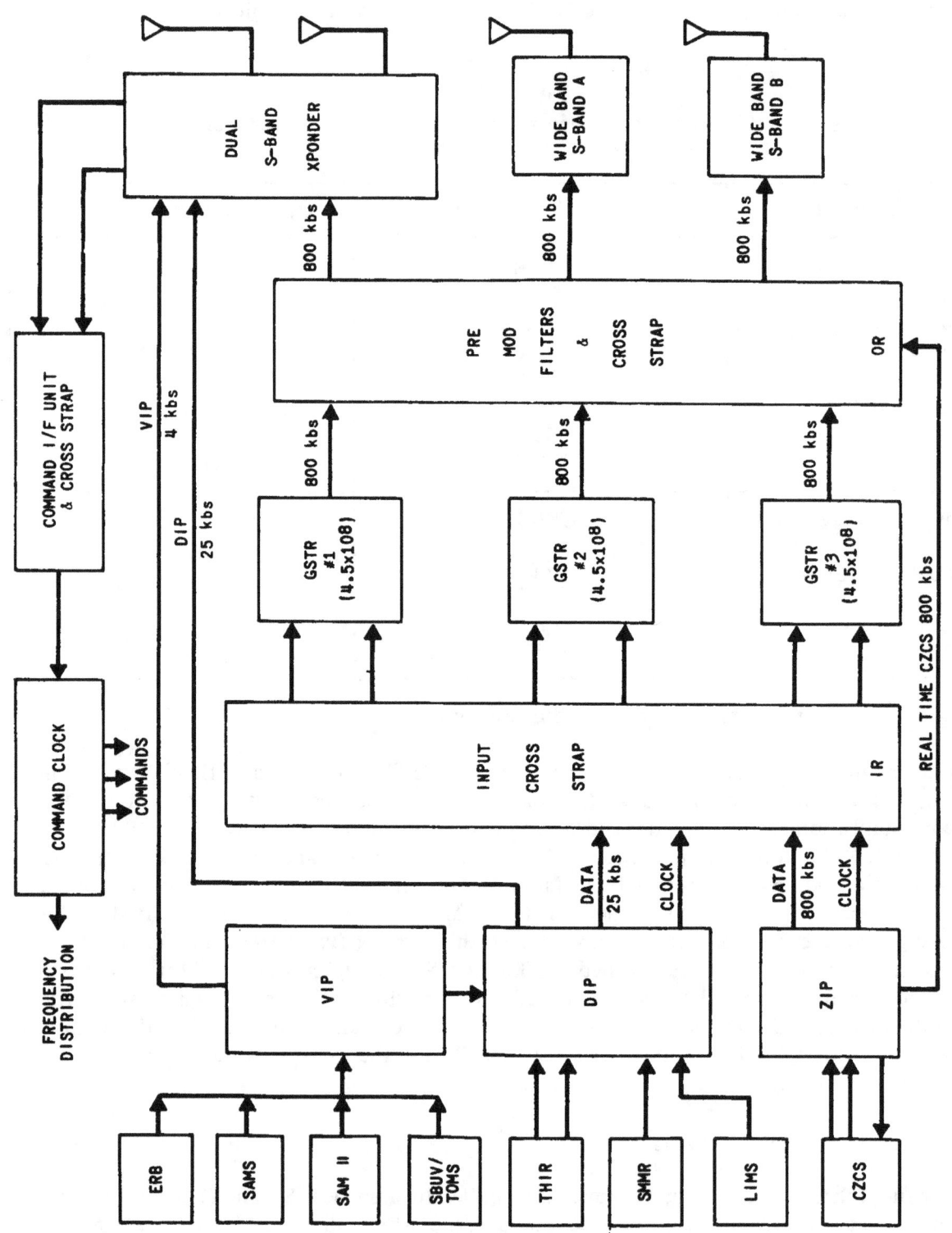

Figure 1-2. Nimbus 7 Spacecraft Data Handling System

2273.5 MHz or the wideband transmitter downlink frequency of 2211.0 MHz. Additionally, any combination of signals may be used to modulate one transponder and one wideband transmitter. One combination is an 800 kbs bi-phase PCM playback from any of the three on board tape recorders. When using this combination the playback data may be either recorded digital information processor data, or Coastal Zone Color Scanner sensor data. Another combination is an 800 kbs bi-phase PCM of real-time CZCS sensor data.

The transponder modulation is baseband PCM/PM multiplexed with the VIP telemetry subcarrier. The wideband transmitter modulation is PCM/FM. The transponder can be phase modulated by ranging tones, by 800 kbs bi-phase PCM data or by 25 kbs bi-phase PCM real-time DIP data on a mutually exclusive basis, selected by command. Simultaneous commanding, range tracking, and VIP telemetry transmission are possible with the selected baseband modulation signal.

1.3.4.2 Data Processing System (on board)

The data processing system on board the Nimbus 7 spacecraft processes analog and digital data in addition to the processing of housekeeping and sensor data. The low-rate housekeeping and sensor data (up to 400 bps) is processed, formatted, multiplexed with medium rate sensor data (2 to 12 kbs), and stored for a full orbit on a multispeed (Goddard Standard) tape recorder at 25 kbs. The high data rate ocean color sensor data (800 kbs) are processed and stored for a portion of the orbit.

Input and output cross strapping provides alternate signal routing in the event a failure occurs with a tape recorder or transmitter. (See Figure 1-2).

The VIP subsystem monitors the observatory housekeeping data and low data rate sensor data using three types of input gates; 576 analog, 320 digital B, and 16 are digital A. It digitizes the analog data and formats the data into a 6400-word major frame (each word contains 10 bits). It outputs these data at 4000 bps, which can be transmitted to the ground in real time or stored on an observatory tape recorder for later playback to the ground. The format consists of an 80-by-80 matrix (major frame) that repeats every 16 seconds. Each row of the matrix contains 80, 10-bit words and is called a minor frame. The VIP output pulses are coherent with the beginning of each major frame and are available to sensors for synchronization to the VIP sampling sequence.

The DIP and ZIP are part of the data processor subsystem (DAPS), which is the primary communication system for observatory data. The flight data-handling equipment (FDHE) is the spacecraft components part of the DAPS. The FDHE also includes the THIR A/D converters, the spacecraft tape recorders, and the input and output routers (IR & OR) for cross-patching major elements. The DIP is a four-channel time-division multiplexer, which accepts synchronous, serial digital data from predetermined sources and combines them into a synchronous, serial output bit stream at a 25 kbs rate. The common denominator used for Nimbus 7 is 2 kbs, therefore, all inputs to the DIP must be an exact multiple of 2 kbs. The resulting 25 kbs output is decommutated on the ground, where each digital input is extracted from the composite and is available in its original format.

The ZIP is a special-purpose processor for handling CZCS sensor data. Specifically, the ZIP multiplexes the six channels of CZCS digital radiometric data, removes nonsensible data, gates in calibration and synchronization data, and compresses the resulting output to a rate compatible with the spacecraft tape recorders and S-band transmission systems.

1.3.4.3 Tape Recorder Subsystem

The Nimbus 7 observatory carries three identical tape recorders, one classified as a redundant unit. Each recorder is capable of recording either DIP or ZIP data, but not both simultaneously. In a normal recording operation, one recorder records 25 kbs data from the DIP for periods of 100 to 257 minutes and then reproduces it in reverse direction at a rate of 800 kbs. Normally, one recorder will be used for recording DIP data and another is used for recording ZIP data. Each recorder has a total record capacity of 305 minutes of DIP data recorded at 25 kbs or 9.56 minutes of ZIP data at 800 kbs.

Data are played back at a rate of 800 kbs for either the DIP or ZIP with a lapse rate of 9.56 minutes for a full tape. Less time is required if less than a full tape has been recorded. A fast rewind mode (three minutes) allows 500 feet of active tape to be moved from beginning of tape (BOT) to end of tape (EOT) at a high speed rate. This is accomplished without record, playback or erasure of previously recorded data. In normal operation, one recorder is played back to a receiving station while the other is recording; thus avoiding loss of data. A summary of operating modes is presented in Table 1-2.

Table 1-2
Tape Recorder Modes Summary

Mode	Data Rate (kbs)	Tape Speed (in./sec)	Max. Continuous Oper. Time (minute)
Record DIP	25.0	0.327	305
Record ZIP	800.0	10.45	9.56
Playback (DIP or ZIP)	800.0	10.45	9.56
Rewind		35.0	3.0

1.3.5 Thermal Control Subsystem

The thermal control subsystem is designed to provide a controlled environment of 25°C (plus or minus 10°C) within the observatory to promote long life for subsystem and instrument components. Thermal control is accomplished by both semipassive (shutter and heaters) and passive (radiators, insulation, and coatings) elements. Shutters are located on most of the peripheral compartments on the sensory ring, and are actuated by fluid-filled bellows assemblies. The assemblies are fastened to a sensor plate which is in contact with the dissipating components that position the shutter blades to the proper heat-rejection level. Heaters are bonded at various locations in the sensory ring to prevent temperatures from falling below minimum levels during extended periods of low equipment duty cycles. The heaters are energized selectively by ground command when the temperature level at these locations falls below a pre-determined value. The upper and lower surfaces of the sensory ring are insulated to prevent gain or loss of heat through those areas. External structure and radiating surfaces are coated to provide the required values of emission and absorption. Passive radiators, coated with a low-absorptivity, high-emissivity finish, are used to assist the shutters in rejecting heat from the sensory ring.

1.4 Data Handling and Processing Complex

The Nimbus Project at GSFC has the responsibility for the initial processing of the observations from all eight instruments on board the Nimbus 7 spacecraft. Data processing for several sensors continues at GSFC. These are processed into archieved tape and film products. Data from some sensors is sent to intermediate processing facilities outside of the GSFC complex. The centers return data to GSFC for final processing and archival.

This procedure is a departure from the traditional method of having the processing of data from each experiment the responsibility of each principal investigator. For the Nimbus Project to meet this new responsibility, a data handling and processing complex was established at the Goddard Space Flight Center and designated the Nimbus Observation Processing System (NOPS). The purpose of NOPS (Figure 1-3) is to organize and oversee the processing of payload data (except SAMS) into scientific investigations. For a graphic view of the relationship between the Nimbus Observation Processing System and the Nimbus Data Applications System (NDAS) see Figure 1-4.

Since data handling and processing are a major task not easily handled in a single facility at GSFC, a plan was devised to distribute the processing among several facilities while converting the computational results into the data products in one facility (GSFC). This plan affords the opportunity to have a broader range of display equipment available for all instrument data products than if the computational and display efforts were both distributed. The specific responsibilities of the NOPS are:

- to calibrate, quality check, and geographically locate the raw sensor observations,

- to convert the observations into meaningful parameters through the application of scientific algorithms,

- to establish a broad data base by correlating observations from related sensors,

- to display the derived parameters in the most useful forms (products) for scientific investigations and correlations,

- to distribute the generated products to Nimbus Experiment Teams (NET) and selected investigators on a limited basis, and

- to distribute archival quality tapes and film products to archive centers for their dissemination to all interested users.

The Information Processing Division (IPD) at GSFC is utilized as the central data products generation and distribution facility. IPD generates contoured maps, cross sections, atmospheric profiles, plots, listings, montages, and images on 16 mm microfilm, 35 mm color slides, 105 mm color film, and 241 mm black and white film. They also generate and distribute a wide variety of magnetic tapes.

The Earth Radiation Budget (ERB), Scanning Multichannel Microwave Radiometer (SMMR), and Solar Backscatter Ultraviolet/Total Ozone Mapping System (SBUV/TOMS) data, receives initial processing in the Science and Applications Computer Center (SACC) at GSFC. ERB processing in SACC is a two part operation; the second step delayed until information derived from the first step is obtained from the National Oceanic and Atmospheric Administration (NOAA). The Temperature

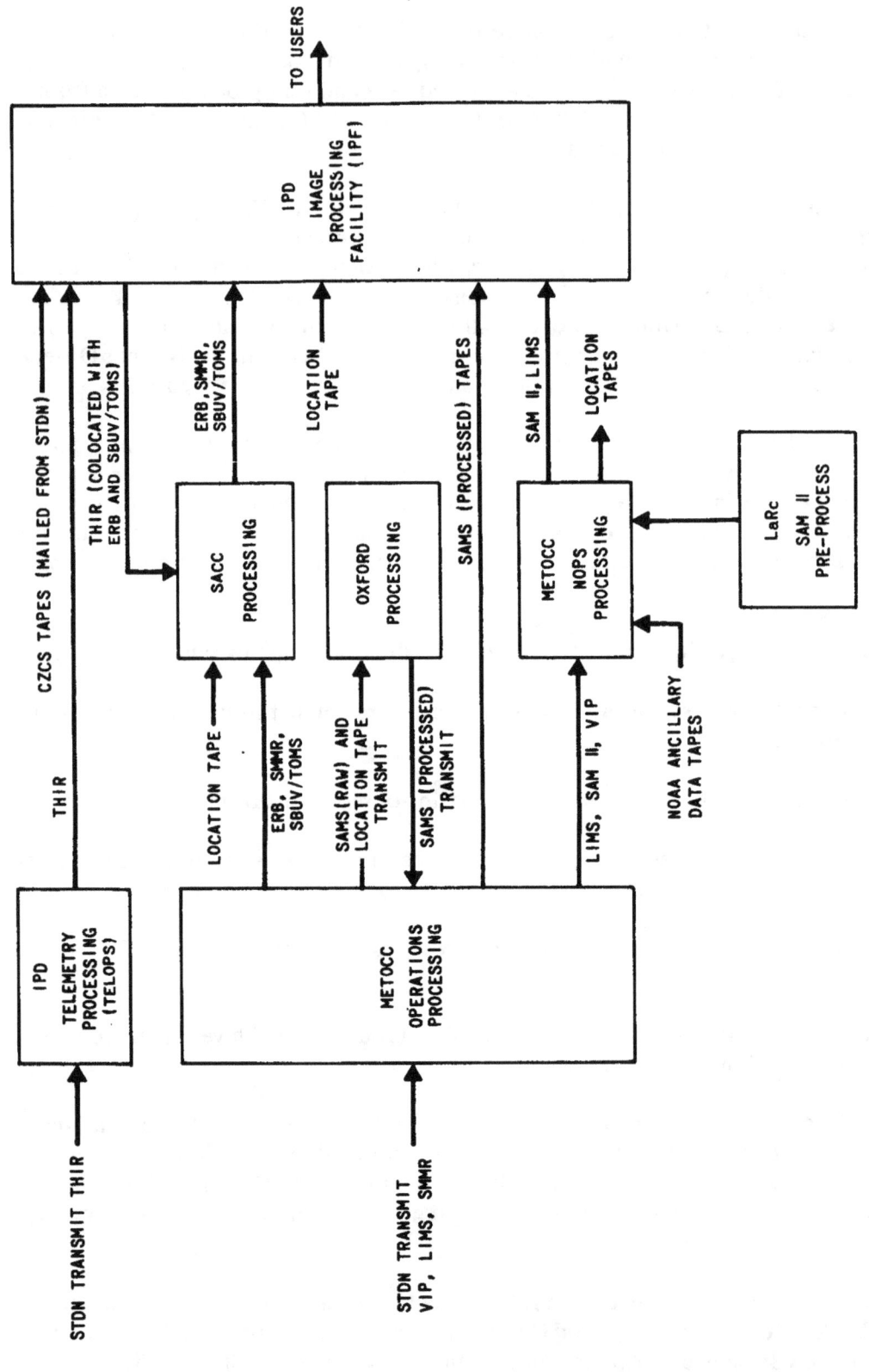

Figure 1-3. NOPS Data Flow and Facility Plan

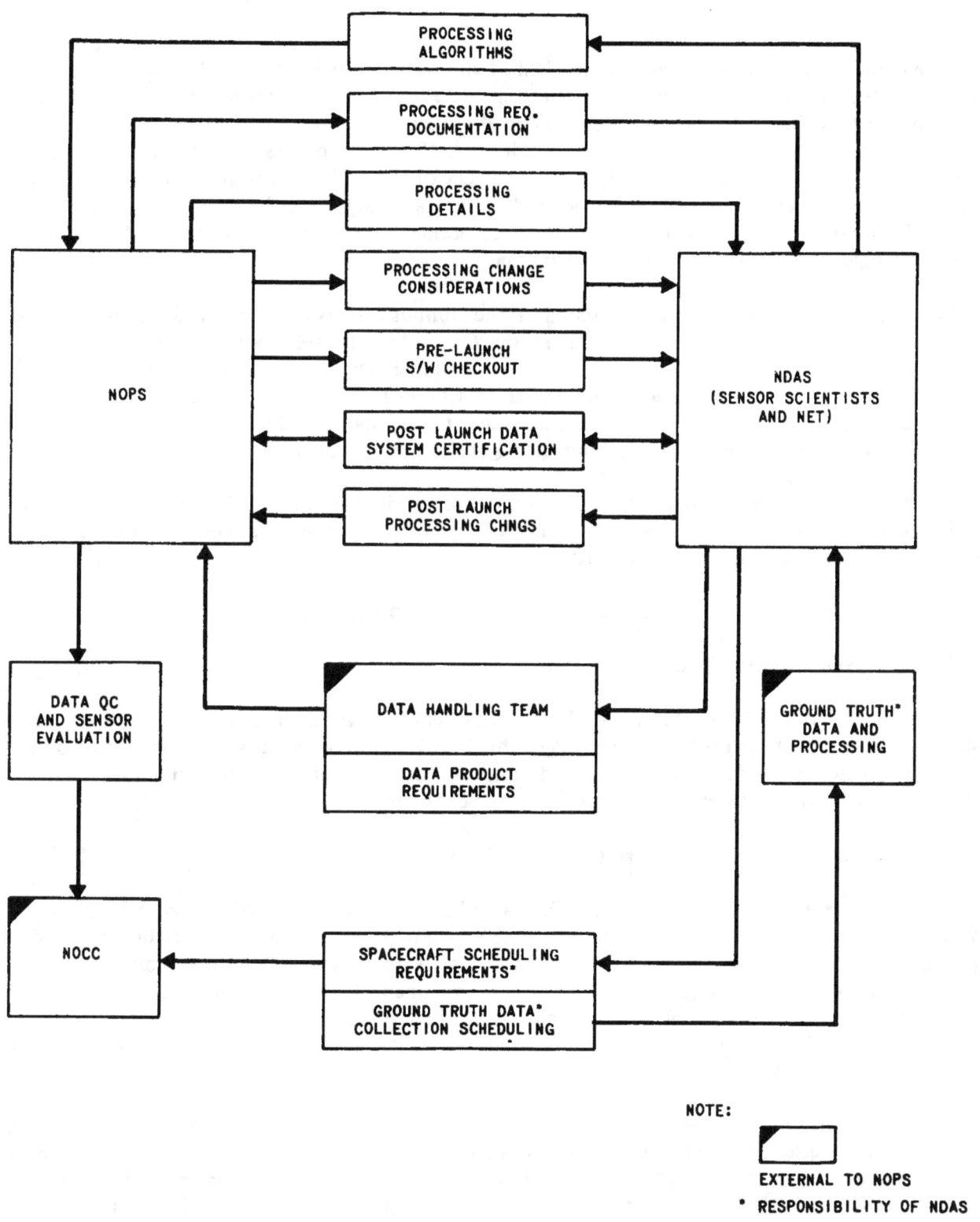

Figure 1-4. Relationship Between the Nimbus Observation Processing System (NOPS) and the Nimbus Data Applications System (NDAS)

Humidity Infrared Radiometer (THIR) and Coastal Zone Color Scanner (CZCS) data, receives initial processing by IPD at GSFC, incorporating the straight-forward calibration science processing requirements within the image generation requirements.

The cloud content in the instantaneous field of view (IFOV) of ERB and SBUV/TOMS are computed from the THIR data and utilized in the respective processing for these instruments. THIR cloud content in the SMMR IFOV's is listed as corroborative information for SMMR data product evaluation. Stratospheric and Mesospheric Sounder (SAMS) data is processed at Oxford, England. SAMS data is sent by data phone lines from GSFC to Oxford. The Limb Infrared Monitoring of the Stratosphere (LIMS) experiment data is processed at the National Center for Atmospheric Research (NCAR) facilities, and the Stratospheric Aerosol Measurement II (SAM II), experiment data is processed at Langley Research Center (LaRC), Virginia.

A common image location program based on the Nimbus 6 ERB program for deriving spacecraft attitude is included in the Meteorological Operations Control Center (MetOCC) processing with the results utilized in the processing for all the instruments. The image location tapes are sent with the user formatted output (UFO) tapes or the sensor data tapes (SDT) to the appropriate facility. See Sections 2 through 9 for details on the processing plans for the individual instruments. These processing plans are based on the currently available algorithms as provided by the sensor scientists.

All initial photographic processing and reproduction services are provided by IPD. IPD provides for the distribution of the data products to the NET members and archival centers. The film and tape archival plan is presented in Section 1.5.

For a composite view of the Nimbus 7 data flow, refer to Figure 1-5 and 1-5a.

1.4.1 Nimbus Data Application System

The function of the Nimbus Data Applications System (NDAS) is to set the requirements for Nimbus 7 data products and the processing algorithms. NDAS evaluates submitted proposals for investigations, disseminates data, administers and monitors contracts related to scientific investigations, and coordinates results with investigators and users. See Figure 1-4.

1.4.2 Nimbus Experiment Teams

There are seven Nimbus Experiment Teams (NET's), one for each of the NASA-provided sensors, plus the United Kingdom team for the Stratospheric and Mesospheric Sounder (SAMS) experiment. The NET's have met at frequent intervals from the initial inception of each committee and will meet through at least one year after launch. Each team consists of five to ten members and is supported by applications scientists and data processing support personnel. Additionally, each NET is also supported by the Nimbus 7 Data Applications System Manager or his appointed representative.

The function of the NET members is to assist and provide advice on all aspects of their respective sensor program and to perform studies or tasks in their areas of expertise during pre-launch and post-launch activities. They determine the principal research and development requirements of each experiment and perform the required tasks commensurate with priorities and available resources.

A complete list of the Nimbus Experiment Team members is given in Appendix B. A summary of the tasks and study areas for each NET are:

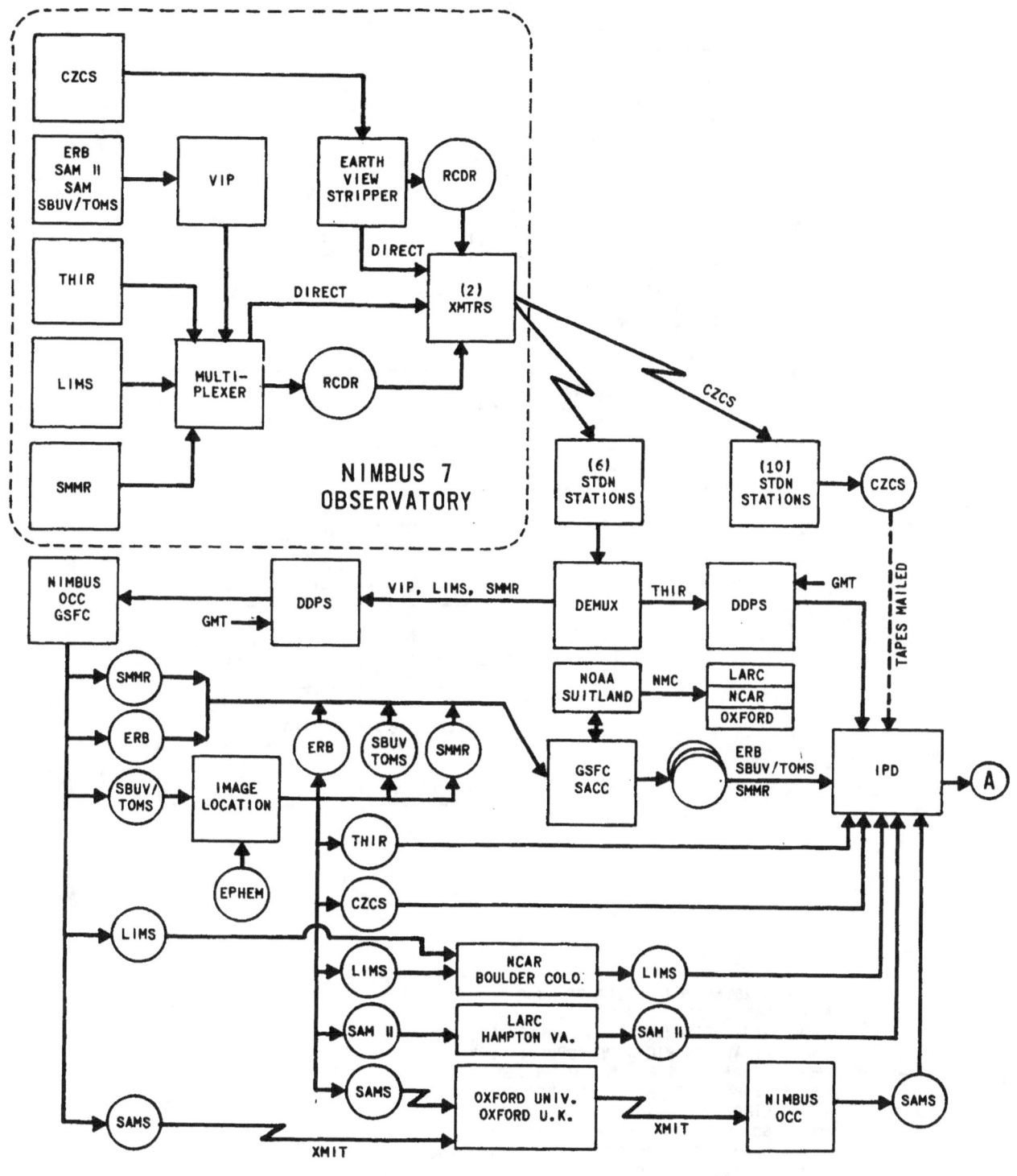

Figure 1-5. Nimbus 7 Data Flow

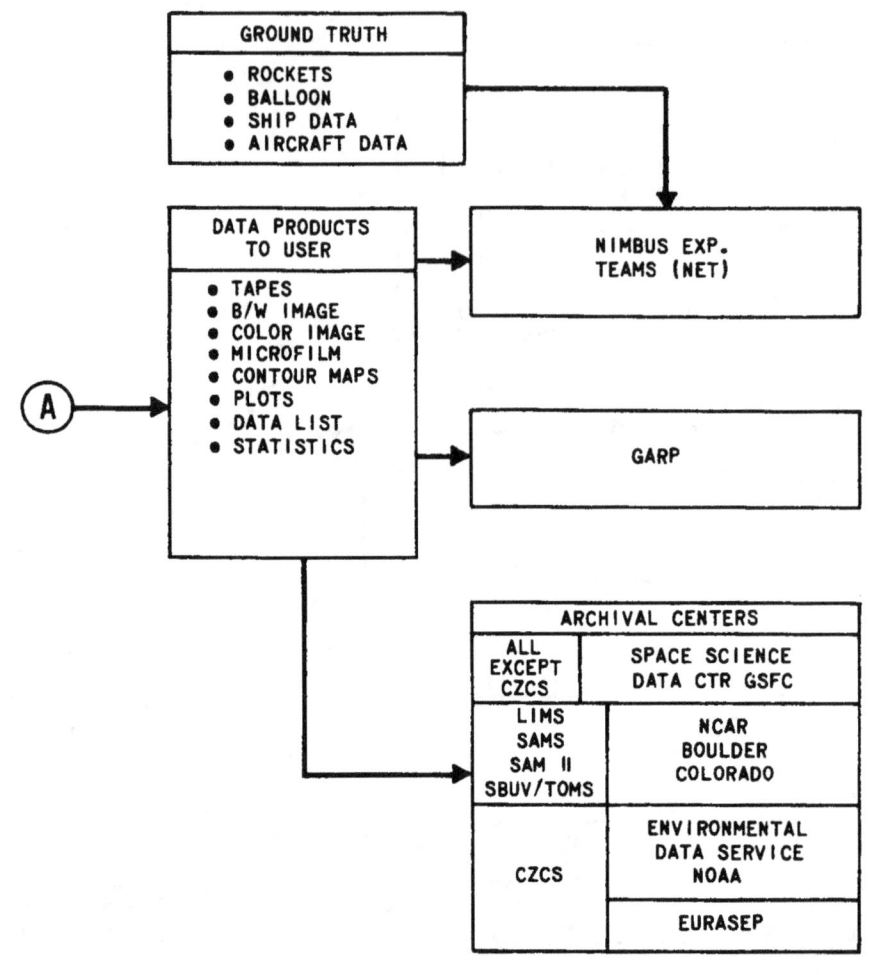

CZCS - COASTAL ZONE COLOR SCANNER
ERB - EARTH RADIATION BUDGET
SAM-II - STRATOSPHERIC AEROSOL MEASUREMENT
SAMS - STRATOSPHERIC & MESOSPHERIC SOUNDER
SBUV/TOMS - SOLAR BACKSCATTER ULTRAVIOLET & TOTAL OZONE MAPPING SPECTROMETER
THIR - TEMPERATURE HUMIDITY INFRARED RADIOMETER
LIMS - LIMB IR MONITORING OF THE STATOSPHERE
SMMR - SCANNING MULTICHANNEL MICROWAVE RADIOMETER
IPD - INFORMATION PROCESSING DIVISION (GSFC)
SACC - SCIENCE AND APPLICATIONS COMPUTING CENTER (GSFC)
VIP - VERSATILE INFOR. PROCESSOR
DDPS - DIGITAL DATA PROCESSING SYSTEM
GARP - GLOBAL ATMOSPHERIC RESEARCH PROGRAM
NMC - NATIONAL MET. CENTER
RCDR - RECORDER
OCC - OBSERVATORY OPERATION CONTROL CENTER
NCAR - NATIONAL CENTER FOR ATMOSPHERIC RESEARCH

Figure 1-5a. Nimbus 7 Data Flow (Continued)

- to develop, use and test processing and science algorithms,

- to define the general content of the film and tape product,

- to verify sensor calibration and performance,

- to participate in the planning for data acquisition, time schedule sharing of sensor operations,

- to certify the quality of data output products by comparison measurements with other data (ground truth measurements), and

- to perform initial post-launch experiment investigations and issuing appropriate reports and publications.

1.5 Data Archiving

The Nimbus 7 archival data products distribution is accomplished after the individual Nimbus Experiment Teams have validated the data. This validation process is expected to take from two to six months after launch, depending on the experiment and the type of anomalies that may arise during the initial checkout phase.

Listed in Table 1-3 are the film data products available to all users. Table 1-4 provides the same information for tape data products. All products, except for CZCS, are archived by the National Space Science Data Center (NSSDC). CZCS data are archived by the Environmental Data Information Service (EDIS). The addresses of these agencies are:

- Environmental Data Information Service
  World Weather Building
  Room 606
  Camp Springs, Maryland 20733

- National Space Science Data Center
  Goddard Space Flight Center
  Code 601
  Greenbelt, Maryland 20771

In addition to the film and tape data products, EDIS will publish a CZCS catalog listing all available CZCS data. NSSDC will publish a meteorological catalog listing data from all meteorological satellites including tape and film output products from Nimbus 7. To obtain copies of these catalogs, write to NSSDC or EDIS.

All requests from foreign researchers for Nimbus 7 data archived and available through NSSDC must be specifically addressed to:

       Director, World Data Center A for Rockets and Satellites
       Code 601, Goddard Space Flight Center
       Greenbelt, Maryland 20771, U.S.A.

Table 1-3
Film Data Products Available Through EDIS and NSSDC.

Archival Data Center	Sensor	Reproducible Copy Is:	User Copy Will Be:
EDIS	CZCS	2nd generation 241 mm (9.5") black and white negative transparency.	a. 241 mm black and white positive or negative transparency. b. 241 mm black and white positive print.
NSSDC	ERB, SAM II, LIMS, SBUV/TOMS, SAMS	2nd generation 16 mm negative (black background) transparency.	a. 16 mm positive transparency. b. 241 mm hard copy (in limited quantity).
	TOMS (montage)	2nd generation 241 mm black and white negative transparency.	a. 241 mm black and white positive or negative transparency. b. 241 mm black and white positive print.
	SMRR (all data)	2nd generation 35 mm color positive transparency.	a. 35 mm color slide. b. color prints — maximum size 203 mm x 254 mm (8" x 10").
	THIR (montage)	2nd generation 241 mm black and white negative transparency.	a. 241 mm black and white positive or negative transparency.

When ordering data from either the NSSDC or the World Data Center, a user should specify why the data are needed, the subject of his work, the name of the organization with which he is connected, and any government contracts he may have for performing his study. Of course, each request should specify the experiment data desired, the day and area of interest, plus any other information that would facilitate the handling of the data request. Requests for specific tape types, as listed in Table 1-4, should specify the tape specification (last column in Table 1-4). This number references a tape specification document describing the record and file content and word format of each tape type. A user receives a tape specification document for each requested tape type.

A user requesting data on magnetic tapes should provide additional information concerning his plans for using the data, e.g., what computers and operating systems will be used. In this context, the NSSDC is compiling a library of routines which can unpack or transform the contents of many of the data sets into formats which are appropriate for the user's computer. NSSDC will provide, upon request, information concerning its services.

When requesting data on magnetic tape, the user must specify whether he will supply new tapes prior to the processing, or return the original NSSDC tapes after the data have been copied.

Table 1-4
Tape Types Available Through EDIS and NSSDC

Archival Data Center	Sensor	Tape Name	Tape Quantity per Year	PDFC	Tape Spec. No.
EDIS	CZCS	CRCST	5500	ZB	T749021
		CAT	12	ZC	T749031
NSSDC	ERB	MATRIX	12	AA	T134031
		MAT	365	AC	T134081
		SEFDT	12	AD	T134021
		ZMT	2	AE	T134091
	LIMS	MATRIX-M	14	EA	T564041
		MATRIX-C	14	EB	T564081
		PROFILE-R	7	EC	T564111
		PROFILE-I	21	ED	T564071
		RAT	210	EE	T564011
		IPAT	105	EF	T564021
		MAT	70	EG	T564051
		CAT	70	EH	T564091
		SMAT	7	EI	T564101
		SCAT	7	EM	T564121
	SAMS	MATRIX	24	HA	T884011
		RAT	180	HC	T884041
	SAM II	MATRIX	4	DA	T454021
		PROFILE	12	DB	T454011
		RDAT	12	DC	T454041
		BANAT	12	DD	T454051
	SBUV/TOMS	MATRIX	24	FA	T634071
		MONTAGE	52	FC	T634081
		RUT-S	26	FD	T634111
		OZONE-S	12	FE	T634041
		OZONE-T	180	FF	T634091
		ZMT	2	FH	T634061
		RUT-T	120	FJ	T634121
	SMMR	MAP-30	12	BD	T234051
		MAP-LO	12	BE	T234101
		MAP-SS	12	BF	T234111
		PARM-30	60	BG	T234041
		PARM-LO	30	BH	T234121
		PARM-SS	30	BI	T234131
		TAT	183	BJ	T234021
		CELL-ALL	61	BK	T234071
	THIR	CLDT	730	ID	T344011
		CLE	53	IE	T343031
		CLT	104	IF	T343041

SECTION 2

THE COASTAL ZONE COLOR SCANNER (CZCS) EXPERIMENT

by

Dr. Warren Hovis
National Oceanic and Atmospheric Administration
National Environmental Satellite Service
Room 0135, FOB 4
Washington, D.C. 20233

2.1 Introduction

The Coastal Zone Color Scanner (CZCS) is the first instrument devoted to the measurement of ocean color and flown on a spacecraft. Although instruments on other satellites have sensed ocean color, their spectral bands, spatial resolution, and dynamic range were optimized for land or meteorological use. In the CZCS, every parameter is optimized for use over water to the exclusion of any other type of sensing. The signal-to-noise ratios in the spectral channels sensing reflected solar radiance are higher than those required in the past. These ratios need to be high because the ocean is such a poor reflecting surface that the majority of the signal seen by the reflected energy channels at spacecraft altitudes is backscattered solar radiation from the atmosphere rather than reflected solar energy from the ocean. The CZCS thermal channel utilizes the 10.5 μm to 12.5 μm region used on many other thermal mappers. This CZCS channel is unique, however, since it is registered with the reflected solar energy bands and has the same spatial resolution.

The data processing techniques for the CZCS are also unique in that off-setting is used to enhance contrasts over the ocean and remove much of the effect of the backscattered atmosphere. Attempts will be made to process the data into derived products such as pigment concentration and diffuse attenuation coefficient prior to distribution to users. The archived magnetic tapes contain both calibrated radiances and equivalent blackbody temperatures, plus the derived products, so a user with a large computer facility would be able to utilize a more complicated algorithm than that used in production and processing at GSFC. A user without such a facility can utilize the derived products provided by NASA.

The CZCS is a conventional multi-channel scanning radiometer utilizing a rotating plane mirror at a 45 degree angle to the optic axis of a Cassegrain telescope. The rotating mirror scans 360 degrees, however, only ±40 degrees of data centered on the spacecraft nadir is collected for ocean color measurements. During the rest of the scan, the instrument acquires a view of deep space and of internal instrument sources for calibration of the various channels. The radiation collected by the telescope is divided into two portions by a dichroic beam splitter. One portion is transmitted to a field stop that is also the entrance aperture of a small polychromator. The radiant energy entering the polychromator is disbursed and reimaged in five wavelengths on five silicon detectors in the focal plane of the polychromator. The spectral channels are described in detail in Section 2.3. The portion of the beam reflected off of the dichroic mirror is directed to a cooled mercury cadmium telluride detector sensing in the 10.5 μm to 12.5 μm region. The CZCS utilizes a radiative cooler that cools the mercury cadmium telluride detector to approximately 120 Kelvin during spacecraft flight.

The CZCS is intended primarily as a tool for determining the content of water. It is well known that the content of water, be it organic or inorganic particulate matter or dissolved substances, affects its color. Ocean water, containing very little particulate matter, scatters as a Rayleigh scatterer with the well known deep purple or bluish color of the ocean. As particulate matter is added to the water, the scattering characteristics are changed and the color is changed. Phytoplankton, for instance, have specific absorption characteristics and normally change the water to a more greenish hue although some phytoplankton, such as the various red tide, can change the water to colors such as red, yellow, blue-green, or mahogany. By sensing the color with very high signal-to-noise ratios, the CZCS provides a mechanism for analyzing that color for the content of the water. Inorganic particulate matter in water, such as the terrigenous outflow from rivers, has a different color from organic material typically brownish in color but sometimes varying with red.

2.2 Scientific and Technical Objectives

2.2.1 Scientific Objectives

The scientific objective of the CZCS is to determine the specific nature of the contents of water as quantitatively as possible and to carry out such measurements over large areas in short periods of time in a way not possible with other techniques such as surface ship investigations. Specifically, the CZCS experiment attempts to discriminate between organic and inorganic materials in the water, determine the quantity of these materials in the water sample to the best degree possible and, in certain instances, attempts identification of organic particulates such as discriminating between various types of red tide organisms.

By conducting measurements over a large area in a short period of time, the CZCS allows oceanographers to view the ocean as never seen before from ships. As an example, in one two-minute data segment, the CZCS covers approximately 1.3 million square kilometers of the ocean surface allowing examination, nearly simultaneously, on a scale never before accomplished. Measurements on this scale allow oceanographers to determine such things as the standing stock of phytoplankton and its distribution in various fishing areas and, potentially, to assess the ability of that area to support a standing stock of fish. In addition to examining the existing fisheries, the CZCS will be used to look for new areas of potential fish production around the globe.

2.2.2 Technical Objectives

The technical objective of the CZCS program is to determine if remote sensing of color can be used to identify and quantify material suspended or dissolved in water. If ocean color measurements can be used to derive such products as chlorophyll and sediment concentration, they will guide further development of the ocean color discipline and help to determine if such an instrument is a candidate for operational satellite use in the future.

The algorithms being developed for the derived products from CZCS are the result of the most extensive ocean color measurements ever made and are a considerable step forward from those available in the past. Corrections for such things as atmospheric backscatter and limb brightening are included in the CZCS processing algorithms. The processing goal is to take the observed radiance, determine the radiance that would be seen directly above the ocean surface, and then derive from that radiance, the content of the water below the ocean surface.

2.3 Instrument Description

2.3.1 Operation

The CZCS has considerable flexibility built into it to accommodate a wide range of conditions. The first four spectral bands, for instance, have four separate gains that change, on command, to accommodate the range of sun angles observed during a complete orbit and throughout the various seasons. The gains are changed to utilize the best dynamic range possible without saturating over water targets. Normally, the gain used in the first four channels is determined by the solar elevation angle of the target to be acquired. When a special circumstance is expected, such as a particularly bright material in the water, the gain can be changed to accommodate the special circumstances.

In addition to gain change, the CZCS scan mirror can be tilted from nadir to look either forward or behind the spacecraft line of flight. It can tilt in two degree increments up to twenty degrees in either direction. This feature was built into the instrument to avoid the glint caused by capillary waves on the ocean that would obscure any scattering from below the surface. The angle of tilt of the scan mirror is determined by the solar elevation angle. It is normally tilted to avoid sunlight and would only be commanded to look into the glint for a special sunglint study.

2.3.2 Viewing Geometry

The CZCS is a scanning multi-spectral radiometer with a recorded scan width of 1566 kilometers centered on spacecraft nadir. The scanner actually scans through 360 degrees, but the electronics limit the high data rate sampling to 39.34 degrees about nadir. As discussed in Section 2.3.1 the scanner looks either ahead or behind the spacecraft nadir in increments of two degrees up to twenty degrees to avoid ocean surface glint. The ground resolution of the IFOV is 0.825 kilometer at nadir and degrades somewhat as the instrument scans away from nadir on either side. The viewing geometry of the instrument is illustrated in Figure 2-1.

2.3.3 Channel Characteristics

The CZCS has six spectral bands, five sensing backscattered solar radiance and one sensing emitted thermal radiance. Figure 2-2 illustrates the method by which discrimination of the spectral bands is achieved. The beam is split by a dichroic beam splitter, one portion of the beam going through a set of depolarizing wedges to a small polychromator where the radiance is dispersed and detected by five silicon diode detectors in the focal plane of the polychromator. Radiance in the 10.5 μm to 12.5 μm spectral band is reflected off the dichroic and then imaged onto an infrared detector of mercury cadmium telluride cooled to approximately 120 Kelvin. Table 2-1 shows the center wavelengths, the spectral bandwidths, and the minimum signal-to-noise ratio specified for the instrument at the most sensitive gain setting, that is, the gain setting that would be used for the darkest targets. (Prelaunch tests show the instrument has exceeded the specification for signal-to-noise in every channel). The first four channels were selected to cover specific absorption bands and the so-called hinge point. These channels are meant to look at water only and saturate when the field of view is over most land surfaces and clouds. The spectral response of channels 1 through 5 is illustrated in Figure 2-3.

Channel 5 has the same spectral response as channel 6 of the Landsat multi-spectral scanner series. The gain of channel 5 is fixed and set to produce the same percentage of maximum signal over land targets as the Landsat channel 6. However, the actual radiance for saturation is higher since the Nimbus 7 spacecraft crosses the equator at high noon whereas Landsat crosses the equator at 9:30 a.m. local time.

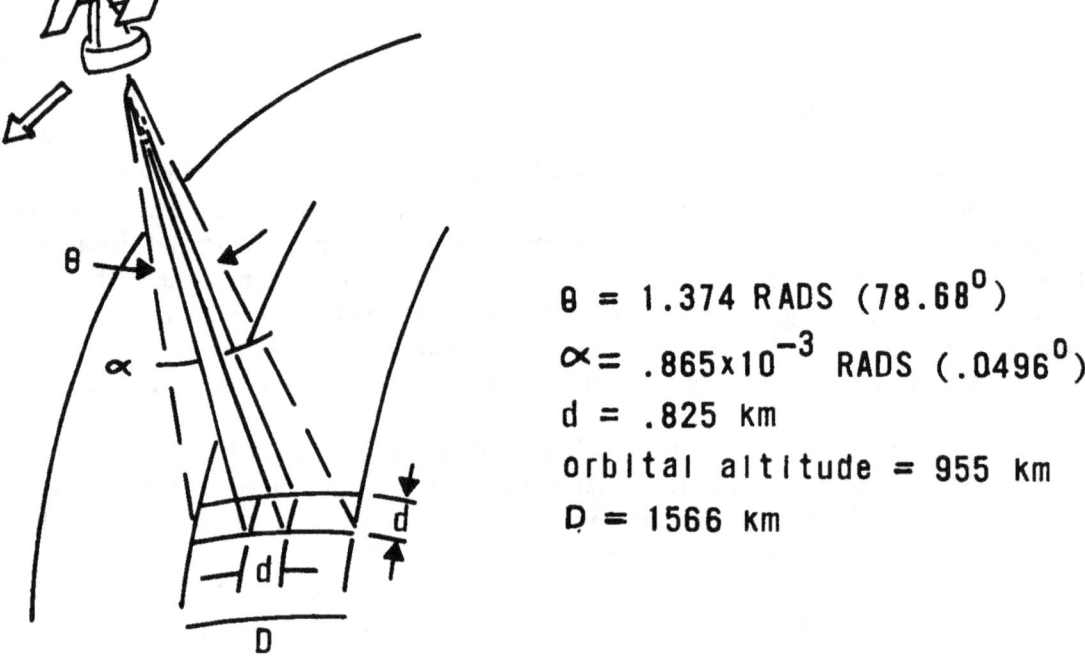

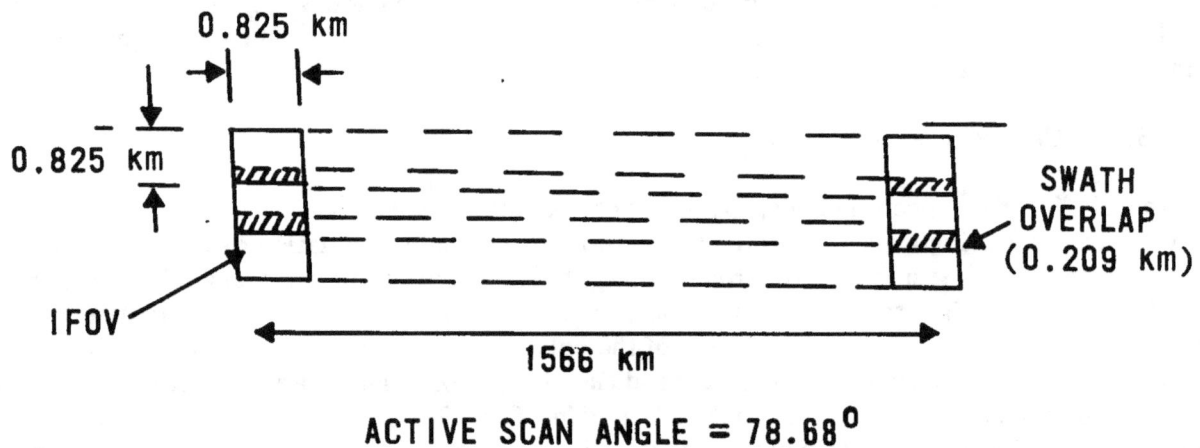

Figure 2-1. CZCS Viewing Geometry and Earth Scan Pattern

The 10.5 μm to 12.5 μm channel measures equivalent blackbody temperature as seen by the sensor with a noise equivalent temperature difference of less than 0.35 Kelvin at 270 Kelvin. Atmospheric interference with this channel, principally from weak water vapor absorption in the 10.5 μm to 12.5 μm region, can produce measurement errors of several degrees. Temperature gradients, however, should be seen quite well because of the extremely low noise equivalent temperature difference of this sensor.

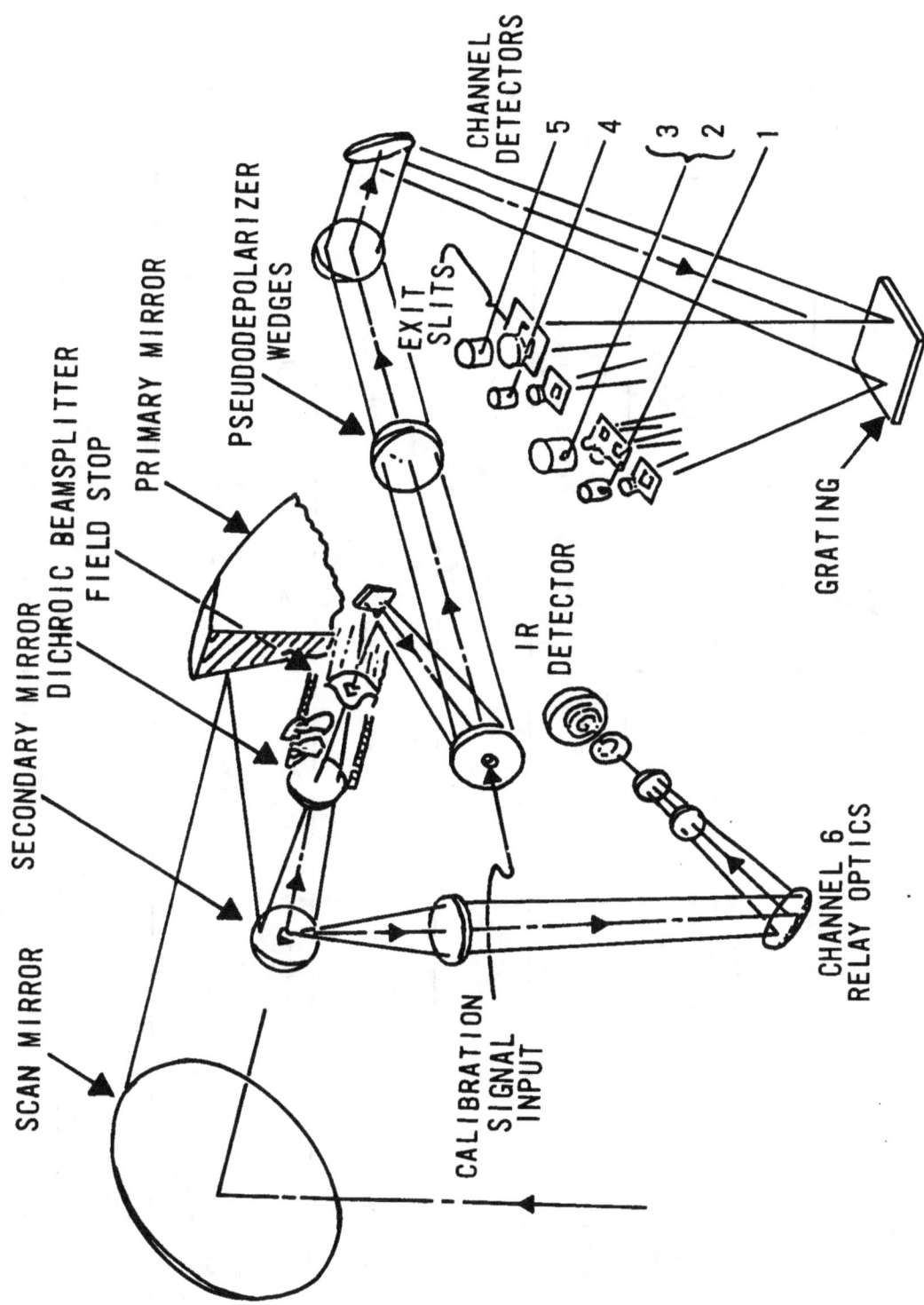

Figure 2-2. CZCS Optical Arrangement

Table 2-1
CZCS Performance Parameters

Performance Parameters	Channels					
	1	2	3	4	5	6
Scientific Observation	Chlorophyll Absorption	Chlorophyll Correlation	Yellow Stuff	Chlorophyll Absorption	Surface Vegetation	Surface Temperature
Center Wavelength λ Micrometers	0.443 (blue)	0.520 (green)	0.550 (yellow)	0.670 (red)	0.750 (far red)	11.5 (infrared)
Spectral Bandwidth Δλ Micrometers	0.433 – 0.453	0.510 – 0.530	0.540 – 0.560	0.660 – 0.680	0.700 – 0.800	10.5 – 12.5
Instantaneous Field of View (IFOV)	0.865 x 0.865 Milliradians (0.825 x 0.825 km at sea level)					
Co-registration at NADIR	<0.15 Milliradians					
Accuracy of Viewing Position Information at NADIR	<2.0 Milliradians					
Signal to Noise Ratio (min.) at Radiance Input $N < (mW/cm^2 \cdot STER \cdot \mu m)$	>150 at 5.41	>140 at 3.50	>125 at 2.86	>100 at 1.34	>100 at 10.8	NETD of 0.220°K at 270°K
Consecutive Scan Overlap	25%					
Modulation Transfer Function (MTF)	1 at 150 km target size, 0.35 min. at 0.825 km target size					

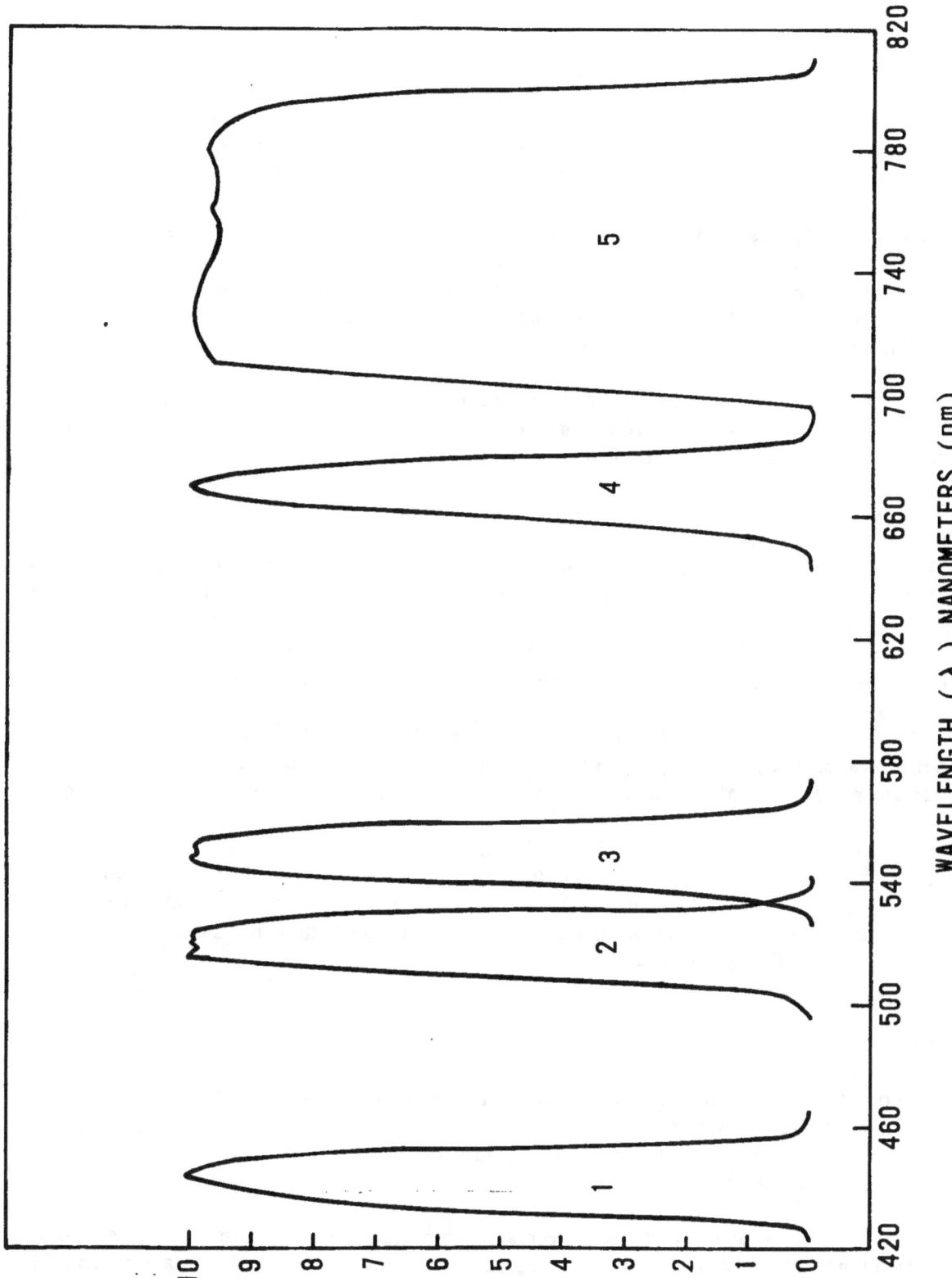

Figure 2-3. CZCS Spectral Response for Channels 1 through 5

2.4 Calibration

2.4.1 Prelaunch Calibration

Prelaunch calibration of the CZCS was achieved utilizing a 76 centimeter diameter integrating sphere as a source of diffuse radiance for channels 1 through 5 and a blackbody source for calibration of channel 6. The integrating sphere was especially constructed for calibration of the CZCS and was, itself, calibrated from a standard lamp from the National Bureau of Standards utilizing a spectrometer and another integrating sphere to transfer calibration from the lamp to the sphere. This same type of sphere has been used in calibrating the multi-spectral scanner for Landsat and will also be used to calibrate the Advanced Very High Resolution Radiometer (AVHRR) for the TIROS-N series that will be flying at approximately the same time as Nimbus 7.

In addition to the sphere and the blackbody, a collimator was also used to calibrate the CZCS in vacuum testing. Calibration was transferred from the primary calibration standard, the sphere and the blackbody, to the collimator using the instrument itself.

2.4.2 In-flight Calibration

In-flight calibration of the CZCS is accomplished for the first five bands by using a built-in incandescent light source. This in-flight calibration source was calibrated using the instrument itself as a transfer against the referenced sphere output. The light source is redundant in the instrument so that in case of failure of one of the lights, another one can be ordered to operate on command. After launch, light calibration source number one will be used routinely, with light source number two tested occasionally to verify its stability.

Channel 6 is calibrated by viewing the blackened housing of the instrument whose temperature is monitored. Deep space is another calibration viewed during the 360 degrees rotation of the scan mirror.

The output from both calibration sources will be monitored during the life of the sensor to determine if any changes in sensitivity occur. If changes in sensitivity are observed, a procedure will be followed to determine if the change is due to a change in sensitivity of the various detectors, channels, or a change in the calibration source itself.

2.5 Operational Modes

Since Nimbus 7 flys from south to north in daylight, the scan mirror is positioned to look behind the satellite when the spacecraft is south of the subsolar point and ahead of the spacecraft when it is north of the subsolar point. Tilt and gain setting information is transmitted with the CZCS data and is part of the data product records.

The CZCS data is transmitted from the spacecraft to ground receiving stations at a rate of 800 kbs either in real time or in playback of the tape recorder. Whenever possible the data is recorded in real time. However, when the satellite is out of the range of tracking stations, the data is recorded on an on board tape recorder. The tape recorded data will normally be played back at the Alaska tracking station. Nine other STDN's also have the capability to receive these playbacks.

To improve the instrument response to ocean color, a DC offset can be inserted into the on board processing of the radiance measured in the first four bands. In this DC offset mode, the entire digital capability of the on board digitizer is utilized to cover approximately the top 30 percent of the signal which contains modulation due to change in ocean color. Since the knowledge of the exact amount of the offset eliminated in the on board processing is always known, it can be reinserted where needed for processing on the ground.

The sensor is turned on in sufficient time prior to collection of data to allow for instrument warm-up and for the sensor to stabilize. Since all channels are calibrated continuously during flight, any effect of turn-on transient should be noticed immediately, but none is expected.

The most important aspect to be understood about the CZCS operation is that the operation is limited due to spacecraft power constraints to approximately two hours per day. Because of the requirement to operate the sensor two hours per day, data must be taken in carefully preselected locations. Minimum on-off data taking time is a two minute segment. Frequently, longer segments are taken — up to a maximum of ten minutes of continuous data.

Interested users are reminded that if they wish to acquire CZCS data over a particular site, they should contact a member of the Experiment Team and inform that member of the location of the site and the dates on which coverage is most highly desired. Even though all of the data is placed in the public archive, there is no guarantee that all areas of the world will be covered. A special effort will be made, however, to cover major oceanographic expeditions where surface truth is being collected by a ship.

The prime operational area for the CZCS are the coastlines of the United States and the Gulf of Mexico. Other areas of coverage are the coast of Europe, including the Mediterranean Sea, the Baltic Sea, the North Sea, the channels between England and the mainland, the Irish Sea, and the test sites designated by the EURASEP Group of the Commission of European Communities. South African NET participation has requested coverage around the southern tip of Africa on both the east and west coasts, and extending toward the Antarctic. In the Antarctic summer following the launch of Nimbus 7, the Scientific Committee on Antarctic Research (SCAR) will conduct a large expedition in Antarctic water. This will also be a prime target area for the CZCS. Other areas, such as the Deep Ocean Mining Experiment Stations (DOMES), and the Antarctic coverage are limited by the extent of the activities such as the time of the DOMES' action, or the availability of sunlight as in the Antarctic where the sensor can only operate usefully during the Antarctic summer. Requests for coverage of other areas have been received from a number of institutions around the world, and every attempt will be made to accommodate requests from other oceanographic institutions, especially when surface truth is being measured by institutions.

All channels of the CZCS instrument operate simultaneously. During daytime operations all six channels provide useful information. If the sensor operates at night, only data from channel 6 is usable.

2.6 Data Processing, Formats, and Availability

2.6.1 Data Processing

The data from the CZCS is transmitted to the ground either in real time, or from tape recorder playback, at a rate of 800 kbs. The data is recorded on magnetic tapes and sent to the IPD at GSFC.

These tapes contain both radiometric information from the imagery and CZCS housekeeping information. IPD uses these data, plus image location tape (ILT) data, to produce the user tapes described in Section 2.6.2 and the images described in Section 2.6.3. After making sufficient tape and film copies for NET users, IPD forwards the tapes and film to the Environmental Data Information Service (EDIS) of NOAA for archiving and reproduction of copies.

At IPD the data are converted from voltages to radiances for bands 1 through 5, and to equivalent blackbody temperature for band 6. This is accomplished by using the calibration curves derived before launch and applying in-flight calibration sources. After calibration, the data is processed using algorithms developed by the CZCS NET to derive products of suspended and dissolved material in the water. As knowledge is gained from the experiment, the algorithms may need to be changed. All algorithms are available to users, and those used to process each tape and image are identified on those products.

2.6.2 Tape Products

The following tape products are produced by IPD and are sent to the EDIS at NOAA for archiving. Brief descriptions of these tapes are as follows:

- CRCST (Calibrated Radiance, Pigment, Diffuse Attenuation Coefficient and Temperature Tape)

  These tapes contain calibrated and located CZCS data from all six channels scan-line-by-scan line with the channels separated, plus derived pigment and diffuse attenuation coefficient parameters, where computed. There is a maximum of three two-minute block (files) of data per tape. Statistical and calibration summaries are at the end of each file.

- CAT (Catalog Tape)

  These tapes contain cataloged information on all images (two-minute files on the CRCST's). Entries are organized chronologically by target area (location).

The form and content of each of these tapes are described in a tape specification document for each tape type. The appropriate document will accompany a tape shipment to a user.

As discussed in Section 2.5, the CZCS is expected to operate a maximum of two hours per day. If the sensors operate for two hours per day for one year approximately 22,000 images and 7,000 magnetic tapes would be generated. Because of weather conditions, principally cloud cover, the CZCS probably will not operate its scheduled two hour period per day as planned. Thus, a more reasonable estimate of total output per year is approximately 12,000 images and 4,000 magnetic tapes.

2.6.3 Film Format

Figure 2-4 is an example of the format for all CZCS images. Each display is produced on 241 mm by 241 mm (9.5 inch) black and white image stock. The title and reference information at the top of each display includes the gain and tilt angle in effect during the scene, and whether the threshold mode of data enhancement was on or off.

Each of the ten chips on a single display has the same latitude and longitude grid around the chip boundaries. Channel displays 1 through 6 show radiances as shades of gray (referenced to the

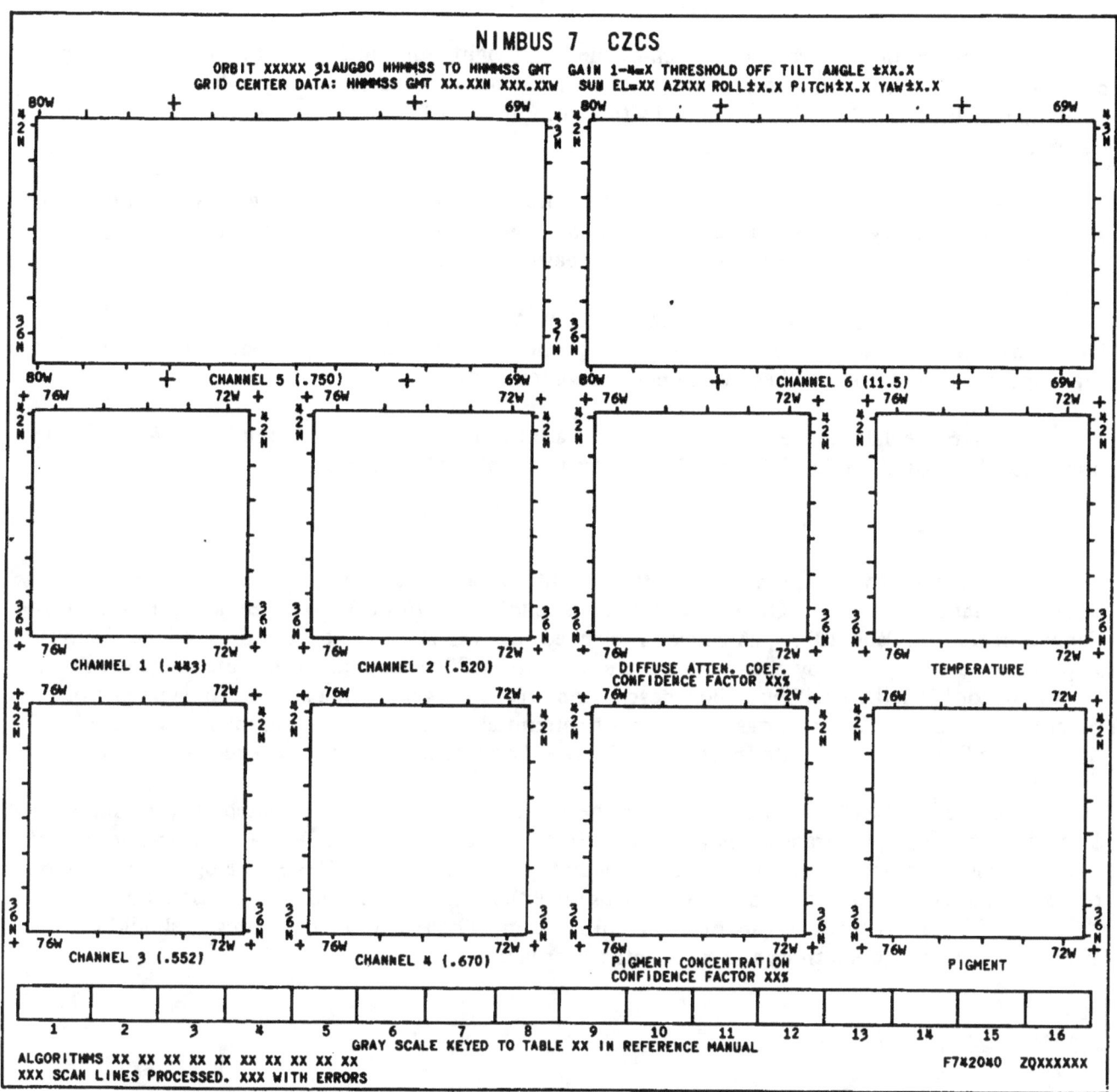

Figure 2-4. CZCS Image Display Format

gray scale at the bottom of the display). The physical parameters of pigment concentration and diffuse attenuation coefficient, if calculated for the scene, are shown as shades of gray (two chips) and as contoured plots (two chips). The top displays (channels 5 and 6) represent the maximum amount of data from a single channel: ±39.36 degrees from nadir for each scan line, and two minutes of data along the orbit track. The bottom eight displays (channels 1 through 4 on the left, plus the four scenes of pigment and diffuse attenuation coefficient parameters) show only the region of best spatial resolution and least geometric distortion, which is within ±20 degrees of nadir. Each scene (chip) is rectified along each scan line and from scan line to scan line so there is an approximately equal scale over each scene.

29

The crosses at the four corners of each of the lower eight chips, and along the top and bottom of the top two chips, are for reference if two or more channels are to be photographically color composited. Cutting apart the chips and aligning two or more sets of crosses provides chip-to-chip registration.

The 16-step gray scale beneath the chips is calibrated in radiances for reference to the images. The values for the gray scale versus radiances for each display are provided in the EDIS catalogs. The appropriate table for each display is identified beneath each gray scale.

The ten algorithms used to generate the ten chips are identified in the lower left corner. The last line in the lower left lists the number of scan lines processed (maximum is 960) and used as input for the chips, and the number of these scan lines containing errors.

The reference data in the lower right corner are the film specification number (F742040) the projection data format code (ZQ), and the film frame number (XXXXXX).

2.6.4   Data Availability

All of the CZCS data is archived with the Satellite Data Services Branch of the Environmental Data Information Service of NOAA. The address is as follows: World Weather Building, Room 606, Camp Springs, Maryland 20233. A catalog is planned that will show the orbital track of the Nimbus 7 spacecraft on a day-by-day basis with the areas where CZCS was operated indicated on the orbital tracks. In addition, there will be a short description on the imagery giving such parameters as cloud cover for each image. The catalogs will be sent to an initial mailing list and will then be available through the Environmental Data Information Service, Satellite Data Services Branch.

The cost of the CZCS data product has not been established, but it is estimated at approximately $3.50 for a photographic transparency and $60.00 for the magnetic tape. All data is available to any user who wishes to purchase it. Data will normally be ordered from the CZCS catalog by specifying the orbit and GMT of the data desired. For users without a CZCS catalog, see Section 1.5 of this document for general tape and film ordering information. The first validated data sets should be available to users between three and six months after launch.

2.7 Planned NET Experiment Investigations and Data Applications

The Nimbus Experiment Team for the Coastal Zone Color Scanner (CZCS) presently plans two major expeditions after launch of Nimbus 7 for validation of the derived product of the CZCS. One expedition will be carried out of the Gulf of Mexico utilizing the research vessel GYRE from Texas A&M University. This expedition will be carried out from approximately mid-October to the first week of November 1978, and will cover various water mass types in the Gulf of Mexico.

The other NET surface validation expedition will be carried out off of Southern California and in the Gulf of California utilizing a research vessel from the Scripps Institute of Oceanography. This expedition will occur sometime after the first year in 1979.

Foreign experiment team members will carry out validation investigations in European waters and off South Africa. The Joint Research Center of the Commission of European Communities will coordinate the activities of the EURASEP group in carrying out surface truth validations in waters around Europe. Information on their planned activity can be obtained from Dr. Bruno Sturm of the Joint Research Center, Ispra, Italy.

The South African experiment team member, Dr. Frank Anderson, Director of the National Research Institute for Oceanology, will coordinate South Africa's efforts in validation measurements made in conjunction with Nimbus 7 overpasses. Their plans have not been finalized and any further information concerning the South African validation efforts should be obtained through Dr. Anderson.

2.8 References and Bibliography

1. Yentsch, C. S., "Limnology and Oceanography," 7,207 (1962).

2. Yentsch, C. S., "Deep-Sea Research," 7.1 (1959).

3. Yentsch, C. S., "The Absorption and Fluorescence Characteristics of Biochemical Substances in Natural Waters," Proc. Symp. on Remote Sensing in Marine Biology and Fisheries, March 1971 (Texas A&M University, College Station, 1971), pp. 75-97.

4. Clarke, G. L., "The Significance of Spectral Changes in Light Scattered by the Sea," Remote Sensing in Ecology (University of Georgia Press, Athens, 1969), Chap. 11.

5. Clarke, G. L., Ewing, G. C. and Lorenzen, C. J., Science 167, 1119 (1970).

6. Hovis, W. A., Forman, M. L. and Blaine, L. R., "Detection of Ocean Color Changes from High Altitude," NASA X-652-73-371, Nov. 1973.

7. Smith, R. C. and Baker, K. S., "The Bio-Optical State of Ocean Waters and Remote Sensing," Scripps Inst. of Oceanography. Ref. 77-2 (1977).

8. Smith, R. C. and Baker, K. S., "Optical Classifications of Natural Waters," Scripps Inst. of Oceanography, Ref. 77-4, (1977).

9. Morel, A., "Analysis of Variation of Ocean Color, Limnology and Oceanography," Vol. 22, No. 4, July 1977.

SECTION 3

THE EARTH RADIATION BUDGET (ERB) EXPERIMENT*

by

H. Jacobowitz and L. L. Stowe
National Environmental Satellite Service
National Oceanic and Atmospheric Administration
Washington, D.C.

and

J. R. Hickey
The Eppley Laboratory
Newport, Rhode Island

3.1 Scientific Objective

The objective of the Earth Radiation Budget (ERB) experiment, a continuation of Nimbus 6 ERB, are 1) to determine over a period of a year, the earth radiation budget on both synoptic and planetary scales by simultaneous measurement of:

- Incoming solar radiation

- Outgoing earth-reflected (shortwave) and earth emitted long wave radiation by:

    a. Fixed wide-angle sampling of these terrestrial fluxes at the satellite altitude.

    b. Scanned narrow-angle sampling of the angular radiance components.

and 2) to develop angular models of the reflection and emission of radiation from clouds and earth surfaces.

Measurements of radiation are obtained in 22 different optical channels. Ten solar channels (labeled 1 through 10c) measure incoming solar radiation. Four earth-looking channels (11 through 14) with fixed wide-angle fields of view measure radiation from the entire earth disc. Eight earth-viewing channels scan from nadir to horizon in several vertical planes with narrow-angle fields of view. Channels (15-18) measure short wavelength radiation while (19-32) measure wavelength radiation. Tables 3-1, 3-2, and 3-3 present the spectral characteristics of solar, wide-angle and narrow-angle channels, respectively.

3.2 Experiment Description

3.2.1 Solar Channels

The ERB experiment measures the incoming solar radiation in ten spectral channels as the satellite orbits over the Antarctic, just before it starts its northward trip on the daylight side of the earth. The spectral intervals monitored by the solar channels are illustrated in Figure 3-1, superimposed on the 1971 standard extraterrestrial NASA curve. These bands were selected to provide measurements

---
*Complete ERB NET membership listed in Appendix B, page B-2.

Table 3-1
Characteristics of ERB Solar Channels

Channel	Sensor (c) Type	Wavelength Limits (μm)	Filter	Solar Irradiance (d) Air Mass Zero (Wm^{-2})	Gain	Noise Equivalent Irradiance (Wm^{-2})
1	N3	0.2 – 3.8	Suprasil W	1370	692.3	1.77 x 10^{-2}
2 (a)	N3	0.2 – 3.8	Suprasil W	1370	685.8	1.77 x 10^{-2}
3	N3	⟨0.2 to⟩ 50	None	1370	607.2	1.43 x 10^{-2}
4	N3	0.526 – 2.8	OG530	970	974.5	1.94 x 10^{-2}
5	N3	0.698 – 2.8	RG695	679	1339.4	1.91 x 10^{-2}
6	N3	0.395 – 0.508	Interference Filter	206	8512.7	3.58 x 10^{-2}
7	N3	0.344 – 0.460	"	166	17964.7	5.73 x 10^{-2}
8	N3	0.300 – 0.410	"	109	26985.3	7.55 x 10^{-2}
9	K2	0.275 – 0.360	"	57	9808.6	0.94 x 10^{-2}
10C (b)	H-F	⟨0.2 to⟩ 50	None	1370	2791.0	2.39 x 10^{-2}

Notes: (a) Channels 1 and 2 are redundant. Channel 1 is normally shuttered and is open periodically to adjust value of Channel 2.
(b) Channel 10C is a self-calibrating cavity channel added to Nimbus 7 and replacing a UV channel on Nimbus 6.
(c) All are types of Eppley wire wound thermopiles.
(d) Values obtained from adjusted Nimbus 6 results.

• The unencumbered FOV for all channels is 10 degrees; the maximum field is 26 degrees for Channels 1 through 8 and 10C. The maximum FOV for Channel 9 is 28 degrees.

Table 3-2
Characteristics of ERB Fixed Wide-Angle FOV Channels

Channel	Wavelength Limits (μm)	Filter	Irradiance Range Anticipated (Wm^{-2})	Approximate Non-Amplified Signal Output (mV)	Amplified Operational Sensitivity (Bits/Wm^{-2})	Noise Equivalent Irradiance (Wm^{-2})
11	<0.2 to >50	None	-200 to +600	-2.1 to 7.6	1.707	6.55 x 10^{-3}
12*	<0.2 to >50	None	-200 to +600	-2.1 to 7.6	1.707	6.55 x 10^{-3}
13	0.2 to 3.8	2 Suprasil W Hemispheres	0 to 450	0 to 5.7	2.276	6.55 x 10^{-3}
14	0.695 to 2.8	RG695 Hemispheres Between 2 Suprasil W Hemispheres	0 to 250	0 to 3.2	4.096	6.65 x 10^{-3}

Notes: *Channels 11 and 12 are redundant channels. Channel 11 has black painted baffles and is used for in-flight calibration of Channel 12. Channel 12 has polished aluminum baffles similar to those on Nimbus 6.
- All channels have type N3 thermopile sensors.
- All channels have an unencumbered FOV of 121 degrees and a maximum FOV of 133.3 degrees. Channel 12 has an additional FOV selection of 89.4 degrees unencumbered, 112.4 degrees maximum.
- Output of these channels is a 3.8 second integral of the instaneous readings.

Table 3-3
Characteristics of ERB Scanning Channels

Channel	Wavelength Limits (μm)	Filter	FOV (Degrees)	Responsivity (V/W RMS/RMS)	Noise Equivalent Radiance (W cm^{-2} sr^{-1})	NEP (W Hz$^{-½}$)
15-18	0.2 to 4.8	Suprasil W	0.25 x 5.12	50	3.7 x 10^{-5}	6.65 x 10^{-9}
19-22	4.5 to 50	Deposited Layers On Diamond Substrate	0.25 x 5.12	50	1.8 x 10^{-5}	1.73 x 10^{-9}

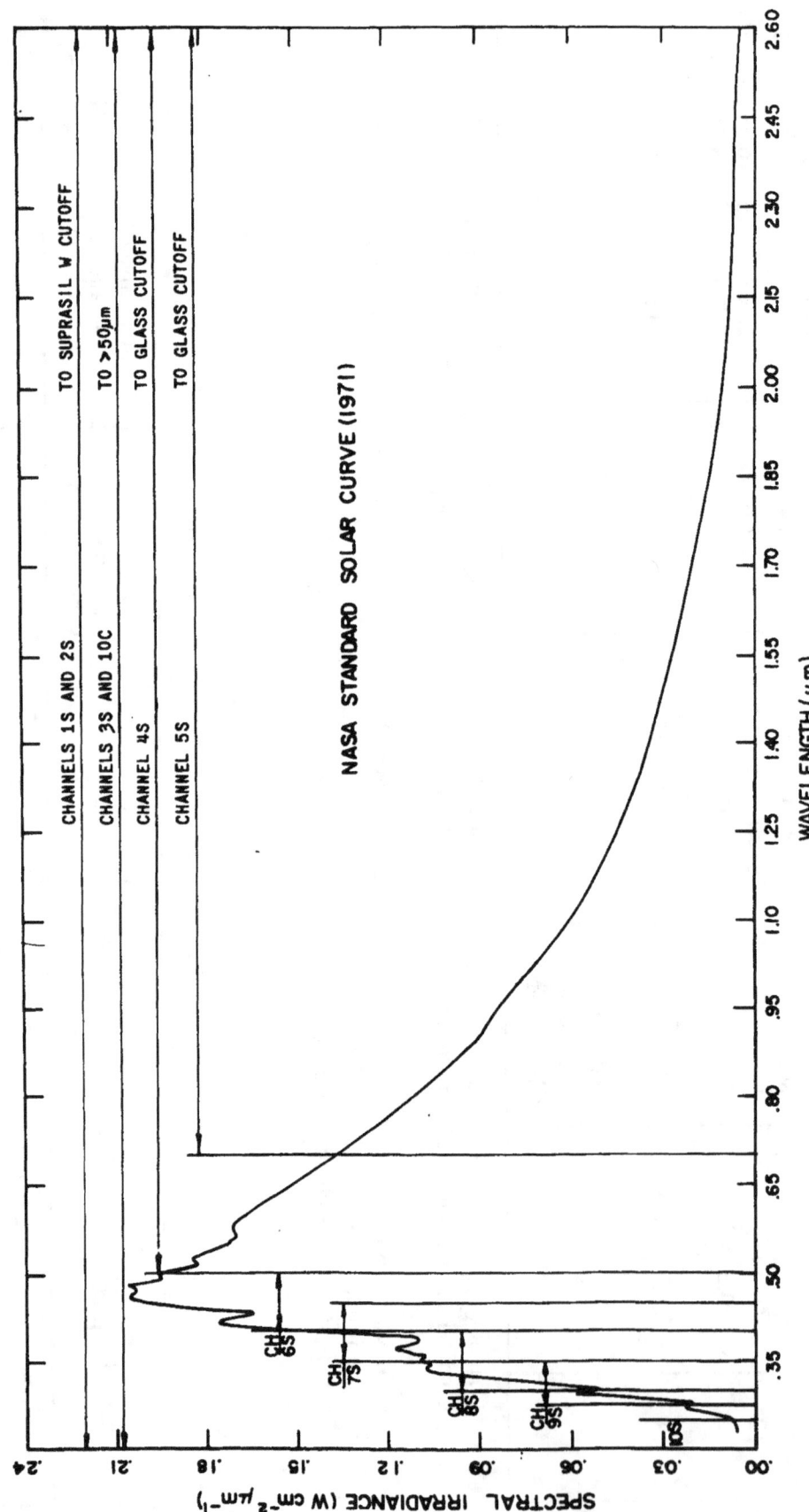

Figure 3-1. Spectral Intervals Monitored by the ERB Solar Channels (With 1971 NASA Standard Extraterrestrial Solar Curve)

of the solar "constant", necessary for earth heat budget computations, and of solar energy in spectral subdivisions in the ultraviolet and visible regions where solar emission variability may occur and where uncertainties exist in present values of the solar emission spectrum. The two "solar constant" channels 3 and 10c measure the entire solar spectrum from <0.2 μm to >50 μm.

3.2.2 Fixed Wide-Angle FOV Channels

Earth-emitted infrared radiation and earth-reflected solar radiation are measured with fixed, wide-angle FOV sensors. The four sensors have unencumbered fields of view of 121 degrees and maximum fields of 133.3 degrees. From the Nimbus 7 orbit altitude of 955 km the earth subtends an angle of 120.8 degrees. This angle is greater than that for the Nimbus 6 ERB because the Nimbus 6 orbit is higher; thereby reducing the angle subtended by the earth. The channel FOV's were not modified for Nimbus 7.

The measurements taken by these channels provide a direct measure of the terrestrial flux passing through a unit area at satellite altitude. An integration of these measurements over the entire globe, together with the solar constant observations, provide a measure of the net radiation balance for the earth-atmosphere system. In principle, the accuracy of this measurement should be compromised only by the diurnal sampling restrictions of the Nimbus sun-synchronous orbit. Measurements of the radiation flux reflected in the shortwave region (0.2 μm to 3.8 μm), in addition to those of the total earth radiation flux (0.2 μm to >50 μm), permit separation of the planetary albedo and long wave flux components of the observed net radiation flux.

An earth flux channel (Channel 14) and a solar flux channel (Channel 5) measure radiation in the 0.698 μm to 2.8 μm interval enabling the planetary albedo to be defined for the spectral subregions $\lambda < 0.695$ μm and $\lambda > 0.695$ μm. These two spectral regions separate the total backscattered radiation into the molecular-plus-aerosol contribution from the aerosol-dominant spectral contribution. This separation is important for assessing the contribution of aerosols to any detectable variations of the earth's planetary albedo.

3.2.3 Narrow-Angle FOV Scanning Channels

The ERB also obtains measurements of the radiance of earth-reflected solar radiation (0.2 μm to 4.8 μm) with Channels 15 through 18 and earth-emitted long wave radiation (5 μm to >50 μm) with Channels 19 through 22. These channels, which have a rectangular IFOV of 0.25 degrees x 5.12 degrees are designed to obtain a large number of angularly independent views of the same geographical area as the Nimbus spacecraft orbits overhead. Characteristic angular distribution models are derived for a variety of reflecting surface conditions from a composite of the scanning channel observations of each area. These models are used with the scanning channel observations to specify radiation budgets on a scale of about 500 km.

3.2.3.1 Scan Geometry and Scan Modes

The basic scan geometry of the ERB is shown in Figure 3-2. The IFOV is stepped at varying rates over each half second measurement to partially maintain a ground resolution of about 150 km from nadir to horizon.

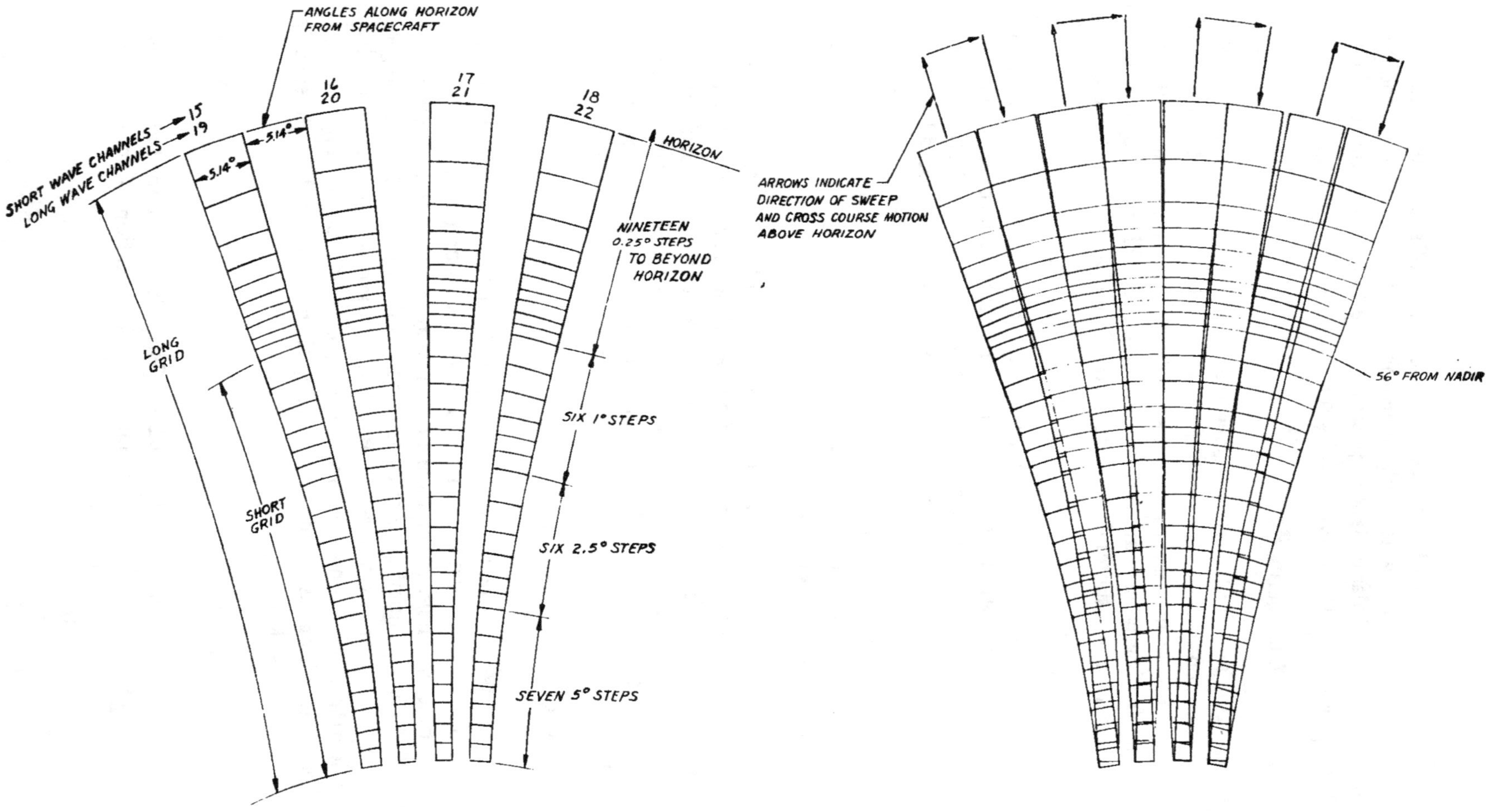

SCAN PATTERN ON EARTH FOR SINGLE SCAN HEAD
SWEEP FROM NADIR TO HORIZON (SPACECRAFT MOTION NEGLECTED).
SHORT GRID LIMITED TO 56° NADIR ANGLE (1250 n mi FROM SSP).
LONG GRID EXTENDS BEYOND HORIZON.
SHORT WAVE AND LONG WAVE CHANNEL FIELDS OF VIEW ARE COINCIDENT.

PATTERN ON EARTH WITH NADIR TO HORIZON AND RETURN SWEEPS SUPERIMPOSED. GIMBAL IS TURNED 6° ABOUT S/C Z AXIS BETWEEN SWEEPS. LONG GRID SHOWN. GIMBAL IS TURNED AT END OF SAMPLE WITH 56° UPPER EDGE DURING SHORT GRID.

Figure 3-2. ERB Scan Grid Earth Patterns

To observe the radiance from various scenes over a wide variety of incident and emerging angles, there are five different scan modes of operation. These routines are schematically illustrated in Figure 3-3. Four scan patterns are a composite of long and short grids shown in Figure 3-2 (a long grid in the forward direction is followed by a short grid in the cross-track direction and then concluded with a long grid in the aft direction). The fifth scan pattern is a composite of scan pattern 3 followed immediately by scan pattern 4. Scan modes 1, 2, 3, and 4 obtain a maximum number of angular independent views of a given geographical area. When the instrument is in one of these four modes of operation, that scan pattern is repeated every 112 seconds or every 700 km along the subpoint track. These four scan modes ensure the ability to obtain numerous observations in the principal plane of the sun, the plane in which the greatest angular variations in reflected sunlight occur. Scan mode 5, which is the normal mode of operation to obtain maximum earth coverage, is repeated every 224 seconds or every 1400 km along the subpoint track.

Figures 3-4 and 3-5 show a complete scan pattern projected on an imaginary sphere coincident with the earth's surface and fixed with respect to the satellite. The solid line with the arrowheads indicates the motion of a point on the earth's surface relative to the imaginary sphere and scan pattern. The small target areas considered for illustration are located at 40°N latitude in Figure 3-4 and

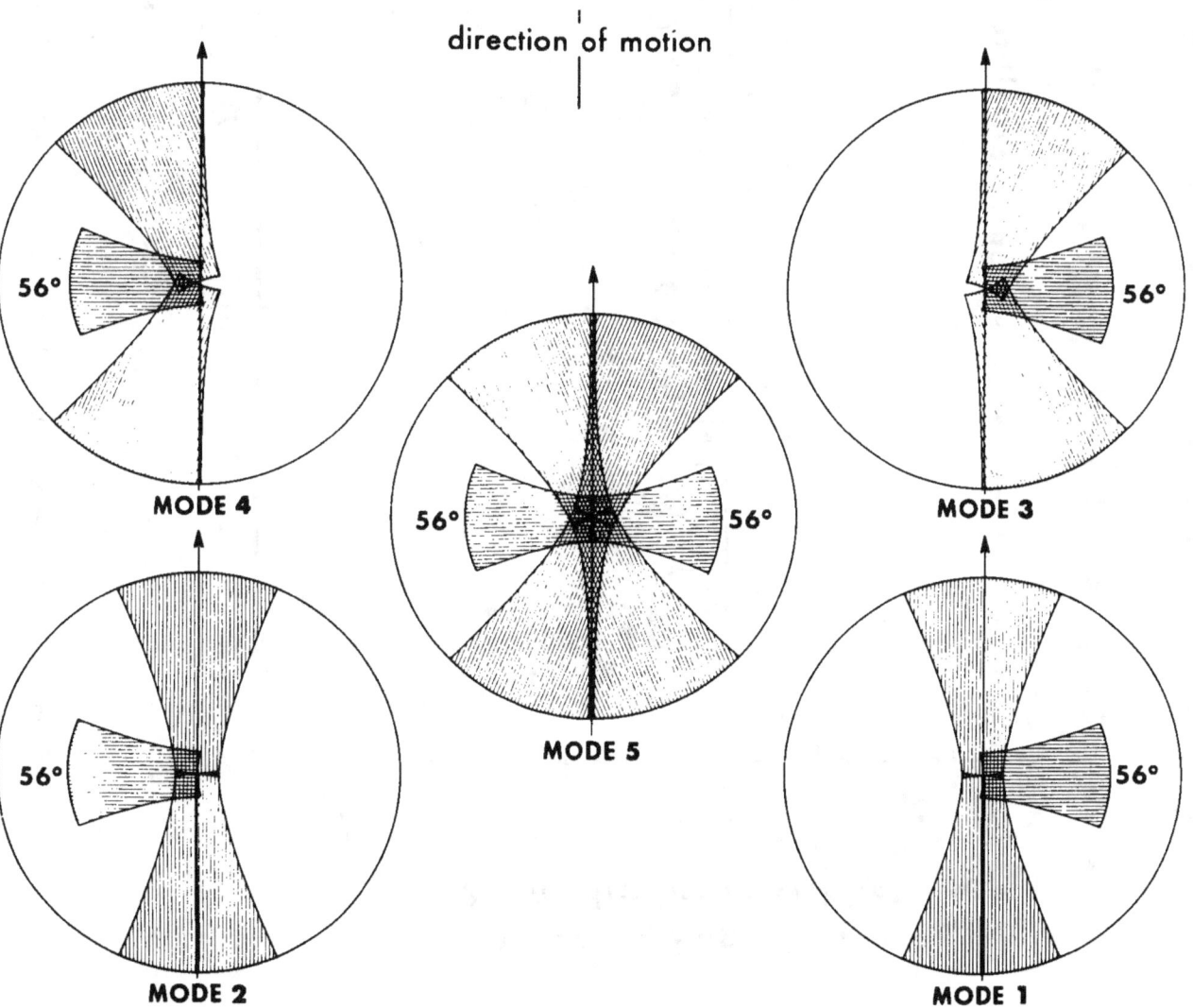

Figure 3-3. ERB Scan Modes

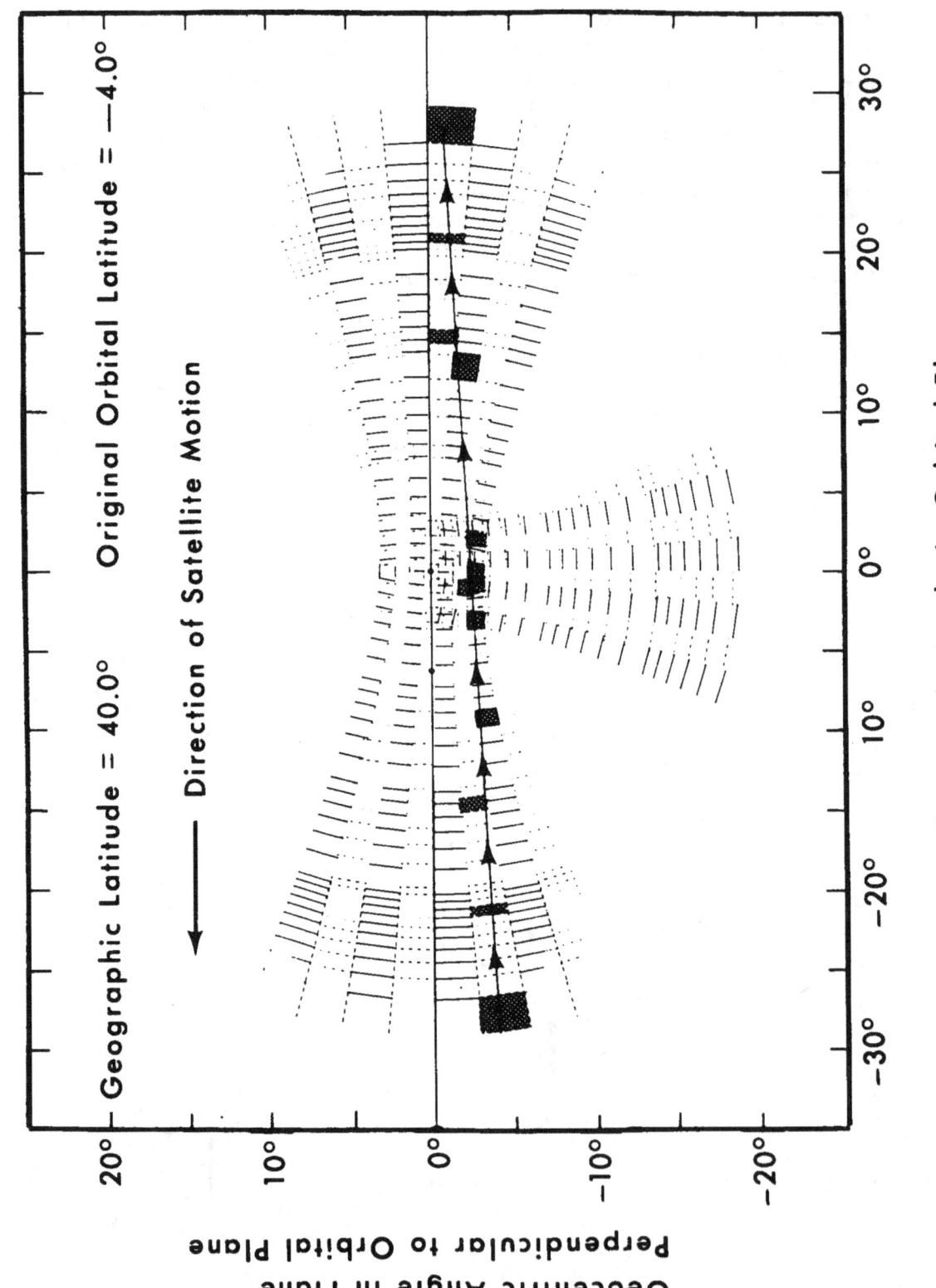

Figure 3-4. ERB Scanning Channel Views of a Geographical Area Near the Subpoint Track

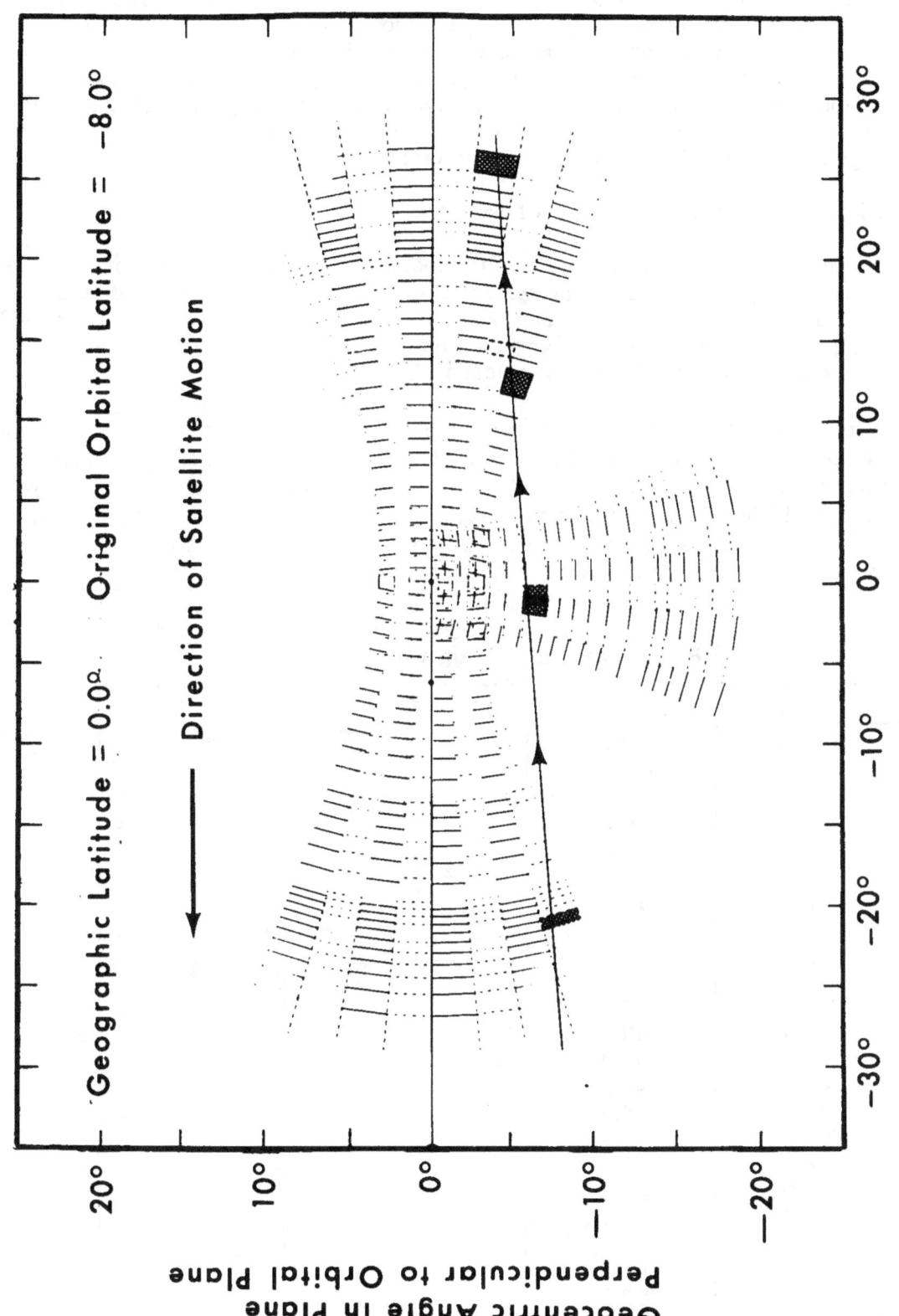

Figure 3-5. ERB Scanning Channel Views of a Geographical Area Away From the Subpoint Track

0° latitude in Figure 3-5. The shaded portions of the scan pattern indicate which FOV's contain the target area. The area is first observed near the forward horizon (in the direction of satellite motion) at a view angle of 58.5 degrees. During succeeding scan patterns, as the satellite approaches the area, the area is viewed at angles of approximately 56, 51, 49, 15, and 0 degrees. As the satellite moves away from the area, radiance observations are made over the other half of the scanning plane at view angles of 15, 40, 51, 47, and 58.5 degrees. Consequently, a fairly complete picture of the angular distribution of radiation emerging from this geographical area in the scanning plane is obtained. Figure 3-5 is presented to illustrate how the side grid scan helps to sense the angular distribution of radiation from geographical areas which are not near the subpoint track.

3.2.3.2 Modeling the Angular Distribution of Radiance

The ERB scanning channel observations are used to model the angular distribution of the radiance reflected and emitted by the earth and atmosphere. The models are required if the directional radiance observations are to be converted into hemispheric fluxes. The scanning portion of the ERB experiment was designed to obtain a number of angularly independent views of the same geographical area so that empirical angular models can be developed.

Initial data processing involves sorting the scanning channel measurements as a function of geographical area. To accomplish this, the earth has been divided into 2070 target areas, each about 500 km square. As a matter of convenience in the data handling, the boundaries of the target areas are chosen to coincide with the latitudes and meridians specified in Table 3-4. Each of these is further subdivided into nine subtarget areas approximately 160 km on a side. Each ERB scanning channel field of view (FOV) is located within one of these subtarget areas.

Table 3-4
ERB Scanning Channel Target Areas

Latitude Limits		Longitude Interval*
Lower Limit	Upper Limit	
0.0	4.5	4.5
4.5	9.0	4.5
9.0	13.5	4.5
13.5	18.0	4.5
18.0	22.5	5.0
22.5	27.0	5.0
27.0	31.5	5.0
31.5	36.0	5.0
36.0	40.5	6.0
40.5	45.0	6.0
45.0	49.5	6.0
49.5	54.0	7.5
54.0	58.5	8.0
58.5	63.0	9.0
63.0	67.5	10.0
67.5	72.0	12.0
72.0	76.5	18.0
76.5	81.0	22.5
81.0	85.5	40.0
85.5	90.0	120.0

*For each latitude band the longitude intervals start at the 0 degree meridian and progress east by the increments listed.

Results of this sorting will be archived on magnetic tape together with the following information concerning the character of each subtarget area: the amount of high, middle, and low clouds; the fraction of land, water, snow, and ice; the topography of the land; and the mean solar zenith and azimuth angles corresponding to the time of observation.

On the basis of this data set, angular distribution models will be developed by grouping systematically the angular data obtained from subtarget areas with similar characteristics over a given time period. Each model will be classified into one of a small number (5 – 10) of categories which sufficiently characterizes the different types of angular distribution found. As a result, one can then define the range of surface type, solar angles, and cloudiness which belong to a given category. Once this is accomplished, the set of angular distributions for each category may then be mathematically described, most efficiently, as a series of several empirical orthogonal functions. Estimation of the flux for any orbital pass over an area with known characteristics then reduces to computing the coefficients of an expansion of the observations in orthogonal functions for the appropriate category and integrating the resulting linear combination of functions over all directions of the upward hemisphere.

3.3 Instrument Description

3.3.1 Physical Layout

The ERB instrument consists of one radiometer unit, as shown in Figure 3-6 with the approximate dimensions of 33 cm x 36 cm x 48 cm and weight of 32.7 kg. All ERB electronics and optics are located within the radiometer unit. The rotatable solar channel assembly is located on the +X surface, facing forward in the direction of spacecraft motion. This assembly can be rotated 20 degrees in 1 degree steps to either side of the spacecraft x-axis in order to acquire an on-axis view of the sun under the expected variation of the satellite orbit plane with respect to the sun. The scanning channel assembly with the gimbal-mounted cylindrical scan head, and the fixed wide-angle earth-flux channels are located on the +Z (nadir) face of the instrument. Adjacent to the solar channel assembly on the +X surface is the diffuser-plate target for in-flight checks on the calibration of the shortwave narrow-angle scanning channels. An earth shine shield has been added to this target on the Nimbus 7 ERB.

3.3.2 Solar Channels

Each of the ten solar channels is an independent, individually replaceable modular element with a mated amplifier as part of the unit. The sensors are advanced versions of wirewound-type thermopiles. There are no imaging optics in the solar channels; only filters, windows, and apertures. No optical amplification is required to maintain high signal-to-noise ratios because the thermopile sensitivities are high and state-of-the-art electronics are used. The spectral intervals of the solar channels have been illustrated in Figure 3-1. Channels 1 and 2 are duplicated, Channel 1 being the reference for Channel 2 for the in-flight calibration program. Channel 1 is normally shuttered. Channels 4 and 5 contain broad bandpass filters with transmittance spectra matching those of the standard Schott glasses, OG530 and RG695, of the World Meteorological Organization. (The RG695 glass is also used in Channel 14, one of the shortwave fixed earth-flux channels.) The filters are protected against particle radiation by 4 mm thick windows or hemispheres of fused silica. The interference filters are deposited on Suprasil W (grade III) fused silica substrates to minimize degradation. The transmittance of a 2 mm thick piece of Suprasil W from 0.2 μm to 5 μm is shown in Figure 3-7. Blocking outside the primary transmission bands, is achieved by interference layers only. No absorbing glasses

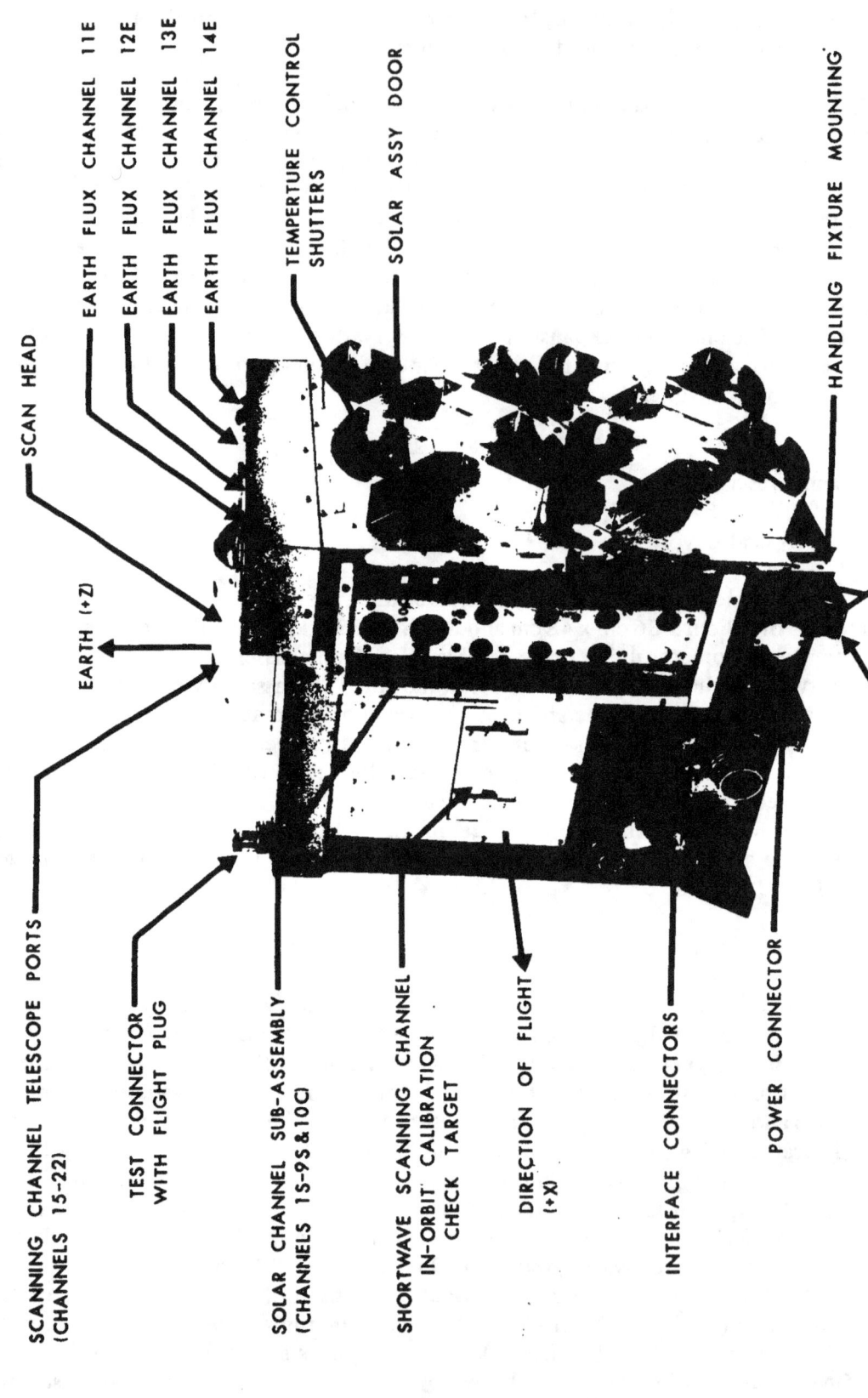

Figure 3-6. ERB Radiometer Unit

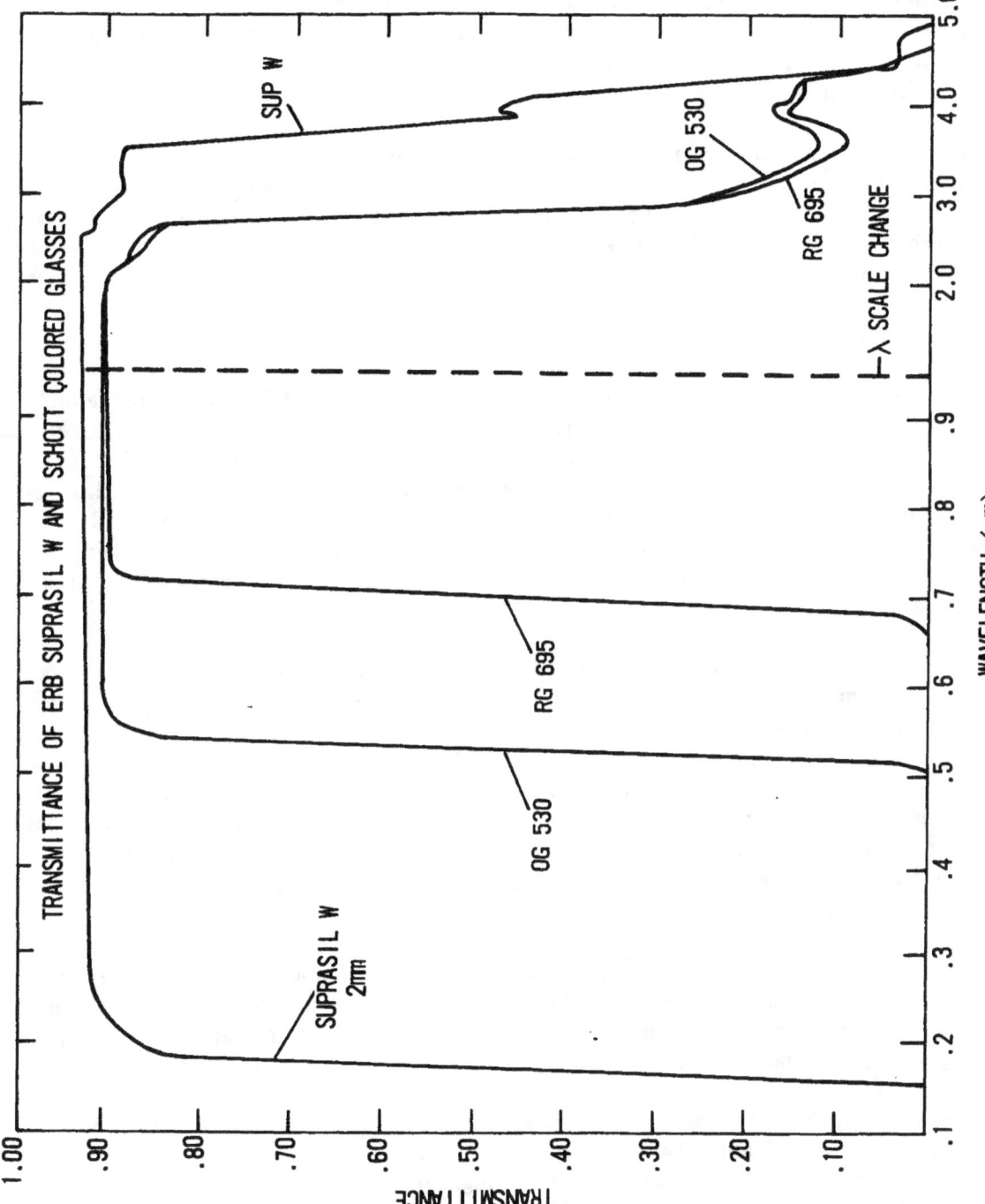

Figure 3-7. Transmittance of ERB Suprasil W and Schott Colored Glasses

are used. The radiation in the 0.2 μm to 0.526 μm, 0.526 μm to 0.695 μm, and 0.20 μm to 0.695 μm is obtained by differential treatment of the channel 4 and 5 data, together with readings obtained from Channel 2. Table 3-1 lists the solar irradiances at the receiver, the FOV, and the measurement type for each channel. Channels 1 through 8 have type N3 thermopiles; Channel 9 has type K2. Channel 10C has a modified model H-F self calibrating cavity element. The cavity is mounted to a circular wirewound thermopile. The electrical heater used for self calibration is energized when the "GO/NO GO" heater command is issued. The thermopile output and the heater voltage and current are then sub-multiplexed into the Channel 10C data stream.

As the satellite comes over the Antarctic region each orbit, the sun is viewed within the unencumbered field for about three minutes. The unencumbered field is that for which the entire sun is contained in the receiver FOV. The solar channels are monitored before and after solar acquisition in order to obtain the space radiation reference (or "zero-level" response). The outputs of the solar channels are sampled once per second.

3.3.3 Fixed Wide-Angle FOV Channels

The fixed wide-angle FOV channels are numbered 11 through 14. The FOV of each channel encompasses the entire earth surface visible each moment from the Nimbus orbit. To allow for the possibility of a small angular misalignment of these channels with respect to nadir, the FOV acceptance angle is slightly larger than that required to view the earth disc. In addition, Channel 12 has an insertable stop so, upon command, it can view slightly less than the entire earth surface. On Nimbus 6 the sun would impinge on these channels twice per orbit at the day/night and night/day transitions at an angle of about 58.5 degrees. This will not happen for Nimbus 7 unless the spacecraft is pitched up or down, beyond normal operating limits.

Channel 11 is a duplicate of 12 and is used only occasionally as a calibration check of Channel 12. For Nimbus 7 the Channel 11 baffles have been painted black in order to investigate a "space loading" sensor radiation induced signal offset. These two channels, with no filters or windows, measure the absolute irradiance over the band from 0.2 μm to 50 μm. The earthward-facing surfaces of these channels are highly polished. Each employs a type N3 thermopile with a circular receiver. These are fabricated from aluminized kapton on which a circle of cured Chemglaze Z-306 paint has been applied and then overcoated with 3M-type 401-C10 black velvet optical paint. Channel 11 shares the "reference channel open" command with solar Channel 1 and is opened periodically, only as directed in the in-flight calibration and data validation procedures.

Channel 13, the shortwave (0.2 μm to 3.8 μm) fixed earth-flux channel, is equipped with two hemispheres of Suprasil W (grade III) fused silica. The spectral band matches that of solar Channels 1 and 2. The difference in measured radiation between Channel 11 (or 12 with full field) and Channel 13 is the long wave terrestrial component. Channel 13 is similar to a precision pyranometer.

Channel 14 has a broadband (RG695) filter hemisphere, to match the band of Channel 5. The RG695 hemisphere (partial) of Channel 14 is between two Suprasil W, fused silica hemispheres. The outer one is thick to attenuate particle radiation which might damage the glass. The inner hemisphere is the characteristic IR blocker included in all precision pyranometers. The use of RG695 as a separator of the shortwave irradiance about its cutoff wavelength of approximately 0.7 μm is a common practice in albedo measurements. After proper correction of the measured irradiance values of Channels 13 and 14, the irradiance in the band between 0.20 μm and 0.7 μm is determined. Thus, the primarily scattering and primarily absorbing regions of the shortwave reflected radiation are independently assessed. Table 3-2 lists characteristics of these channels.

3.3.4 Thermopiles in Channels 1 through 14

The detectors for all the solar and fixed wide-angle FOV channels are improved versions of the wirewound-type thermopiles employed in the Eppley -JPL radiometers. Type N3 is used for Channels 1 through 8, and 11 through 14. Those in the earth channels (11 through 14) have a circular blackened area on the receiver, while those in the solar channels 1 through 8 are blackened over the entire square. Channel 9 has a type K2 thermopile. This is larger than the N3 type, but of similar construction. Channel 10C has a circular wirewound model H-F thermopile. The thermopiles are constructed to react to a conductive thermal transient in such a way that both active and reference receivers will respond simultaneously and equally to the temperature offset, thus cancelling any offset in the output signal. Also, the time constants of the actual active and reference couples are matched by position control during the plating operation. The receivers are matched coated and mounted in a manner which assures time constants near balance. Time constants are approximately four seconds for vacuum operation. The modification of the cavity thermopile is that its cold junctions have a flat rather than a cavity receiver. There was not enough room to fit the rear cavity in the volume available for the retrofit. The balance requirement is extremely important for the solar channels because the measurements are made during a thermal period when the satellite crosses the terminator and its front face takes the full heat of the sun. This is very critical for channel 9 (and channel 10 on Nimbus 6) which have K2 thermopiles because of the low radiant input in the ultraviolet bands as opposed to the full solar fulx and because of the required high sensitivity. For this reason K2 sensors must be balanced to within 15 percent to meet the measurement accuracy requirements.

3.3.5 Interference Filters in Channels 1 through 14

Since ideal square wave response characteristics cannot generally be realized, filter factors are calculated to specify the transmittance characteristics of most filters. The filter factor which is the reciprocal of the effective transmittance should exhibit a variability of less than one percent when calculated for all possible extraterrestrial solar curves and for the solar simulator employed during testing. The wavelength limits for the solar channels are given in Table 3-1. The filter factors are given in Table 3-7. Figure 3-8 illustrates the spectral transmittance functions for the broadband filters. Figures 3-8 through 3-11 show the transmittance curves (which correspond to the channel's relative spectral response) for the interference filter on Channels 6 through 9.

3.3.6 Narrow-Angle FOV Scanning Channels

The cylindrical scan head contains four telescopes aligned such that the telescope center lines are 12 degrees apart when projected onto the horizontal plane. The telescopes contain off-axis mirror objectives which are deployed to accept a common chopper interrupting the beams immediately ahead of the field stops. In addition to its time-sharing beam splitter function, the chopper also separates the beam energy into the long wavelength (4.5 μm to 50+ μm) and the short wavelength (0.2 μm to 4.8 μm) relays. Each relay consists of a field stop, spectral filter, cross-axis tilted relay mirror, referred aperture stop, and detector. With the off-axis mirror objective and cross-axis tilted relay mirror tilted equally with respect to the optical axis, the effects of polarization induced by reflection are minimized. The detector is a pyroelectric element immune to solar exposure and intentionally defocussed to provide uniformity in field response.

Table 3-5
Characteristics of the Fixed Channels

Channel No.	λ1 (μm)	λ2 (μm)	FF (Flight)	Filter Factor* (calculation & calibration)	Preliminary Sensitivity for flight data reduction (counts/Wm^{-2})
1	0.18	3.8	None	1.156 & None	1.300
2	0.18	3.8	None	1.156 & None	1.272
3	0.2	50	None	None	1.216
4	0.526	2.8	None	1.255 & 1.087	1.720
5	0.698	2.8	None	1.255 & 1.087	2.424
6	0.395	0.508	None	2.913 & 2.563	6.935
7	0.344	0.460	None	4.642 & 4.452	9.951
8	0.300	0.410	None	5.252 & 5.265	12.72
9	0.275	0.360	None	7.720 & 9.492	30.18
10C	0.2	50	None	None	1.313**
11	0.2	50	None	None	1.495
12	0.2	50	None	None	1.721/1.709 (Wide/Narrow)
13	0.18	3.8	None	1.156 & None	1.939
14	0.698	2.8	None	1.255 & 1.087	4.122

*Filter factors used in calculation of desired gain include effects of fused silica IR blocker. Filter factors for calibration are those of the NIP filter wheel.

**Channel 10C is self calibrating; this is a baseline sensitivity which will be changed as in-flight calibrations are performed.

Note: Sensitivity values will be updated.

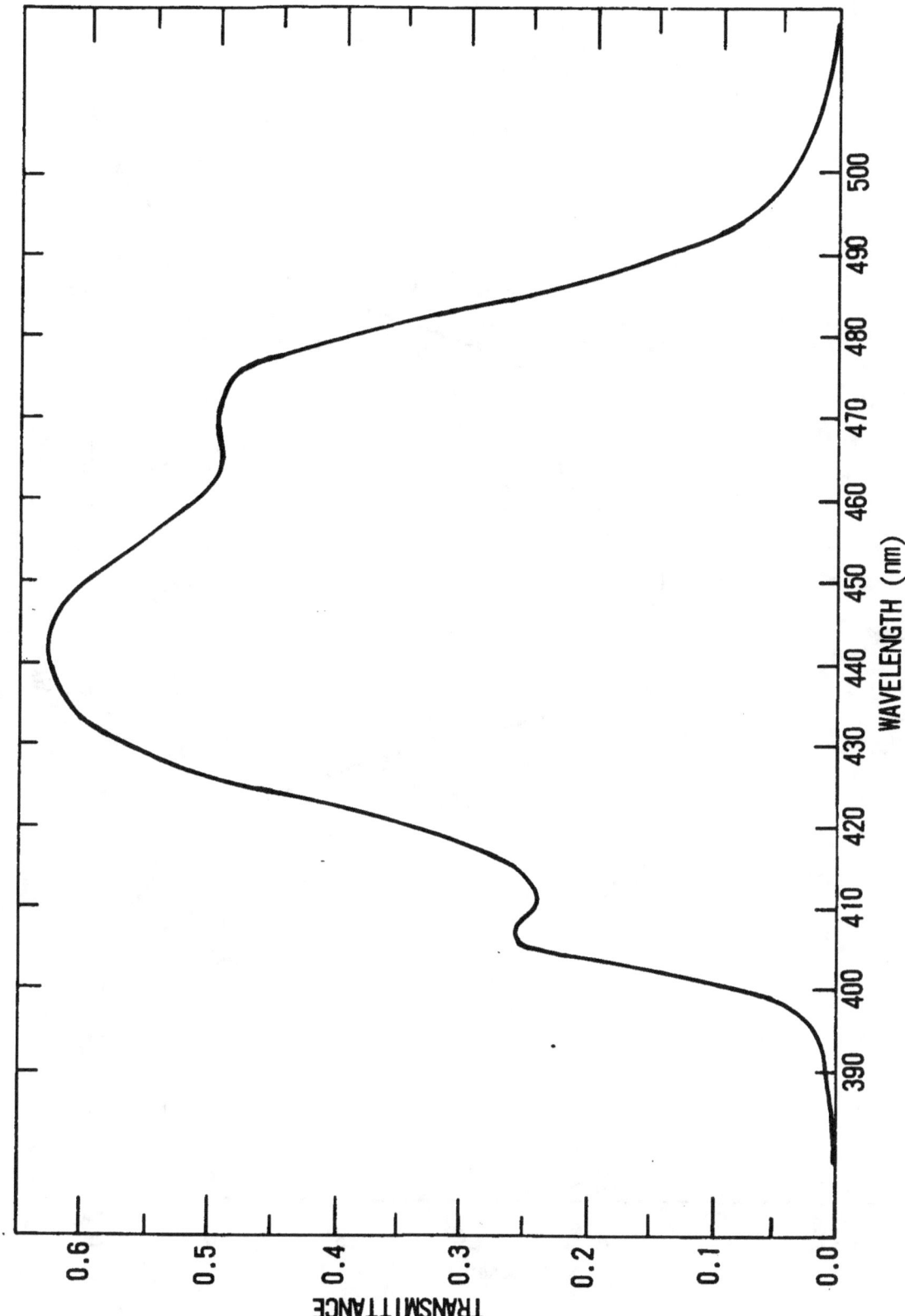

Figure 3-8. Channel 6S ERB Flight Model

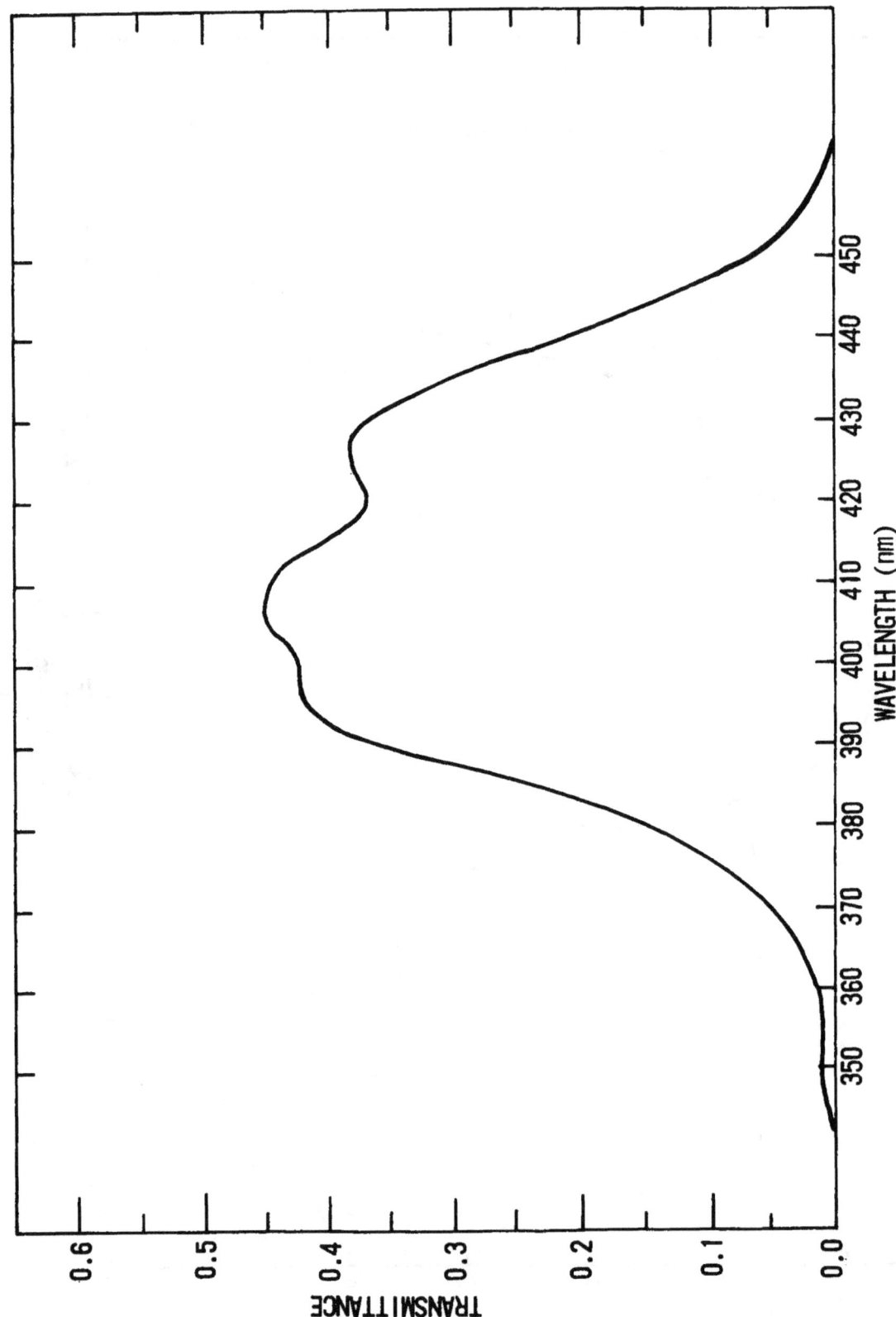

Figure 3-9. Channel 7S ERB Flight Model

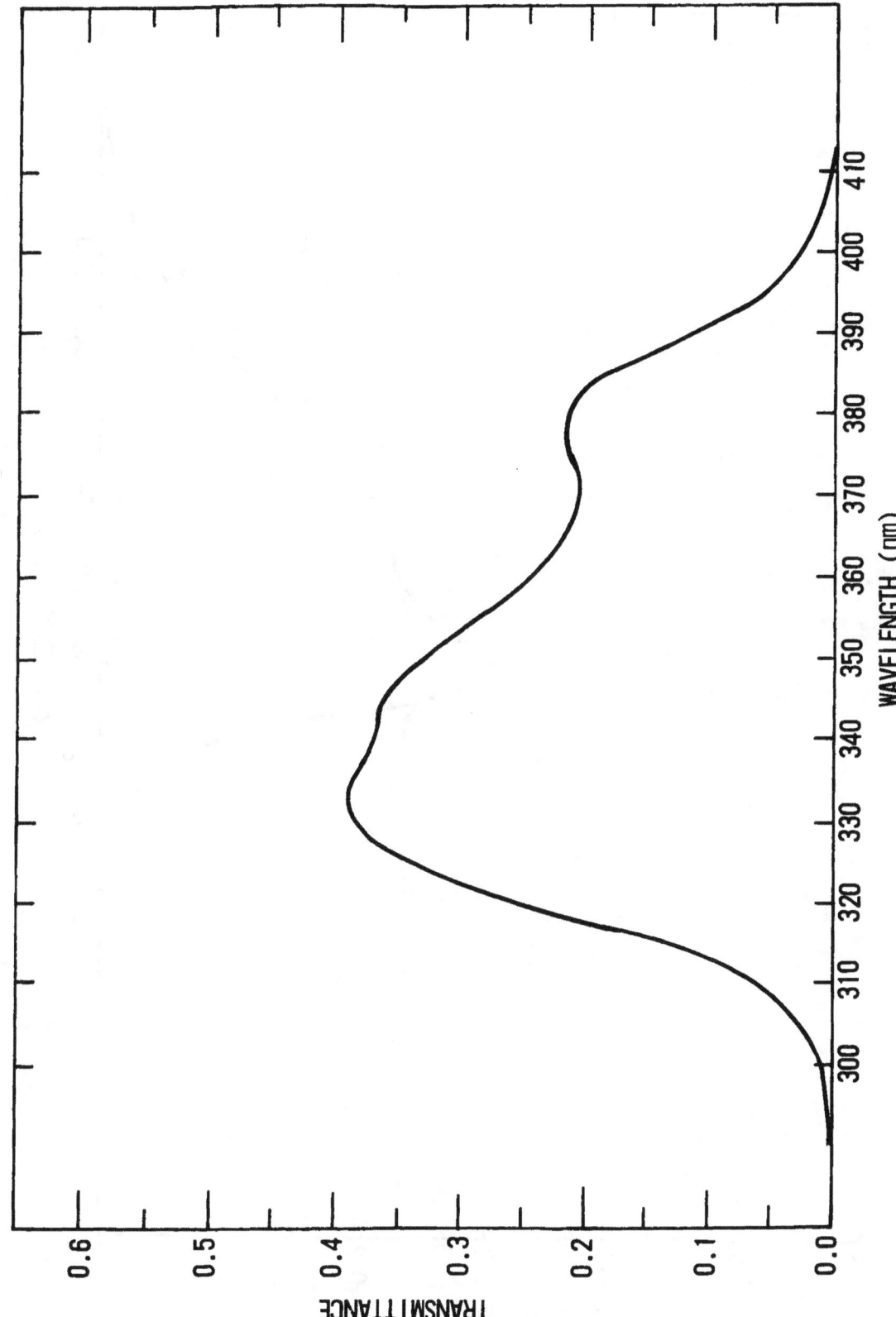

Figure 3-10. Channel 8S ERB Flight Model

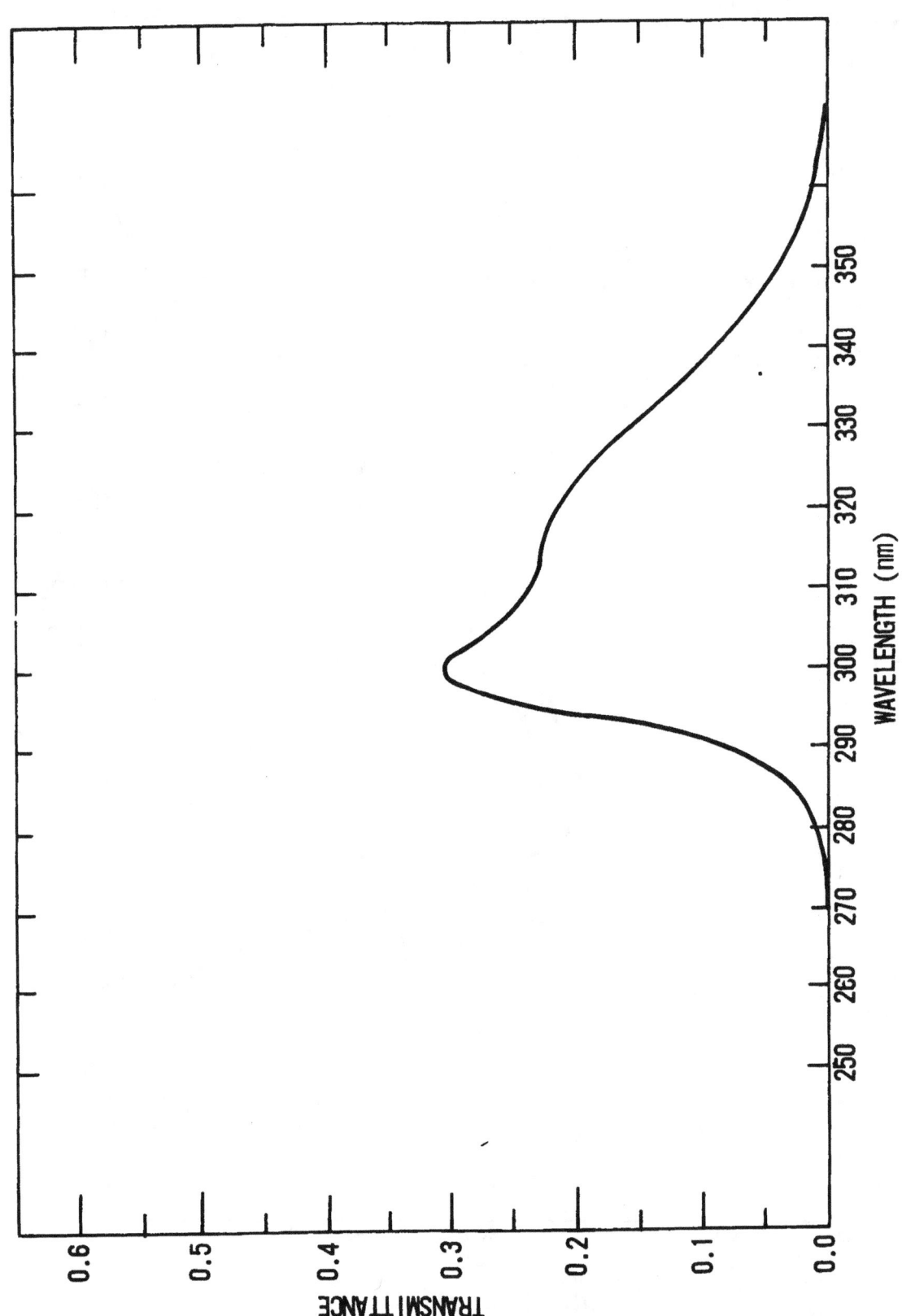

Figure 3-11. Channel 9S ERB Flight Model

The scan head is on a gimbal mounted on the main frame of the radiometer unit. The gimbal arrangement allows the pointing direction of the scan head to be varied within a vertical plane by rotation of the scan head, and within a horizontal plane by rotation of the gimbal. The pointing direction of the scan head is specified by two angles, a and β. The angle a, in the scan plane, is measured between the nadir ($a = 0°$) direction and the curve connecting the upper edges of the FOV's. The angle β, in the plane perpendicular to the scan plane, is measured between the orbital plane and the scan plane. Thus, the scan plane is the XZ plane when $\beta = 0°$. The vertical motion is accomplished with a stepper drive which rotates the scan head in steps of 0.25 degrees. The gimbal rotation is driven by a stepper motor which rotates the gimbal in steps of 0.5 degrees.

The FOV's of the four telescopes are rectangular, 0.25 degrees x 5.12 degrees, and are arranged so that with $a = 60.6$ degrees the upper corners of the FOV's lie along the earth's horizon as shown in Figure 3-2. The narrow-angle (0.25 degrees) side of the FOV is in the direction of vertical (a) motion. The FOV's of the short wavelength channels (15 through 18) are coincident, respectively, with those of the long wavelength channels (19 through 22).

The five scan modes described in Figure 3-3 are based on four gimbal position sequences and one scan head rotation sequence. These sequences are defined in Table 3-6 in terms of VIP major frame count from the start of the basic scan routine, the encoder position, and the angle of rotation. Note that many of the channel output samples result from integration over several (4, 10, or 20) fields of view. The scan head rotation speed is varied to accomplish this so that the sampling frequency of two samples per second is constant.

The scan head may also be commanded to check and calibration positions called "space-look", "longwave check" and "shortwave check".

3.4 Pre-launch Calibration

3.4.1 Solar Channels

The prelaunch calibration for the solar channels consists of a number of absolute intercomparisons and transfer operations. The reference for the absolute calibrations is the new World Radiometric Reference (WRR) scale which is embodied in a number of self-calibrating cavity radiometers. Channel 10C of the Nimbus 7 ERB is itself such a device. This new scale can be referenced to previous scales such as the International Pyrheliometric Scale (IPS 1956).

The four major solar channels (1, 2, 3 and 10C) have been directly intercompared with self calibrating cavity instruments of both the JPL-PACRAD and Eppley model H-F types. The PACRAD employed in this program has been an Eppley manufactured version. The serial number is 11402. This unit has been involved in a number of intercomparisons, including International Pyrheliometric Intercomparison IV (IPCIV).

For transfer operations usually employing a solar simulator as source normal incidence pyrheliometers (NIP) of the ERB reference set are employed. The two devices used for the solar channel intercomparisons bear serial numbers 12016E6 and 12018E6. Both of these are also traceable to the WRR.

When calibrating the filtered channels (4, 5, 6, 7, 8 and 9) the NIP is fitted with a filter wheel containing matching filters to the flight set. The incident irradiance is calculated using the measured irradiance and the appropriate filter factor for the particular filter.

Table 3-6
ERB Scan Head Positions During Scan

A. Alpha Encoder Positions*

Major Frame	Encoder Position
1	0(20)140, 150(10)200, 204(4)224, 225(1)236
2	237(1)243, 3*263, 2*243, 242(-1)224, 220
3	216(-4)200, 190(-10)140, 120(-20)20, 8*0, 20(20)140
4	150(10)200, 204(4)224, 224, 224(-4)200, 190(-10)140, 120(-20)20
5	7*0, 20(20)140, 150(10)200, 204(4)224, 225(1)230
6	231(1)243, 3*263, 2*243, 242(-1)229
7	228(-1)224, 220(-4)200, 190(-10)140, 120(-20)20, 9*0

*Alpha encoder positions (m) represent the number of steps (1 step = 0.25° of rotation) from the reference position (0°) so that the angle a is given (in degrees) by $a = 0.25 \cdot m$ (degrees).

- This sequence of a angles (encoder positions) is the same for all scan routines.

- The notation $n_1 (n_2) n_3$ is used for an arithmetic progression n_1, $n_1 + n_2$, $n_1 + 2 \cdot n_2$, ..., n_3. For example 0(20)140 can be written 0, 20, 40, 60, ..., 140, and 220(-4)200 is the same as 220, 216, 212, ..., 200.

- The notation $n_1 * n_2$ means n_1 occurrences of the value n_2.

B. Beta Encoder Positions[+]

Major Frame	Scan Mode	Encoder Positions	Major Frame	Scan Mode	Encoder Positions
1	1	16*512	1	3	16*557
2	1	4*512, M, 11*524	2	3	4*557, M, 11*569
3	1	9*524, 3*M, 4*692	3	3	9*569, 3*M, 4*692
4	1	6*692, M, 9*704	4	3	6*692, M, 9*704
5	1	3*M, 13*872	5	3	3*M, 13*827
6	1	7*872, M, 8*884	6	3	7*827, M, 8*839
7	1	12*884, 4*M	7	3	12*839, 4*M

Table 3-6 (Continued)

Major Frame	Scan Mode	Encoder Positions	Major Frame	Scan Mode	Encoder Positions
1	2	16*524	1	4	16*479
2	2	4*524, M, 11*512	2	4	4*479, M, 11*467
3	2	9*512, 3*M, 4*344	3	4	9*467, 3*M, 4*344
4	2	6*344, M, 9*332	4	4	6*344, M, 9*332
5	2	3*M, 13*164	5	4	3*M, 13*209
6	2	7*164, M, 8*152	6	4	7*209, M, 8*197
7	2	12*152, 4*M	7	4	12*197, 4*M

+ Beta encoder positions (n) represent the number of steps (1 step = 0.5° of rotation) from the reference position (n_o = 518) so that the angle β is given (in degrees) by $\beta = 0.5 \ (n - n_o)$

• The letter M in the table means that the scan head is in motion so that the encoder readout may not be repeated exactly at those times.

The ERB reference sensor model (RSM), which is a duplicate of the flight instruments relative to the solar channels, has been employed as a transfer and checking device throughout the Nimbus 6 and Nimbus 7 calibration programs. This device is being maintained in order to trace calibrations as required. All vacuum calibrations of the Nimbus 6 and 7 ERB solar channels are referencable through the RSM as well as many of the calibrations performed at atmospheric pressure.

It is reiterated here that Channel 10C of ERB on Nimbus 7 is a self calibrating cavity device which does not rely on transfer calibration. There is no equivalent to channel 10C in the RSM.

The solar channels are not calibrated during thermal vacuum testing of the spacecraft. Their calibrations are checked during an ambient test after the thermal vacuum testing. Final calibration values for the solar channels are expressed in units of Counts/Watt/meter^{-2} (C/Wm^{-2}) relating the on-sun signal output to the incident extraterrestrial solar irradiance in the pertinent spectral band of the channel.

3.4.2 Fixed Wide-Angle FOV Channels

There are longwave and shortwave calibrations of Channels 11 and 12. The longwave calibrations are performed during thermal vacuum testing with a special blackbody source named the total earth-flux channel blackbody (TECB). The source is a double cavity blackbody unit designed for calibrating Channels 11 and 12 after they are mounted on the ERB radiometer unit. It operates over a temperature range of 180 K to 390 K with an apparent emissivity under test conditions in vacuum of 0.995 or greater. Temperatures are measured and controlled to an accuracy of 0.1°C during these calibrations. These calibrations are performed during both Instrument and Spacecraft level testing. The entire FOV of the channels is filled by the TECB including the annular ring which normally views space in the angular element between the unencumbered and maximum FOV's. For Nimbus 7 Channel 12 was also calibrated for shortwave response by normal incident irradiation by the solar simulator while the instrument was in vacuum. The reference NIP was employed as the transfer standard during this calibration.

Channels 13 and 14 are calibrated for response within their respective spectral bands only. These tests were performed in the same manner as the shortwave calibration of Channel 12. For Channel 14 the reference NIP is fitted with a matching RG695 filter (as for Channel 5) in order to isolate the radiation to its proper spectral band.

An angular response scan is performed on each wide FOV channel in order to relate the normal incidence calibrations described above to the overall angular response of the channels.

3.4.3 Shortwave Scan Channels

These channels are calibrated for radiance response by viewing a diffuse target. Three methods have been employed. These are: the viewing of a smoked magnesium oxide (or Barium sulphate) plate which is irradiated by the solar simulator, exposure in a diffuse hemisphere illuminated internally by tungsten lamps, and viewing a diffusing sphere from outside. The last method employs the "Hovis Sphere" as source. For methods 1 and 3 the reference instrument is a high sensitivity NIP calibrated in terms of radiance. The second method employs a pyranometer as reference instrument. Differences in the results obtained by the various methods are still under investigation. The sensitivity values selected for use are an average of methods 1 and 3. Unfortunately these tests can only be performed at atmospheric pressure while the scan channel performance is superior in vacuum.

Another calibration of these channels is the in-flight check target. With the channels in the shortwave check position (viewing the scan target) the instrument is irradiated by the solar simulator beam. This test is performed at normal incidence when the instrument is in vacuum. The reference is one of the reference NIP's. In air the instrument is similarly calibrated at a number of angles both in elevation and azimuth to obtain the angular characteristics necessary for the reduction of in-flight shortwave check operations.

3.4.4 Longwave Scan Channels

These channels are calibrated in vacuum at both the instrument and spacecraft level thermal vacuum tests. The sensors view a special blackbody source called the longwave scanning channel blackbody (LWSCB) which has a separate cavity source for each channel. The procedure is straightforward and covers a range of temperatures covering the complete range of in-flight measurement possibilities.

3.5 In-flight Calibration

In-flight calibration of the main solar channels may be referred to Channel 10C, a self-calibration channel using the cavity heater activated by the GO/NO GO heater command. In addition the degradation of Channel 2 is checked by the occasional exposure of its duplicate Channel 1. Channels with filters do not have a direct method of optically checking their calibration but must rely on whatever correlations are made with the main channels.

All the thermopile channels are equipped with a GO/NO GO heater. It is used to check for response during prelaunch activities to assure that the channels are functioning. The heater can be used in flight as a rough check for all except Channel 10C. Also, Channels 1 through 14 are equipped with an electrical calibration which inserts a precision voltage at the input to the entire signal conditioning stream. While the electronic calibration cannot be used to infer sensor or optics changes, it insures prevention of misinterpretation of an electronic problem.

Channel 12 relies on the stability of the normally shuttered matching Channel 11. Channels 13 and 14 have no inherent in-flight calibration capability. They rely on occasional looks at the sun during the spacecraft transitions to aid in assessment of drift or degradation. For this mission, a spacecraft pitch maneuver is required. For Nimbus 6, this calibration was available for each orbit.

Channels 15 through 18 are checked using the shortwave scan channel check target as previously described. They also view space as a "zero radiation" reference to evaluate offset.

The long wave scan channels can view space or the on board reference blackbody. They share the only true in-flight calibration capability with Channel 10C.

Additional checks are possible by comparing the wide FOV channels with the integral values obtained by the corresponding scan channels for appropriate scenes.

3.6 Data Processing, Products, and Availability

3.6.1 Data Processing

The raw ERB telemetry data and location information are processed onto magnetic tapes by the MetOCC at GSFC and sent to SACC. SACC generates tapes described in Section 3.6.2. IPD uses the MATRIX and TABLES tapes to make the 16 mm microfilm displays. Duplicate copies of the MAT, SEFDT, and ZMT are sent to NET members requesting these tapes. All original tapes and microfilm are then sent to NSSDC where tape copies are made available to users.

3.6.2 Tape Products

The following tapes are produced by SACC, used by IPD, and then sent to the NSSDC for archiving. Brief descriptions of these tapes are as follows:

- MAT (Master Archiving Tape)

  Contains calibrated and raw digital data values for all channels, plus values of temperature monitoring, orbit, attitude, and DSAS data.

- MATRIX (Mapped Data Matrix Tape)

  Contains daily, six day, monthly, and three month world grids of data, and polar stereographic map matrices of derived parameters. (This tape is utilized in generating the contoured map microfilm displays.)

- SEFDT (Solar and Earth Flux Data Tape)

  Contains up to 30 days of solar data (Channels 1 through 10) and earth flux data (Channels 11 through 14) stripped from the MAT.

- TABLES (ERB tables for microfilm display)

  Contains data for production of all ERB tables on microfilm listed in Table 3-7.

- ZMT (Zonal Means Tape)

  Contains in computer-compatible format the tabular listings of solar irradiance, zonally averaged insolation, logitudinal and latitudinal averages of earth flux, albedo, and net radiation.

The form and content of each of these tapes is specified in a tape specification document for each tape type.

The appropriate documents will accompany a tape shipment to a user. See Section 1.5 of this document for details.

3.6.3 Display Products

There are 26 different types of ERB map products and 11 different types of tabular listings produced on 16 mm microfilm. Table 3-7 lists the titles of all these displays and the film specification number pertaining to each title. Interpreting the fourth digit from the left in each film specification number gives the frequency of production of each display.

        XXX1XX = produced every day;
        XXX4XX = produced every six days,
        XXX7XX = produced every month, and
        XXX8XX = produced every three months.

All ERB map displays have the same format. Figure 3-12 is an example. The only differences are in data content within the maps (the display title identifies the data content), and in the time span of the display.

One northern hemisphere and one southern hemisphere polar stereographic map are at the top with a Mercator map immediately below. The Mercator map (with an equatorial scale equal to the equatorial scale of the polar maps) provides overlapping (and redundant) coverage between the two polar maps. Contouring information for all three maps is beneath the Mercator map. The "on-off-cycle" scale identifies the days of data contributing to the contoured values. If a day is "filled in" the instrument was on, data was collected, interpreted, and used in the contouring. If a day is not filled in, the instrument was either off or the data was unuseable for some reason. Title and reference information at the bottom are mostly self-explanatory. The right half of the last line, however, requires explanation. These items are: the physical tape number the data is stored on (TXXXXX), the algorithm reference number used in processing the data (ALGO XXX), the film specification number (F133XXX), the project data format code (AA), and the film frame number (XXXXXX).

Examples of the ERB table displays are given in Figures 3-13 through 3-19. Examples of the NFOV observations (last three titles in Table 3-6) are not shown as they are identical to the WFOV observations displays. Interpretation of the last line on the table displays is the same as for the maps. The only difference is that the PDFC is AB for ERB tables.

3.6.4 Data Availability

The ERB experimental data consisting of the magnetic tapes described in Section 3.6.2 and the 16 mm microfilm maps and tables listed and illustrated in Section 3.6.3 are archived at the NSSDC. These data will be available to the archive center three to six months after launch. Users requesting ERB data should read Section 1.5 of this document for general tape and film ordering information.

Table 3-7
ERB Microfilm Products Titles and Specification Numbers

Film Spec Number	Film Product Title
	MAPS
133701	L.W. TERRESTRIAL FLUX FROM WFOV OBSERVATIONS – ASCENDING NODE
133801	
133702	L.W. TERRESTRIAL FLUX FROM WFOV OBSERVATIONS – DESCENDING NODE
133802	
133703	L.W. TERRESTRIAL FLUX FROM WFOV OBSERVATIONS – ASC+DSC NODE
133803	
133704	DATA POPULATION OF WFOV OBSERVATIONS – ASCENDING NODE
133804	
133705	DATA POPULATION OF WFOV OBSERVATIONS – DESCENDING NODE
133805	
133406	DATA POPULATION OF WFOV OBSERVATIONS – ASC + DSC NODE
133706	
133806	
133707	EARTH ALBEDO FROM WFOV OBSERVATIONS (0.2 – 4.0 μm)
133807	
133708	EARTH ALBEDO FROM WFOV OBSERVATIONS (0.7 – 3.0 μm)
133808	
133709	EARTH ALBEDO FROM WFOV OBSERVATIONS (0.2 – 0.7 μm)
133809	
133410	NET RADIATION FROM WFOV OBSERVATIONS
133710	
133810	
133711	L.W. TERRESTRIAL FLUX FROM NFOV OBSERVATIONS – ASCENDING NODE
133811	
133712	L.W. TERRESTRIAL FLUX FROM NFOV OBSERVATIONS – DESCENDING NODE
133812	
133713	L.W. TERRESTRIAL FLUX FROM NFOV OBSERVATIONS – ASC + DSC NODE
133813	
133714	DATA POPULATION OF NFOV OBSERVATIONS – ASCENDING NODE
133814	
133715	DATA POPULATION OF NFOV OBSERVATIONS – DESCENDING NODE
133815	

Table 3-7 (Continued)

Film Spec Number	Film Product Title
	MAPS
133416	DATA POPULATION OF NFOV OBSERVATIONS – ASC + DSC NODE
133716	
133816	
133717	EARTH ALBEDO FROM NFOV OBSERVATIONS
133817	
133718	MINIMUM EARTH ALBEDO FROM NFOV OBSERVATIONS
133818	
133419	NET RADIATION FROM NFOV OBSERVATIONS
133719	
133819	
133720	PERCENTAGE AREA OF CLOUDINESS FROM THIR 11.5 μm DATA
133721	NORMALIZED DISPERSION OF L.W. TERRESTRIAL FLUX FROM WFOV OBSERVATIONS – ASCENDING AND DESCENDING NODE
133722	NORMALIZED DISPERSION OF EARTH ALBEDO FROM WFOV OBSERVATIONS
133723	NORMALIZED DISPERSION OF NET RADIATION FROM WFOV OBSERVATIONS
133724	NORMALIZED DISPERSION OF L.W. TERRESTRIAL FLUX FROM NFOV OBSERVATIONS – ASCENDING AND DESCENDING NODE
133725	NORMALIZED DISPERSION OF EARTH ALBEDO FROM NFOV OBSERVATIONS
133726	NORMALIZED DISPERSION OF NET RADIATION FROM NFOV OBSERVATIONS
	TABLES
136160	MEAN NORMALIZED SOLAR IRRADIANCE (and) ZONALLY AVERAGED INSOLATION
136460	
136760	
136860	
136461	NET RADIATION FROM WFOV OBSERVATIONS (meridional)
136761	
136861	
136462	NET RADIATION FROM NFOV OBSERVATIONS (meridional)
136762	
136862	

Table 3-7 (Continued)

Film Spec Number	Film Product Title
	TABLES
136763	TERRESTRIAL FLUX FROM WFOV OBSERVATIONS
136863	
136764	TERRESTRIAL FLUX FROM NFOV OBSERVATIONS (meridional)
136864	
136765	EARTH ALBEDO FROM WFOV OBSERVATIONS
136865	
136766	EARTH ALBEDO FROM NFOV OBSERVATIONS
136866	
136767	MONTHLY STATUS AND CALIBRATION SUMMARY
136468	NET RADIATION FROM NFOV OBSERVATIONS (zonal)
136768	
136868	
136769	EARTH ALBEDO FROM NFOV OBSERVATIONS (zonal)
136869	
136770	TERRESTRIAL FLUX FROM NFOV OBSERVATIONS (zonal)
136870	

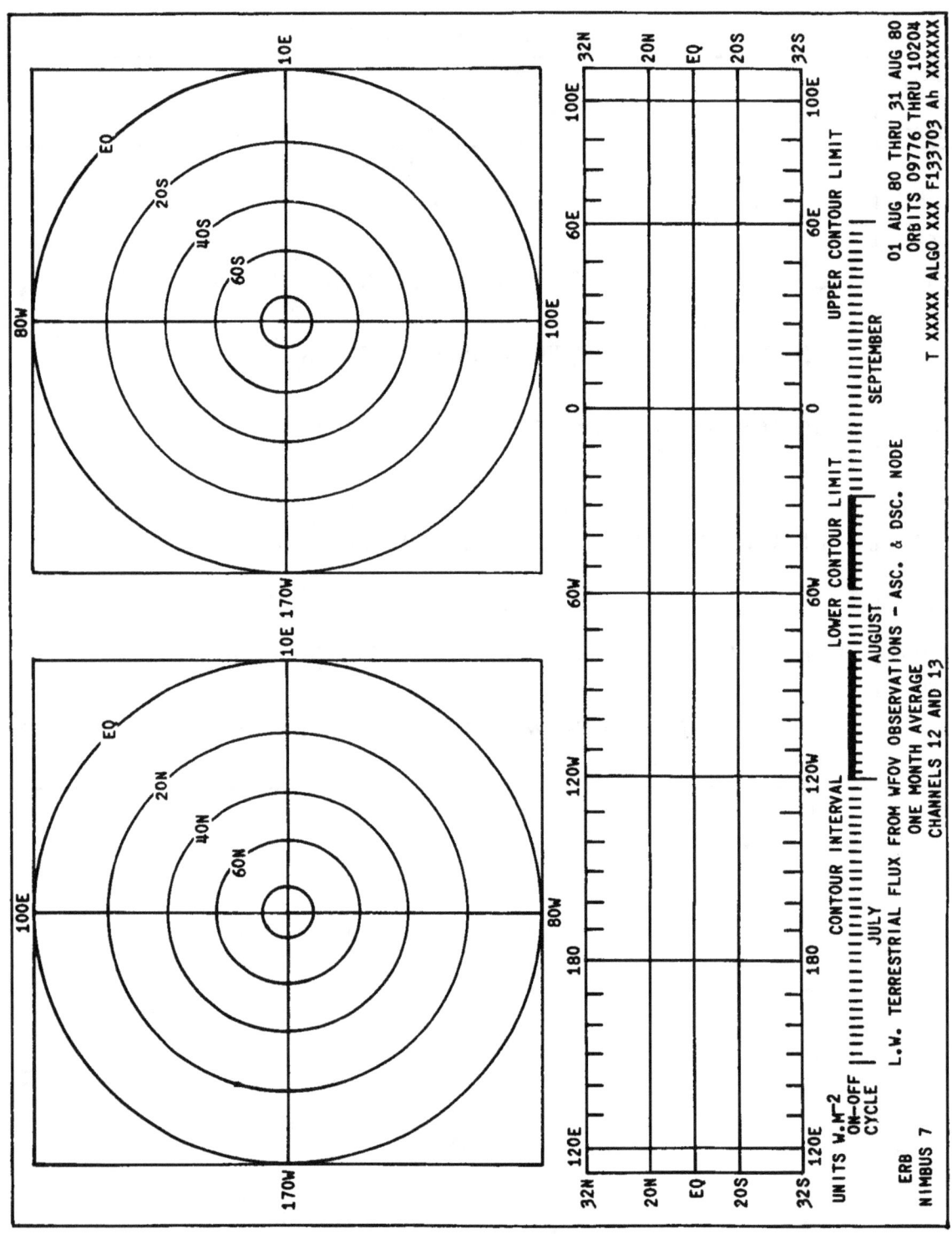

Figure 3-12. ERB Microfilm Map Display Format

MEAN NORMAL SOLAR IRRADIANCE

ONE MONTH AVERAGES

CHANNELS 1 THRU 10C

ERB
NIMBUS 7

01 AUG 80 THRU 31 AUG 80
ORBITS 09776 THRU 10202

EARTH-SUN DISTANCE (A.U.) MIN = X.XXXX   MAX = X.XXXX

CHANNEL NUMBER	SPECTRAL BAND (MICROMETERS)	MEAN IRRADIANCE (W/M*2)	STANDARD DEVIATION	RANGE (MAX.)	DELTA MEAN
1	0.18-3.8	XXXX.X	X.XX	XX.XX	X.XX
2	0.18-3.8	↓	↓	↓	↓
3	0.2-50+				
4	0.530-2.8				
5	0.695-2.8				
6	0.395-0.508	XXX.XX			
7	0.344-0.460	↓			
8	0.390-0.410				
9	0.275-0.360	XXXX.X	X.XX	XX.XX	X.XX
10C	0.2-0.50+	↓	↓	↓	↓

ZONALLY AVERAGED INSOLATION

ONE MONTH AVERAGES

CHANNELS 2, 3, 4, 5, AND 10

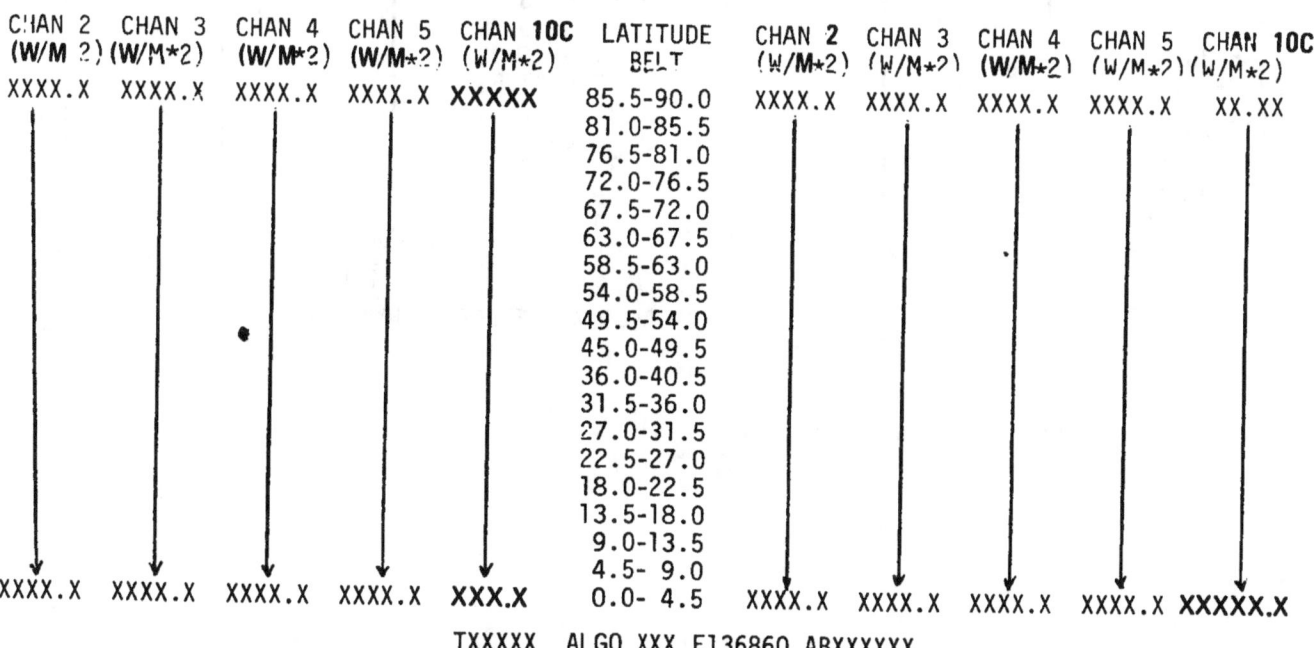

Figure 3-13. Mean Normal Solar Irradiance Display Format

EARTH ALBEDO FROM WFOV OBSERVATIONS
ONE MONTH MEAN ZONAL AVERAGES
CHANNELS 2,5,13 AND 14

ERB   01AUG80 THRU 31AUG80
NIMBUS 7   ORBITS 09776 THRU 10204

LATITUDE BELT	0.2-3.8 μm ALBEDO (%)	0.7-2.8 μm ALBEDO (%)	0.2-0.7 μm ALBEDO (%)	LATITUDE BELT	0.2-3.8 μm ALBEDO (%)	0.7-2.8 μm ALBEDO (%)	0.2-0.7 μm ALBEDO (%)
85.5-90.0N	XX	XX	XX	85.5-90.0N	XX	XX	XX
81.0-85.5N				81.0-85.5S			
76.5-81.0N				76.5-81.0S			
72.0-76.5N				72.0-76.5S			
67.5-72.0N				67.5-72.0S			
63.0-67.5N				63.0-67.5S			
58.5-63.0N				58.5-63.0S			
54.0-58.5N				54.0-58.5S			
49.5-54.0N				49.5-54.0S			
45.0-49.5N				45.0-49.5S			
40.5-45.0N				40.5-45.0S			
36.0-40.5N				36.0-40.5S			
31.5-36.0N				31.5-36.0S			
27.0-31.5N				27.0-31.5S			
22.5-27.0N				22.5-27.0S			
18.0-22.5N				18.0-22.5S			
13.5-18.0N				13.5-18.0S			
9.0-13.5N				9.0-13.5S			
4.5- 9.0N				4.5- 9.0S			
0.0- 4.5N	XX	XX	XX	0.0- 4.5S	XX	XX	XX

TXXXXX ALGO XXX F136765 ABXXXXXX

Figure 3-14. Earth Albedo From WFOV Observations Display Format

EARTH ALBEDO FROM NFOV OBSERVATIONS
ONE MONTH MERIDIONAL VARIATIONS OVER LATITUDE BELTS
CHANNELS 2 AND 15 THRU 18

ERB 01AUG80 THRU 31AUG80
NIMBUS 7 ORBITS 09776 THRU 10204

ALBEDO (%)

LONGITUDE ZONE	LAT 18N-13.5N	LAT 13.5N-09N	LAT 09N-4.5N	LAT 4.5N-00	LAT 00-4.5S	LAT 4.5S-09S	LAT 09S-14.5S	LAT 13.5S-18S
00-09E	XX	XX	XX	XX	XX	XX	XX	XX
09-18E								
18-27E								
27-36E								
36-45E								
45-54E								
54-63A								
63-72E								
72-81E								
81-90E								
90-99E								
99-108E								
108-117E								
117-126E								
126-135E								
135-144E								
144-153E								
153-162E								
162-171E								
171-180								
180-171W								
171-162W								
162-153W								
153-144W								
144-135W								
135-126W								
126-117W								
117-108W								
108-99W								
99-90W								
90-81W								
81-72W								
72-63W								
63-54W								
54-45W								
45-36W								
36-27W								
27-18W								
18-09W								
09-00	XX	XX	XX	XX	XX	XX	XX	XX

TXXXXX ALGO XXX F136766 ABXXXXXX

Figure 3-15. Earth Albedo From NFOV Observations Display Format

NET RADIATION FROM WFOV OBSERVATIONS
ONE MONTH MEAN ZONAL AVARAGES
CHANNELS 2, 3, 12, 13

ERB
NIMBUS 7

01 AUG 80 THRU 31 AUG 80
ORBITS 09776 THRU 10204

LATITUDE BELT	NET RADIATION (W/M*2)	LATITUDE BELT	NET RADIATION (W/M*2)
85.5-90.0N	$\pm$XXX	85.5-90.0S	$\pm$XXX
81.0-85.5N		81.0-85.5S	
76.5-81.0N		76.5-81.0S	
72.0-76.5N		72.0-76.5S	
67.5-72.0N		67.5-72.0S	
63.0-67.5N		63.0-67.5S	
58.5-63.0N		58.5-63.0S	
54.0-58.5N		54.0-58.5S	
49.5-54.0N		49.5-54.0S	
45.0-49.5N		45.0-49.5S	
40.5-45.0N		40.5-45.0S	
36.0-40.5N		36.0-40.5S	
31.5-36.0N		31.5-36.0S	
27.0-31.5N		27.0-31.5S	
22.5-27.0N		22.5-27.0S	
18.0-22.5N		18.0-22.5S	
13.5-18.0N		13.5-18.0S	
9.0-13.5N		9.0-13.5S	
4.5- 9.0N		4.5- 9.0S	
0.0- 4.5N	$\pm$XXX	0.0- 4.5S	$\pm$XXX

TXXXXX ALGO XXX F136761 ABXXXXXX

Figure 3-16. Net Radiation From NFOV Observations Display Format

NET RADIATION FROM NFOV OBSERVATIONS
ONE MONTH MERIDIONAL VARIATIONS OVER LATITUDE BELTS
CHANNELS 2, 3 AND 15 THRU 22

ERB 01 AUG THRU 31 AUG 80
NIMBUS 7 ORBITS 09776 THRU 10204

NET RADIATION (W/M*2)

LONGITUDE ZONE	LAT 18N-13.5N	LAT 13.5N-09N	LAT 09N-4.5N	LAT 4.5N-00	LAT 00-4.5S	LAT 4.5S-09S	LAT 09S-13.5S	LAT 13.5S-18S
00-09E	±XXX	±XXX	±XXX	±XXX	±XXX	±XXX	±XXX	±XXX
09-18E								
18-27E								
27-36E								
36-45E								
45-54E								
54-63A								
63-72E								
72-81E								
81-90E								
90-99E								
99-108E								
108-117E								
117-126E								
126-135E								
135-144E								
144-153E								
153-162E								
162-171E								
171-180								
180-171W								
171-162W								
162-153W								
153-144W								
144-135W								
135-126W								
126-117W								
117-108W								
108-99W								
99-90W								
90-81W								
81-72W								
72-63W								
63-54W								
54-45W								
45-36W								
36-27W								
27-18W								
18-09W								
09-00	±XXX	±XXX	±XXX	±XXX	±XXX	±XXX	±XXX	±XXX

TXXXXX ALGO XXX F136762 ABXXXXXX

Figure 3-17. Net Radiation From NFOV Observations Display Format

TERRESTRIAL FLUX FROM WFOV OBSERVATIONS
ONE MONTH MEAN ZONAL AVERAGES
CHANNELS 12 AND 13

ERB   01AUG80 THRU 31AUG80
NIMBUS 7   ORBITS 09776 THRU 10204

LATITUDE BELT	DAYTIME FLUX (W/M*2)	NIGHTTIME FLUX (W/M*2)	DAY & NIGHT FLUX W/M*2)	LATITUDE BELT	DAYTIME FLUX (W/M*2)	NIGHTTIME FLUX (W/M*2)	DAY & NIGHT FLUX (W/M*2)
85.5-90.0N	XXX	XXX	XXX	85.5-90.0S	XXX	XXX	XXX
81.0-85.5N	↓	↓	↓	81.0-85.5S	↓	↓	↓
76.5-81.0N				76.5-81.0S			
72.0-76.5N				72.0-76.5S			
67.5-72.0N				67.5-72.0S			
63.0-67.5N				63.0-67.5S			
58.5-63.0N				58.5-63.0S			
54.0-58.5N				54.0-58.5S			
49.5-54.0N				49.5-54.0S			
45.0-49.5N				45.0-49.5S			
40.5-45.0N				40.5-45.0S			
36.0-40.5N				36.0-40.5S			
31.5-36.0N				31.5-36.0S			
27.0-31.5N				27.0-31.5S			
22.5-27.0N				22.5-27.0S			
18.0-22.5N				18.0-22.5S			
13.5-18.0N				13.5-18.0S			
9.0-13.5N				9.0-13.5S			
4.5- 9.0N	↓	↓	↓	4.5- 9.0S	↓	↓	↓
0.0 4.0N	XXX	XXX	XXX	0.0- 4.5S	XXX	XXX	XXX

TXXXXX ALGO XXX F136765 ABXXXXXX

Figure 3-18. Terrestrial Flux From WFOV Observations Display Format

TERRESTRIAL FLUX FROM NFOV OBSERVATIONS
ONE MONTH MERIDIONAL VARIATIONS OVER LATITUDE BELTS
CHANNELS 19 THRU 22

ERB 01AUG80 THRU 31AUG80
NIMBUS 7 ORBITS 09776 THRU 10204

DAY PLUS NIGHT FLUX (W/M*2)

LONGITUDE ZONE	LAT 18N-13.5N	LAT 13.5N-09N	LAT 09N-4.5N	LAT 4.5N-00	LAT 00-4.5S	LAT 4.5S-09S	LAT 09S-13.5S	LAT 13.5S-18S
00-09E	XXX	XXX	XXX	XXX	XXX	XXX	XXX	XXX
09-18E								
18-27E								
27-36E								
36-45E								
45-54E								
54-63E								
63-72E								
72-81E								
81-90E								
90-99E								
99-108E								
108-117E								
117-126E								
126-135E								
135-144E								
144-153E								
153-162E								
162-171E								
171-180								
180-171W								
171-162W								
162-153W								
153-144W								
144-135W								
135-126W								
126-117W								
117-108W								
108-99W								
99-90W								
90-81W								
81-72W								
72-63W								
63-54W								
54-45W								
45-36W								
36-27W								
27-18W								
18-09W								
09-00	XXX	XXX	XXX	XXX	XXX	XXX	XXX	XXX

TXXXXX ALGO XXX F136764 ABXXXXXX

Figure 3-19. Terrestrial Flux From NFOV Observations Display Format

SECTION 4

THE LIMB INFRARED MONITOR OF THE STRATOSPHERE (LIMS) EXPERIMENT

by

J. M. Russell*
Langley Research Center, Hampton, Virginia 23665

and

J. C. Gille*
National Center for Atmospheric Research**, Boulder, Colorado 80303

4.1 Introduction

The Limb Infrared Monitor of the Stratosphere (LIMS) experiment is being conducted to determine global scale vertical distributions of temperature and several gases involved in the chemistry of the ozone in the stratosphere. This will allow detailed study of the chemistry, especially of the nitrogen compounds related to ozone distribution, as well as determinations of geostrophic winds and transport of trace constituents. Profiles of ozone (O_3), nitrogen dioxide (NO_2), nitric acid (HNO_3), water vapor (H_2O), and temperature are determined with high vertical resolution from the lower stratosphere (≈ 10 km) to the lower mesosphere (≈ 65 km). This list of measurements includes the catalyst in the chain of NO_x reactions that destroy ozone (NO_2) as well as the end product responsible for removal of NO_x from the stratosphere (HNO_3). These data are determined by inverting measured limb radiance profiles obtained by LIMS, an infrared multispectral scanning radiometer. Measurements are made in each of six spectral regions: one in the 9.6 μm O_3 band, one in the 6.3 μm NO_2 band, one in the 6.2 μm H_2O band, one in the 11.3 μm HNO_3 band, and two in the 15 μm band of CO_2.

This experiment is a follow-on to the successful Limb Radiance Inversion Radiometer (LRIR) experiment flown on Nimbus 6 (Reference 1) to measure O_3, H_2O, and temperature. The LIMS instrument is identical to the LRIR in many respects but is significantly different in that two detectors were added to the focal plane array and five parameters rather than three are being measured. The horizon scan rate was also decreased from one degree per second to one quarter degree per second to provide improved signal-to-noise performance. These changes facilitate measurement of constituents with small signals (e.g. NO_2, HNO_3) and allow extension of the measurements to lower and higher altitudes.

A programmed scanning mirror in the radiometer causes the field of view of the six detectors to make coincident vertical scans across the earth's horizon. (See Figure 4-1 for description of limb-viewing geometry). The data from these scans are stored on a tape recorder for later transmission to the ground. During data reduction the measured limb radiance profiles from the carbon dioxide channels are operated on by inversion algorithms to determine the vertical temperature distribution. This inferred temperature profile, together with the radiance profiles in the other channels, are then used to infer the vertical distribution of the trace constituents.

---

*Co-team Leader of the LIMS Nimbus Experiment Team – other NET members are listed in Appendix B.
**Sponsored by the National Science Foundation.

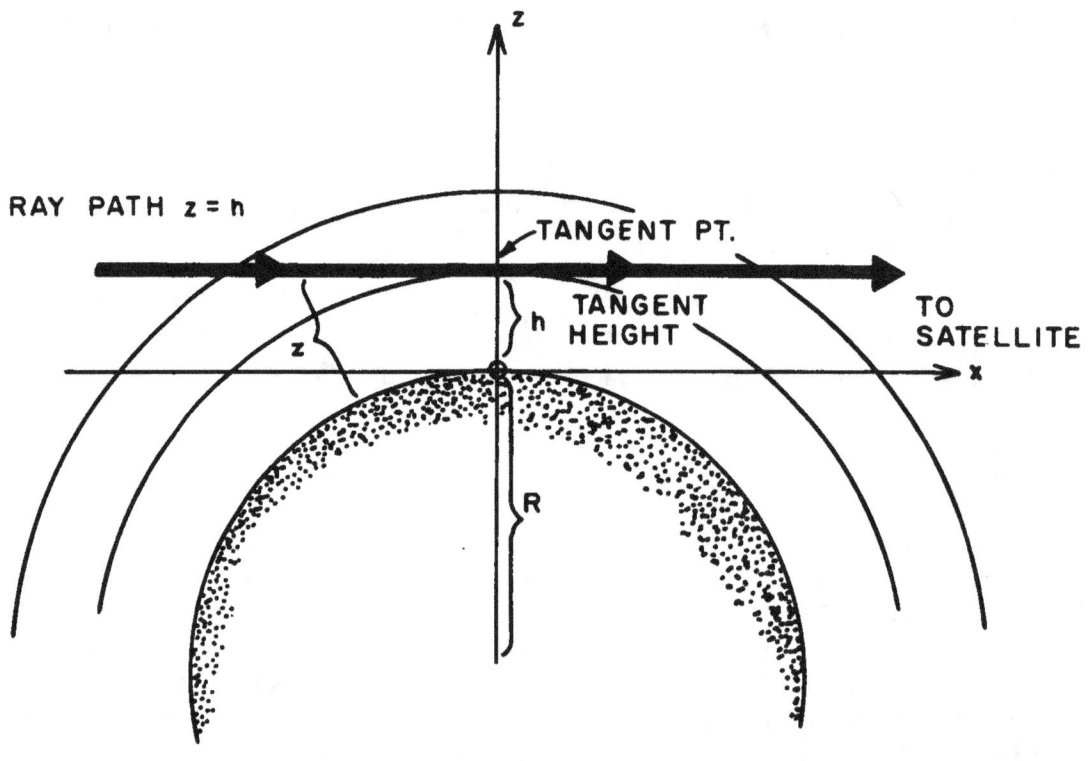

Figure 4-1. LIMS Viewing Geometry

4.2 Scientific and Technical Objectives

4.2.1 Scientific Objectives

The objective of the LIMS program is to obtain the global distributions of temperature, ozone, and three other trace gases related to ozone chemistry, and to apply these observations (in conjunction with results from other Nimbus 7 sensors) to the solution of upper atmosphere chemistry, radiation, and dynamics.

The limb scanning technique received its first satellite-borne test with the Limb Radiance Inversion Radiometer (LRIR) on Nimbus 6. LRIR successfully measured the stratospheric O_3 concentration and temperature field over a seven month period beginning in June 1975. Figures 4-2 and 4-3 show sample LRIR vertical profiles of temperature and O_3 as compared to near simultaneous measurements made by in situ rocket soundings. Using sequential estimation techniques (Rodgers, 1976) it is possible to fourier analyze large time ($\approx$ weeks) and spatial (4° latitude circles from 64°S to 84°N) sets of data to produce global maps of the measured fields as a function of time as in Figure 4-4. The LIMS uses the same techniques to produce standard output mapped products for temperature, O_3, H_2O, HNO_3, and NO_2.

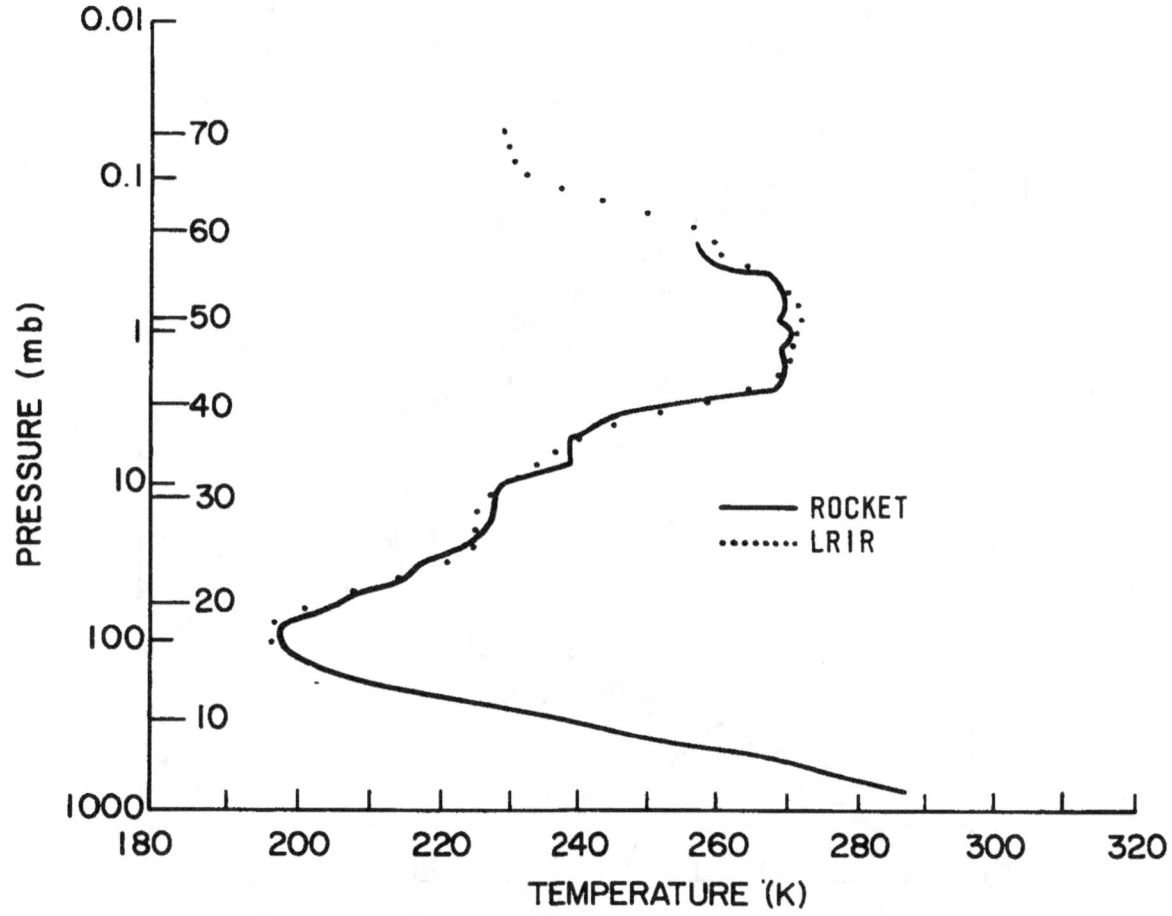

Figure 4-2. LRIR Temperature Retrieval (Antiqua, Dec. 17, 1975) Compared with Rocket Results

The ability to monitor the time variations of the global distributions of temperature and trace gases in the 10-70 km region is important to the understanding of interrelationships between dynamics and chemistry. Chemistry determines in large measure the global stratospheric temperature distribution, which is, itself, responsible for the global wind fields. The winds, in turn, transport the chemical species, continuing the cycle.

Some specific scientific objectives for the LIMS experiment are:

- To understand the chemistry and transport of stratospheric NO_x including sources and sinks, and possible feedbacks upon the dynamics

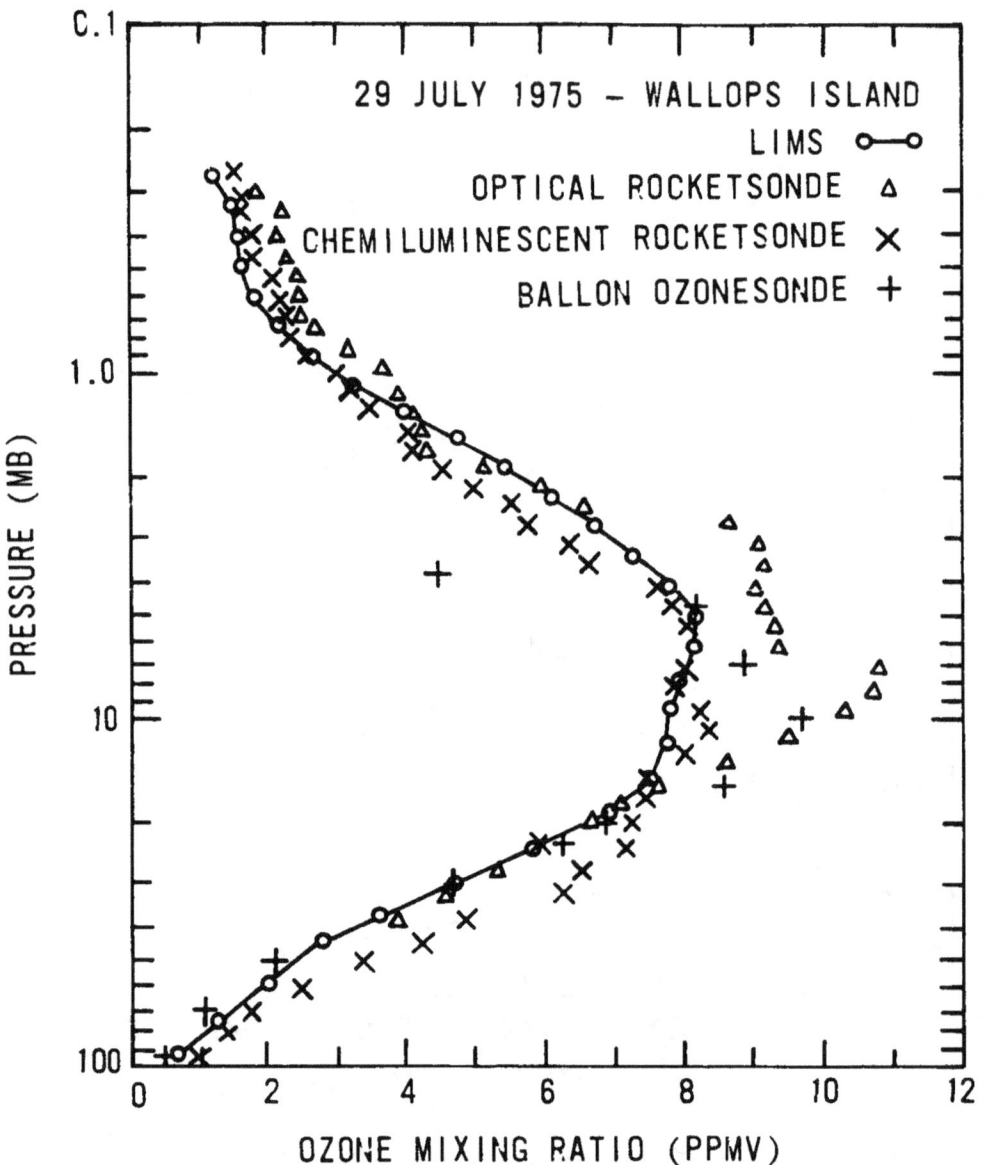

Figure 4-3. LRIR Ozone Retrieval Compared with Rocket Results

- To utilize the capabilities of the LIMS for initial analyses of a variety of stratospheric dynamic phenomena, including planetary waves, equatorial waves, and high frequency waves, plus global modes

- To determine the relationship between temperature and O_3 mixing ratio and its dependence on height, latitude, and season in the photochemically controlled altitude regions

- To determine the response of LIMS measured fields to perturbations within the earth's geophysical environment due to solar and cosmic ray variability, etc., and how these changes might correlate with temperature

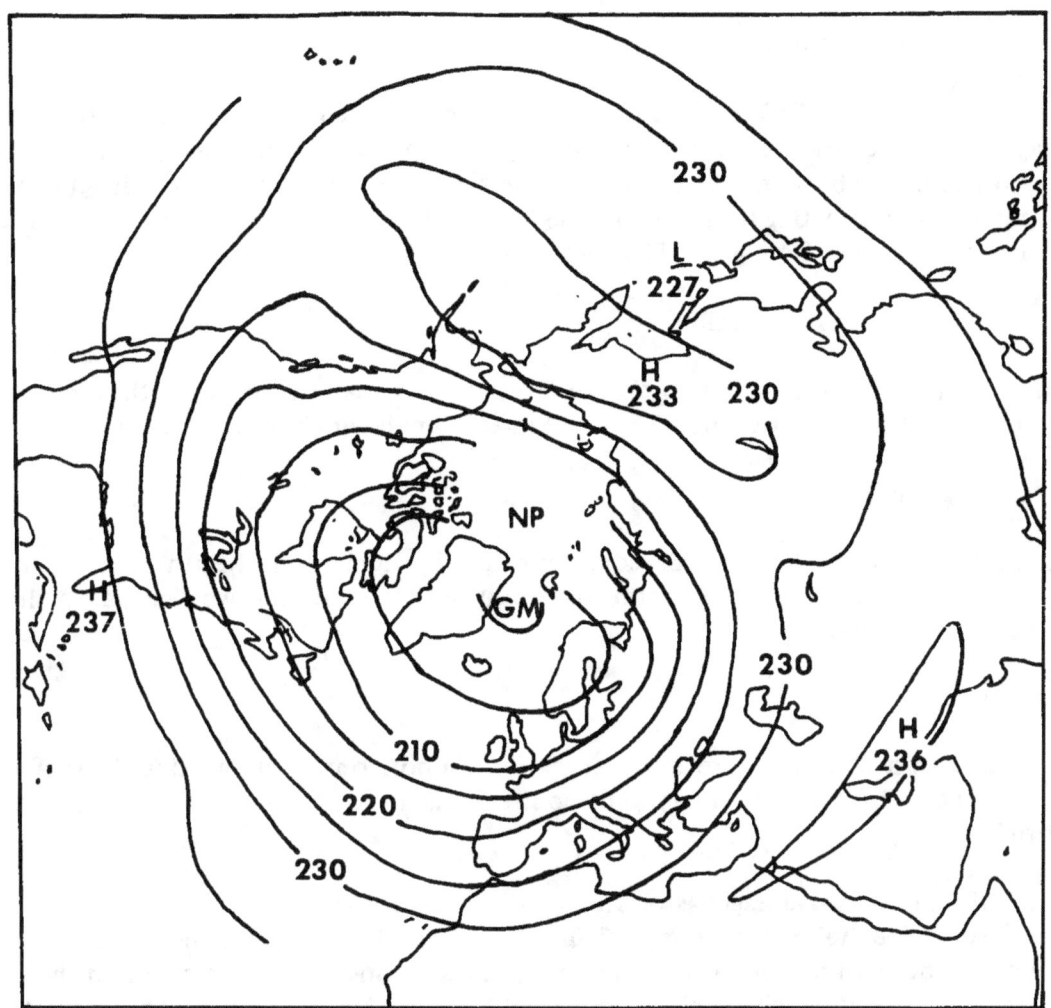

Figure 4-4. LRIR Temperature Map Results for 10 mb Level-Descending Node

- To determine the temporal and spatial distributions of transports of chemical constituents, heat, momentum, energy, and potential vorticity by geostrophic and possibly ageostrophic motions

- To prepare a climatology of the northern and southern hemispheres, containing statistical information on the temporal and spatial variability of the stratospheric variables as measured by LIMS

- To study day-night difference in trace gas concentrations (especially NO_2 and O_3) and temperature and to use these data in testing photochemical models and assessing the magnitude of tidal motions

- To utilize the improved constituent profiles in radiative energy budget calculations of the middle atmosphere

- To employ lower stratospheric LIMS data in a study of stratospheric-troposphere exchange, with particular emphasis on O_3 injections and tropopause folding

- To search for evidence of secondary maxima in the O_3 profiles and correlate with meteorological phenomena

4.2.2 Limb Scan Features

Refering to Figure 4-1, LIMS receives infrared radiation emitted by the atmosphere along a ray path that may be identified by the height (tangent height) or point (tangent point) closest to the surface. The atmosphere may be scanned by sweeping the view direction from tangent heights below the horizon (ray paths intersecting the surface) to high altitudes. The following advantageous features of limb scanning are apparent from a consideration of Figure 4-1.

- High inherent vertical resolution

  For geometric reasons, a small portion of the signal originates from below the tangent height, and most of the signal originates from a 4 km to 5 km layer above the tangent height.

- Zero background

  For heights above the horizon, all radiation received originates in the atmosphere, and all variations in signal are due to the atmosphere since the radiation is viewed against the cold background of space.

- Large opacity

  There is at least 60 times more emitting gas along a horizontal path grazing the surface than there is in a vertical path to the tangent point. Thus, the atmosphere can be sampled to high altitudes.

There are, of course, disadvantages associated with these features. The long paths mean that even for rather transparent spectral regions, it is difficult to see the solid surface of the planet. A cloud along a path acts as a body of infinite opacity, and may cause a considerable alteration in the emerging radiation. For the earth's atmosphere, clouds are usually below the tropopause. This suggests that reliable operation be limited to the upper troposphere and above. Even this area will be subject to occasional clouds.

The radiative transfer equation for a non-scattering atmosphere is local thermodynamic equilibrium may be written as

$$I_i(h) = \int_{-\infty}^{\infty} B_i(T) \frac{d\tau_i(h,x)}{dx} dx \tag{1}$$

where I is the observed radiance at tangent height h and spectral interval i, B is the Planck blackbody function, T is temperature, x the distance coordinate along the ray path, with the origin at the tangent point and positive toward the satellite (located at $+\infty$), and $\tau(h,x)$ the mean transmission in the spectral interval along the path with tangent height h from point x to the satellite.

The temperature inversion problem is to determine B and therefore T from measurements of I, assuming that $d\tau/dx$ is known. The latter requires that the distribution of the emitting species be known, which in practice means that radiation from CO_2, a uniformly mixed gas, is measured. In the limb problem $d\tau/dx$ is also crucially affected by the atmospheric structure.

In the case of the constituent inversion problem, the solution to the temperature inversion problem is utilized with the constituent limb radiance profile to determine the gas concentration as a function of altitude (References 2 and 3). In Equation (1), B is known from the temperature solution. The constituent concentration is determined as an implicit function of transmission.

4.2.3 Technical Objectives

The LIMS experiment is an important next step in an evolving technology for study of the atmosphere using limb sounding. The experiment follows and advances research capabilities demonstrated with the successful flight of LRIR on Nimbus 6. A specific objective of LIMS is to show the capability for remote sensing of tenuous trace species (e.g. HNO_3, NO_2, H_2O) with high vertical resolution. Features of the LRIR instrument such as internal reflections that led to spurious signal levels have been substantially reduced in the LIMS instrument. Advances in detector performance have led to improved signal-to-noise performance. These improvements will be tested in the LIMS experiment to show the potential of the limb emission sounder for future long-term observations of minor trace gases. Algorithms have been developed to perform rapid inversions of limb radiance data and to invert signals from two gases simultaneously where spectral overlap occurs. The LIMS experiment allows these algorithms to be tested, verified, and refined for future application.

4.3 Instrument Description

4.3.1 LIMS Operation

The LIMS instrument is shown in Figures 4-5 and 4-6. It consists of two electronic units and a frame housing assembly (FHA). The FHA is divided into two sections consisting of the solid cryogen

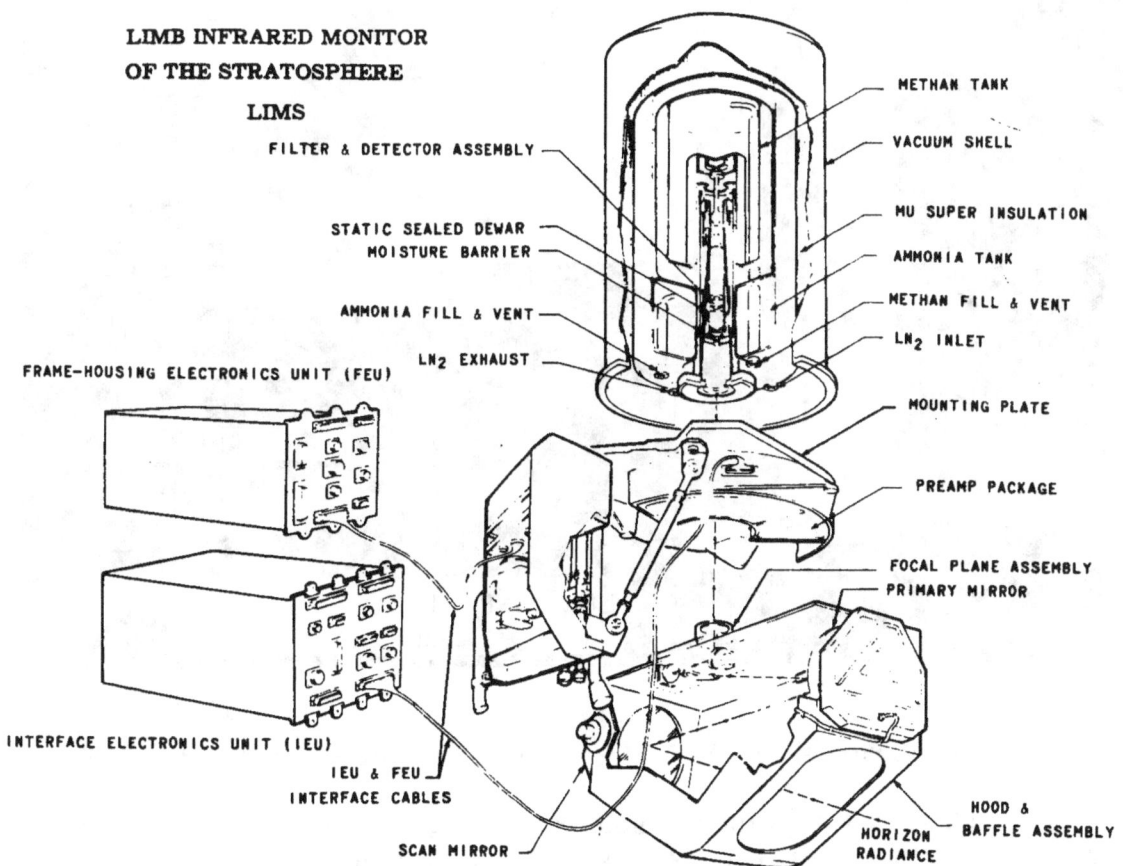

Figure 4-5. LIMS Instrument Configuration

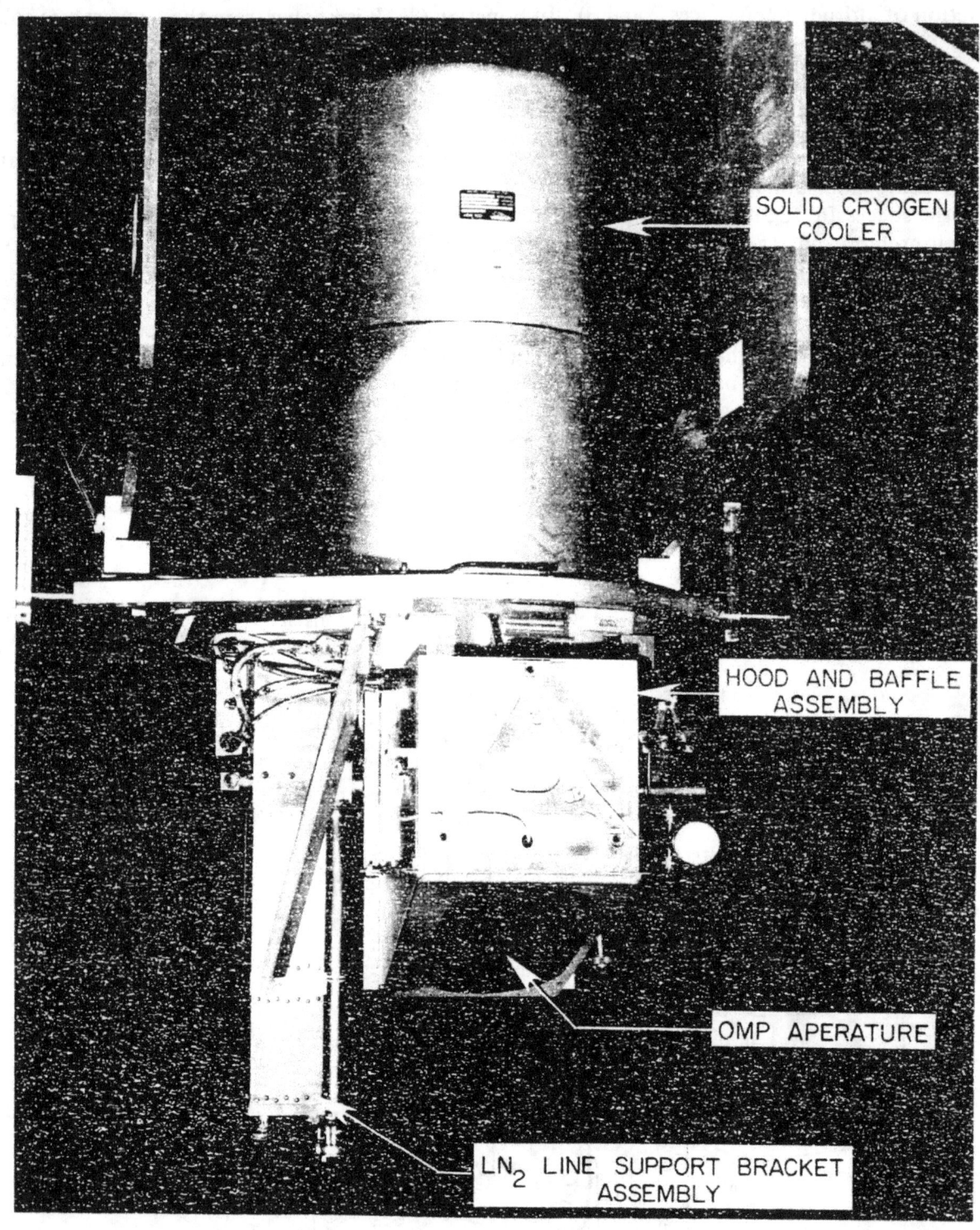

Figure 4-6. LIMS Proto-Flight Model Instrumentation Installed on Mounting Yoke

package (SCP) located above the instrument mounting plate and the optical mechanical package (OMP) located below the mounting plate. The OMP contains the primary optics of the instrument and the SCP contains the six spectral detectors.

The earth's limb energy is directed through the hood and baffle assembly to the scan mirror as shown in Figure 4-7. The beam is directed from the scan mirror (M1) to an 18 cm off-axis parabolic mirror (M2) and then through the focal plane subassembly containing a secondary off-axis parabolic mirror (M3). The secondary parabolic recollimates the energy into a folding mirror (M4) and into the detector capsule assembly (DCA). The CdTe lens at 300°K focuses the beam from the OMP primary optics to a thermally baffled secondary field stop at the 142°K static sealed dewar (cooled by cryogenic package). A 0.32 cm thick window of Irtran VI behind this stop on the dewar allows energy into the DCA. Then back-to-back parabolas (M7 and M8) focus to the detector with a final correction from an f/3 to an f/1 system by an Irtran V1 lens. The parabolic mirrors behind the Irtran V1 window operate at 152°K and the final Irtran V1 lens operates at 64°K. The focused energy is directed onto an array of HgCdTe detectors mounted to a cold finger imbedded in the inner CH_4 stage of a solid cryogen cooler (64°K). The cooler outer stage is NH_3.

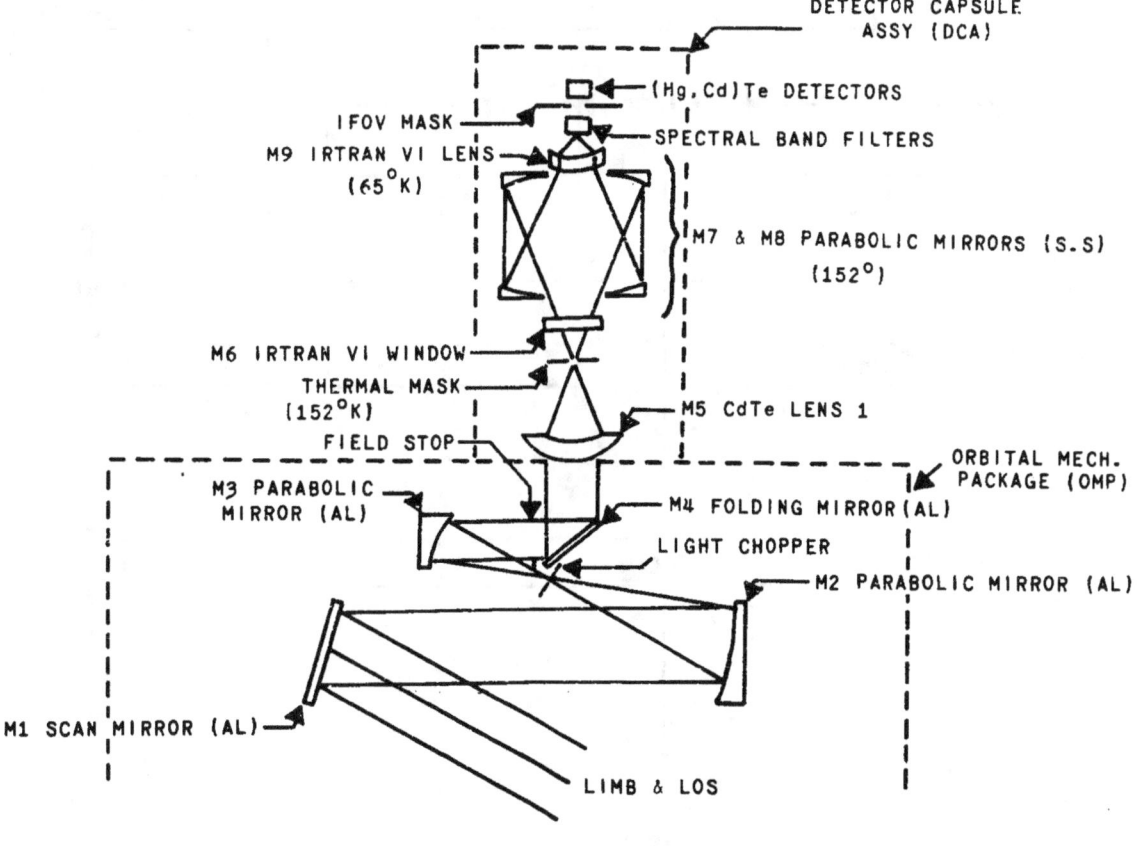

Figure 4-7. LIMS Optical Schematic

The optical views of the LIMS channels and their locations relative to the limb scene are shown in Figure 4-8. The angular resolution (in milliradians) of each detector and their angular positions relative to each other are indicated. When the entire view is projected through the atmosphere at the limb, perpendicular to a vertical plane at the tangent point, the sampled area encompasses a rectangle that is about 56 km in the horizontal by 61 km in the vertical. Within this area the O_3, HNO_3 and the two CO_2 channels each cover areas of about 20 km in horizontal and 2 km in vertical extent. These four channels are positioned symmetrically about the center of the LIMS field of view as shown in Figure 4-8. The H_2O and NO_2 channels cover a larger area — about 30 km by 4 km.

The scan mirror sweeps the horizon image across the detector array over an angular distance of ±6° centered about the optical axis. This axis is oriented at a nominal depression angle of ~29° (~40 km altitude point) during the mission. Two scan cycles are completed in the acquisition scan model (Figure 4-9) during which the instrument senses the angular position of the peak radiance in the narrow band CO_2 channel. This information is used to center the scan (at the 40 percent of peak

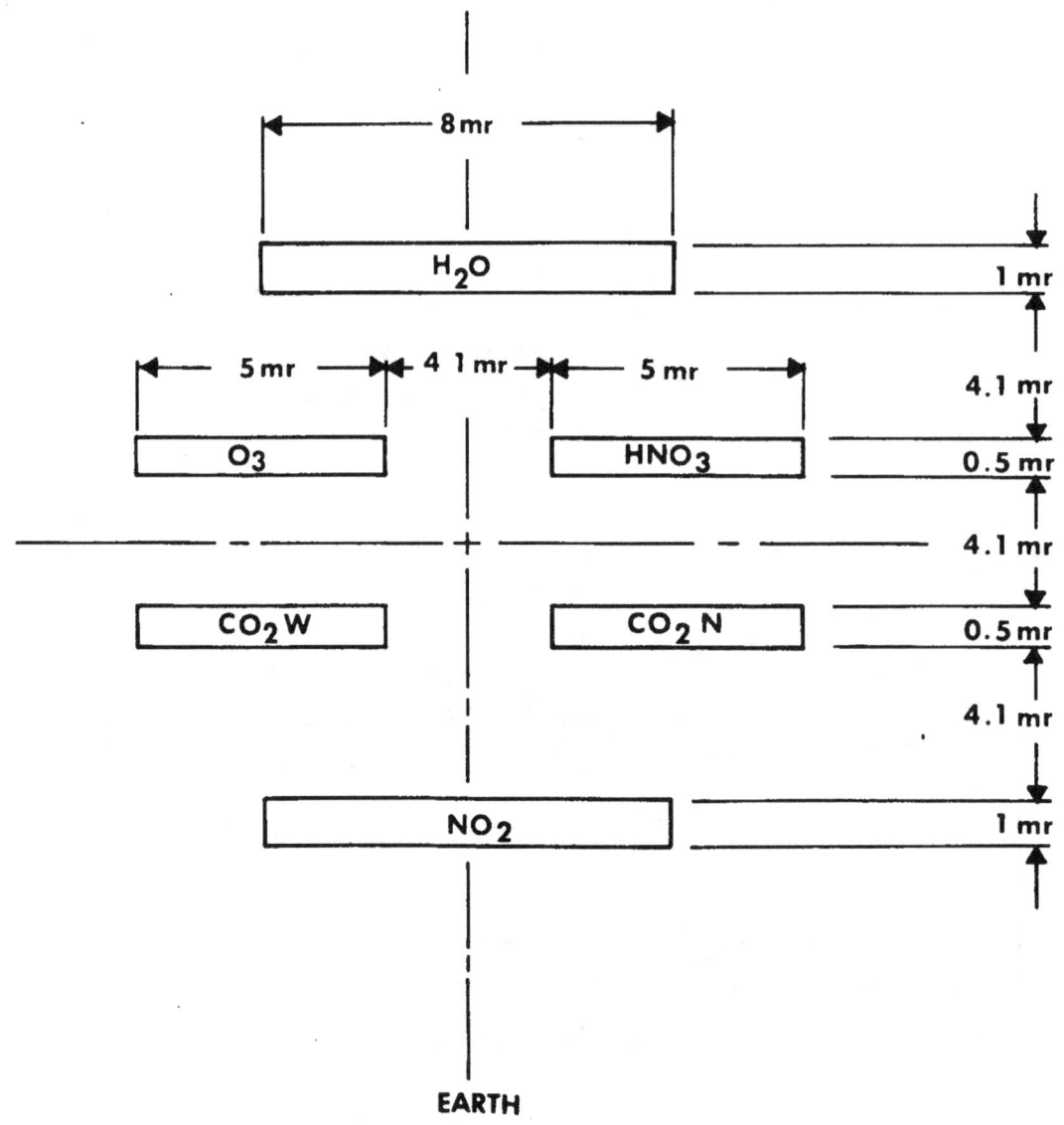

Figure 4-8. LIMS Instantaneous Fields of View

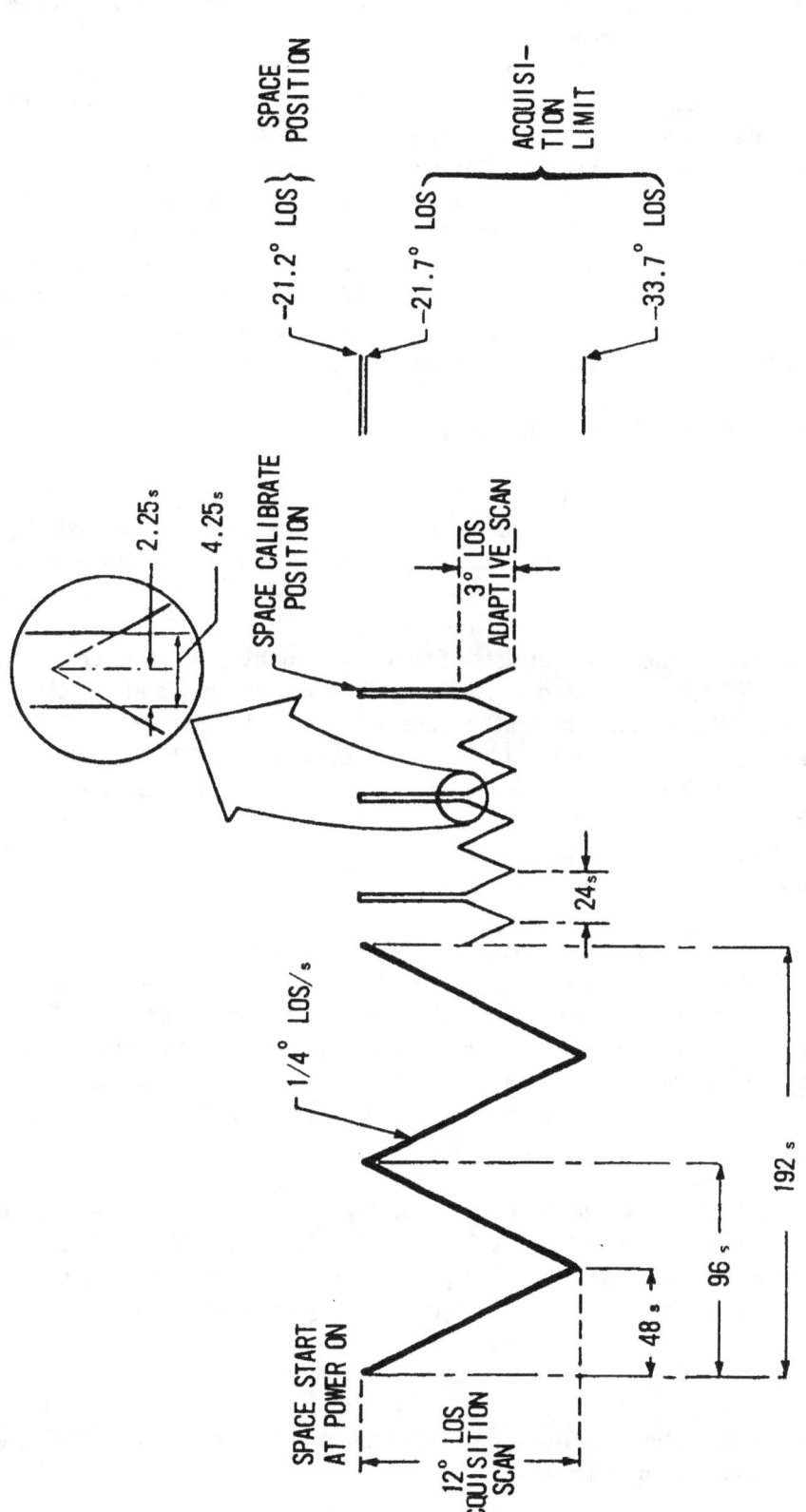

Figure 4-9. LIMS Scan Mode Wave Form

radiance point) for the adaptive scan mode which provides a scan +2° (up) and -1° (down) about the track point. The scan mirror rotates above the scan center by 8 degrees on every other scan cycle in order to view the cold of space for calibration, and it rotates further to an angle of 30 degrees on the same cycle to view in-flight calibration source.

The frame housing electronics unit (FEU in Figure 4-5) conditions radiance signals, controls the scan drive mechanism, and contains the command relay and command logic assemblies. The interface electronics unit (IEU in Figure 4-5) interacts with spacecraft systems and performs the functions of signal processing, control, power conversion, and digitizing of electronic signals. The system block diagram is shown in Figure 4-10. Thermal energy from the horizon focused onto the detector array produces electrical signals from the detectors. These signals are synchronously detected, preamplified, sampled, digitized, multiplexed, and routed to the spacecraft serial data stream for onboard recording and subsequent telemetry transmission to a ground station. The spacecraft system provides timing and command signals to LIMS for data reduction and command functions.

4.3.2 LIMS Location and Limb Viewing Geometry

The information presented in Figure 4-11 illustrates the principal axes of Nimbus 7 and the location of the LIMS relative to these axes and the sensory ring (a). Also shown are associated positions of the optical axis of the radiometer relative to the local horizontal (b) and to the spacecraft heading (c).

As mentioned in Section 4.3.1, the FHA consists of two components; the cryogen cooler (SCP) and radiometer optics (OMP). The SCP is located to the rear of the sensory ring and directly along the roll axis of the satellite. The OMP is situated beneath the sensory ring with the attached SCP projected up through the ring on the inside. The LIMS line of sight is positioned 33.5 degrees to the left of the negative roll axis (-X) in the X-Y plane and 29 degrees below the horizontal of the sensory ring. This positioning is necessary to shield the radiometer from direct solar radiation and to position the mirror scan some 30 to 40 km above the earth's horizon. The mirror in the OMP scans "up and down" through a small angle just above the earth's surface.

It should be emphasized that LIMS views neither the local nadir nor in the orbital plane of the spacecraft. As a result, the LIMS scan track, when projected to earth, is parallel and to the right of the subpoint track relative to direction of flight. When the satellite is near the north pole, the LIMS views across the pole. Near the south pole the LIMS views equatorward of the subpoint latitude position. With this geometric configuration, a greater density of observations are provided in the northern hemisphere as compared to the southern hemisphere. Thus, LIMS data coverage extends from about latitude 84°N to 64°S.

To better understand the LIMS coverage, computer portrayals of successive sub-tangent-point tracks for both hemispheres are presented in Figures 4-12 and 4-13. Note in Figure 4-12 that successive ascending and descending node tracks provide equal spacing in mid-latitudes of the Northern Hemisphere, providing good spatial coverage in this geographical region of the earth.

4.3.3 LIMS Channel Characteristics

· The optical, noise equivalent radiance (NEN) and instantaneous field of view (IFOV) characteristics of the six LIMS channels are given in Table 4-1.

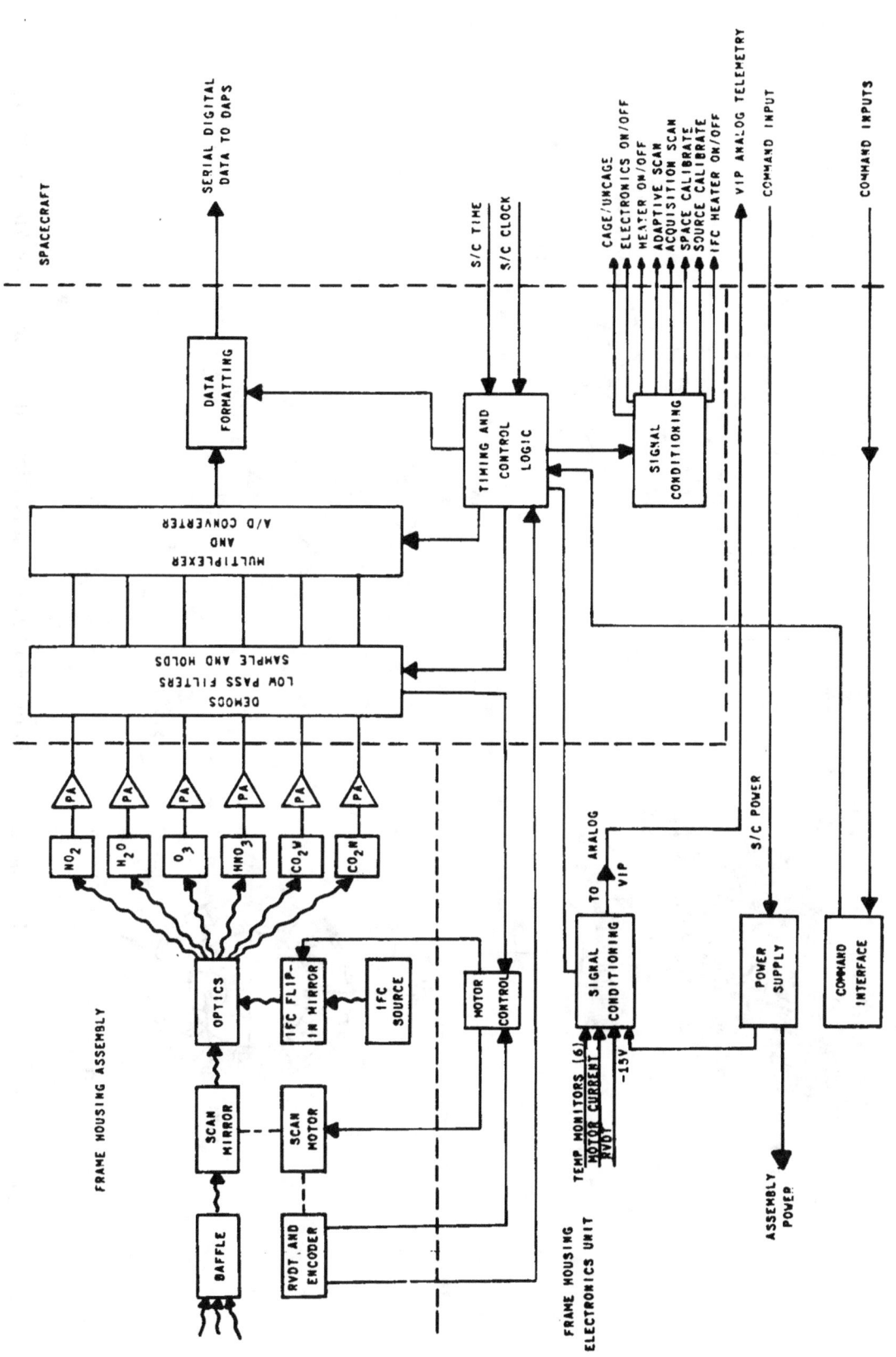

Figure 4-10. LIMS System Block Diagram

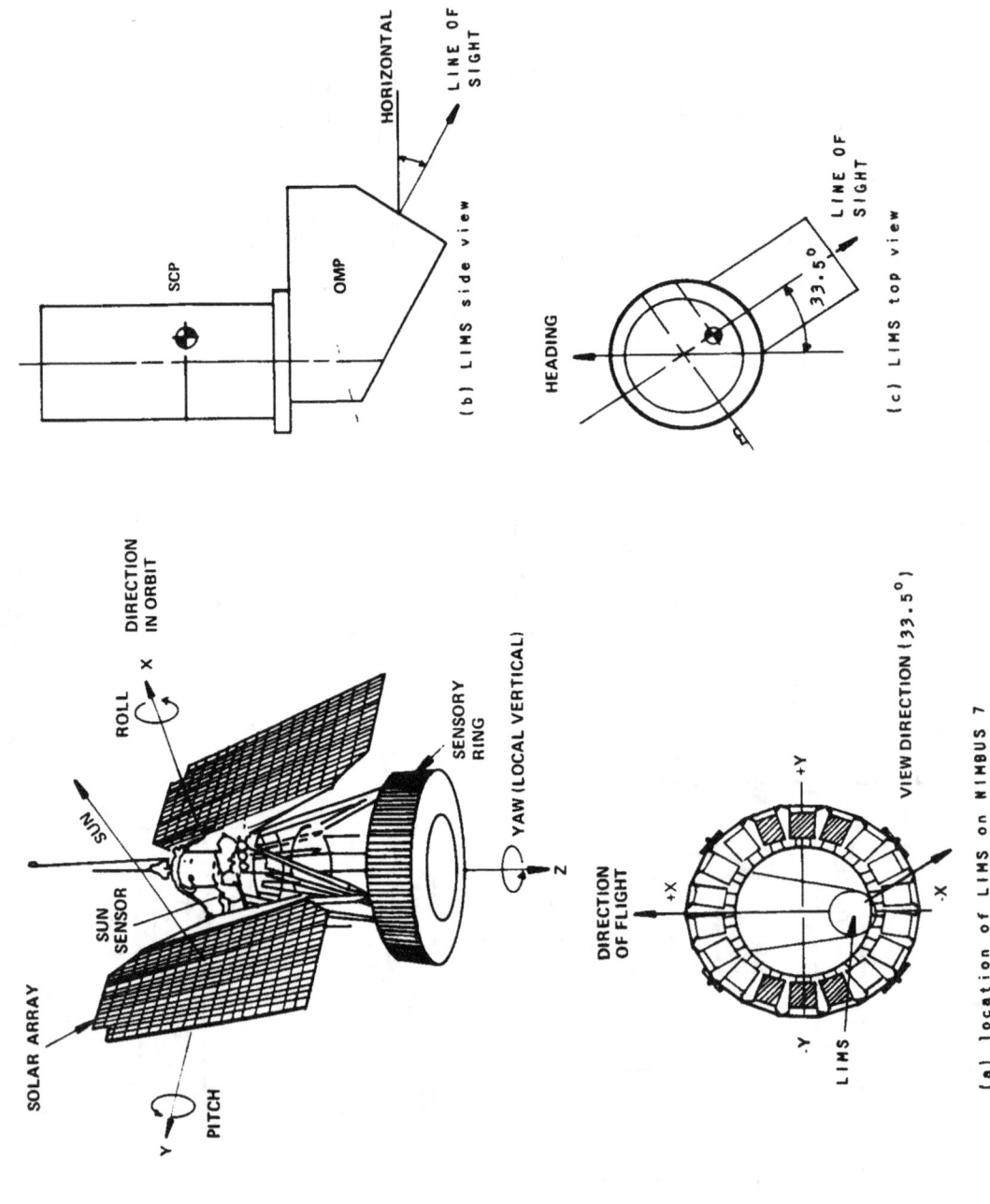

Figure 4-11. Location of the LIMS Instrumentation on the Nimbus 7 Sensory Ring and Associated Viewing Geometry

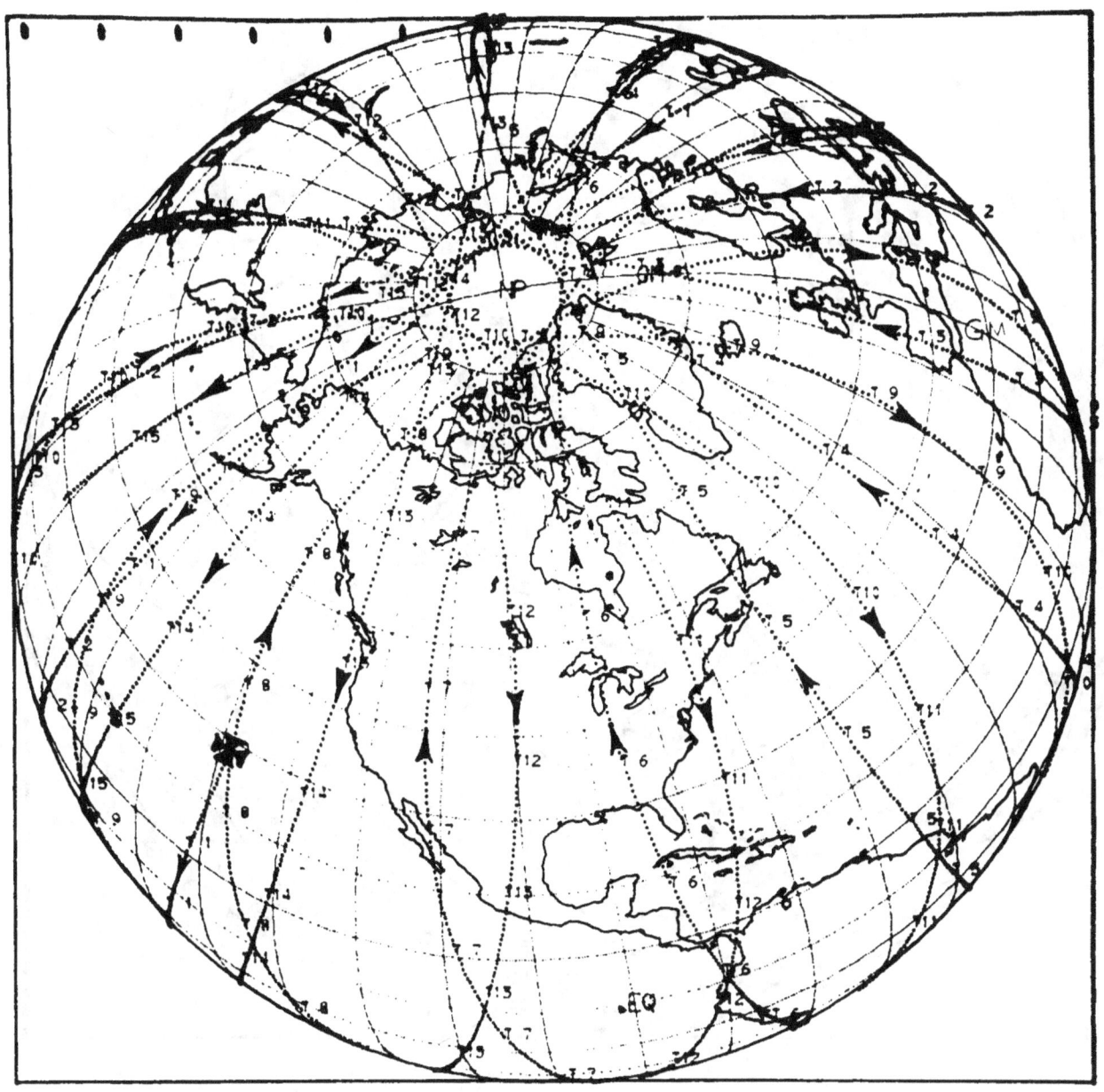

Figure 4-12. LIMS Sub-Tangent Point Tracks for the Northern Hemisphere

The noise performance was measured after completion of all environmental testing of the spacecraft. Channels 5 and 6, the wide (CO_2W) and narrow (CO_2N) carbon dioxide channels, provide information on the temperature structure of the atmosphere. Channels 1 through 4 provide data on the constituents NO_2, H_2O, O_3, and HNO_3, respectively.

The location and width of each channel was optimized to provide the highest signal-to-contaminant-gas-noise ratio. This optimization process was especially important for the NO_2 channel which is located in the center of a strong H_2O band at 6.2 μm. The region of strongest atmospheric NO_2 emission was selected (e.g. 6.1 μm) and the filter cut-off wave length was varied on either side of the central emission to arrive at the best location. A similar procedure was followed for the 9.6 μm (O_3), 6.2 μm (H_2O), 11.3 μm (HNO_3), and 15 μm (CO_2) channels. The spectral bandwidth was set as wide as possible in order to maximize the energy collected. Because of this, it was possible

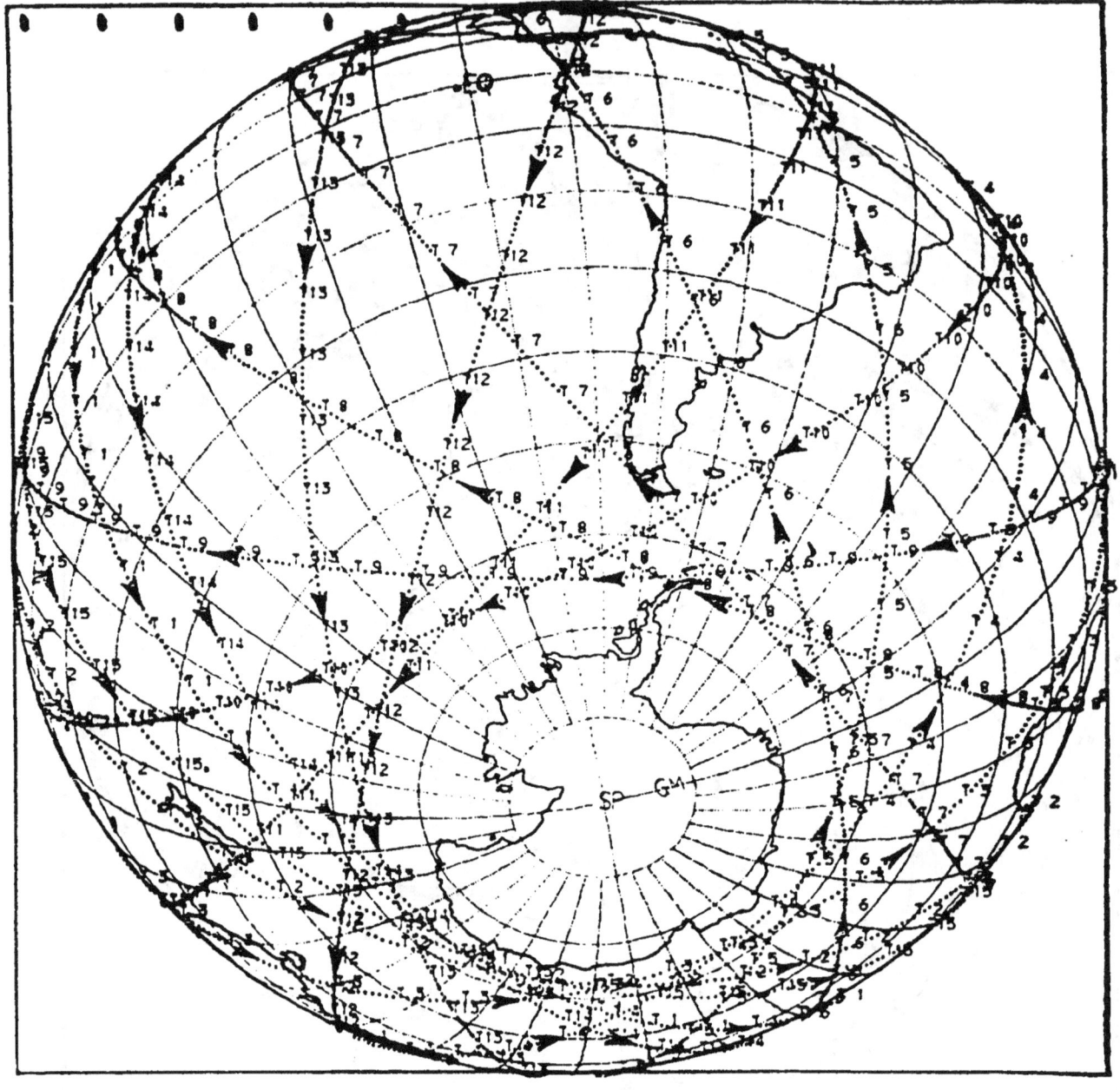

Figure 4-13. LIMS Sub-Tangent Point Tracks for the Southern Hemisphere

to reduce the IFOV's in Channels 3 through 6 to near the diffraction limit thereby providing good vertical resolution during a limb scan. The ozone spectral bandwidth was broadened over the LRIR ozone channel in order to include the more weakly absorbing wings of the ozone band. When these spectral regions are included, it becomes possible, in the absence of clouds, to measure the ozone profile down to the ground level. The locations of the two CO_2 bands are not symmetrical about the 15 μm band center, but are shifted to the short wavelength (higher wavenumber) side of the band. Because each band provides a limb radiance profile of different character, both are utilized in the process of obtaining temperature inversion solutions. Good data can be obtained solely from the CO_2W channel when statistical inversion techniques are applied to the limb radiance data. However, temperature accuracy degrades somewhat.

Table 4-1
Optical Characteristics of LIMS Channels

No.	Gas	Band Pass 5% Relative Response Points	Band Pass 50% Relative Response Points	Field of View		Noise Equivalent Radiance NEN (Watt/m²-n)
				Vertical	Horizontal	
1	NO_2	1560 – 1630 cm^{-1} (6.4 – 6.1 µm)	1580 – 1613 cm^{-1} (6.3 – 6.2 µm)	3.6 km	28 km	0.0006
2	H_2O	1370 – 1560 cm^{-1} (7.3 – 6.4 µm)	1396 – 1527 cm^{-1} (7.2 – 6.5 µm)	3.6 km	28 km	0.0021
3	O_3	926 – 1141 cm^{-1} (10.8 – 8.8 µm)	947 – 1103 cm^{-1} (10.6 – 9.1 µm)	1.8 km	18 km	0.0035
4	HNO_3	844 – 917 cm^{-1} (11.8 – 10.9 µm)	859 – 900 cm^{-1} (11.6 – 11.1 µm)	1.8 km	18 km	0.0015
5	CO_2W	579 – 755 cm^{-1} (17.3 – 13.2 µm)	595 – 739 cm^{-1} (16.8 – 13.5 µm)	1.8 km	18 km	0.005
6	CO_2N	637 – 673 cm^{-1} (15.7 – 14.9 µm)	645 – 673 cm^{-1} (15.5 – 14.9 µm)	1.8 km	18 km	0.0013

4.4 Calibration

The LIMS has undergone preflight calibrations in the laboratory under simulated flight environments to determine its geometric, spectral and system response to known magnitude and positions of radiance sources. These calibration procedures were performed to an absolute accuracy approaching one percent.

4.4.1 Preflight Calibration

The preflight calibrations included the following procedures:

- Calibrations of the encoder used to measure angular position of the plane mirror in scan space

- Determination of the spectral response of each LIMS channel

- Determination of the field-of-view response of each LIMS channel in the scan plane

- Determination of the response of the LIMS system to a variable blackbody radiance source for:

    (a) eight points over the dynamic range of all channels

    (b) three different environmental temperatures of the spacecraft

- Determination of the optical and radiometric characteristics of the in-flight calibration source (IFC) for all LIMS channels.

The first procedure establishes where the instrument is viewing at any time in relative scan space. The second and third procedures are important in determining what is being observed optically across the band pass of each channel and how the limb radiance is being averaged spatially.

Primary calibration (the fourth procedure) determines the response of all LIMS channels to varying radiance sources over the expected dynamic range of limb radiances and for the range of possible spacecraft temperatures anticipated during the mission. This calibration procedure establishes the functional relationship (nearly linear) between instrument response and scene radiance. Once this response is established for the operating environment of the satellite, it is assumed to hold throughout the mission. A large scene radiance (the IFC source) and a zero scene radiance (space view) are then used during flight to determine absolute radiance values.

The last procedure is critical to the absolute accuracy of the flight data. The IFC source provides a signal that establishes system response to a large (warm) radiance source. Absolute accuracy is possible only if the characteristics of the IFC are known and its temperature is monitored accurately.

4.4.2 In-flight Calibration

During flight the scan mode automatically sequences the LIMS field of view to focus the detectors on deep space and then on the in-flight calibration source operating at 310 K. Each position, space and IFC, is viewed for two seconds. The sequence is repeated every second scan cycle. This provides a two point instrument gain determination. The IFC becomes the tie point to the ground calibration. Assuming the instrument linearity characteristics do not change, the IFC and space data coupled with the ground calibration curve can be used to remove any small residual nonlinearity effects and provide a continuous updated absolute calibration.

4.5 LIMS Operational Modes

The operational scan modes of LIMS were described in Section 4.3.1. The acquisition mode will rarely be used under normal circumstances. Almost all data will be measured with the instrument in the adaptive scan, limb track mode.

During acquisition, radiation from the narrow band CO_2 channel is peak detected during the 12-degree acquisition scan. Once the limb is acquired, the scan mirror subsystem switches to the adaptive scan mode during which time limb radiance profiles are recorded over a ±1.5 degree scan about the limb centerline (a nominal 45 km altitude in the atmosphere). From a tangent point at the earth's horizon, the range of altitude of the adaptive scan is between −45 km to 135 km. The limb track position is updated after each scan through the limb.

The six LIMS channels measure radiance profiles simultaneously. As shown previously in Section 4-3, however, the NO_2 data is offset by about 14 km from the O_3 and HNO_3, and the latter by about 14 km from the two CO_2 channels, and so on. The sub-tangent point moves a distance of about 22 km in the time it takes for the top channel (H_2O) to scan across the same altitude point as the bottom channel (NO_2).

The LIMS lifetime is presently estimated at seven months and is limited by the life of the solid cryogen cooler. On-off sequencing causes no reliability problems in the instrument but can have an important effect on the usefulness of LIMS results for some investigations. Fourier analysis of the data to study wave propagation modes is one of these. Data gaps up to one day can be tolerated with an acceptable loss of information, but gaps no greater than one-half day are more desirable. Random rather than regular spacings of the gaps is preferred in order to avoid a sampling bias. Data is recorded continuously while LIMS is on. The expected daily LIMS coverage was shown in Section 4.3.1. The latitude range from 64 degrees south to 84 degrees north is scanned on each orbit. Normal operations provide vertical concentration profiles at about one degree latitude intervals along each tangent point track.

4.6 Data Processing, Products, and Availability

4.6.1 Data Processing

The raw telemetry data for LIMS are processed at GSFC and written on computer-compatible tapes. These tapes are mailed to the National Center for Atmospheric Research (NCAR) in Boulder, Colorado for processing into archival tape products of radiance observations and derived atmospheric variables as listed in Section 4.6.2 and illustrated in Figure 4-14. These tapes are returned to IPD at GSFC. IPD copies some tapes and sends them to NET users. IPD uses other tapes as input to generate the LIMS microfilm displays. All original tapes and microfilms are then sent to NSSDC for archiving. NSSDC makes copies of these products for users.

4.6.2 Tape Products

The following tapes are produced by NCAR, used by IPD, and then sent to the NSSDC for archiving. Brief descriptions of these tapes are as follows:

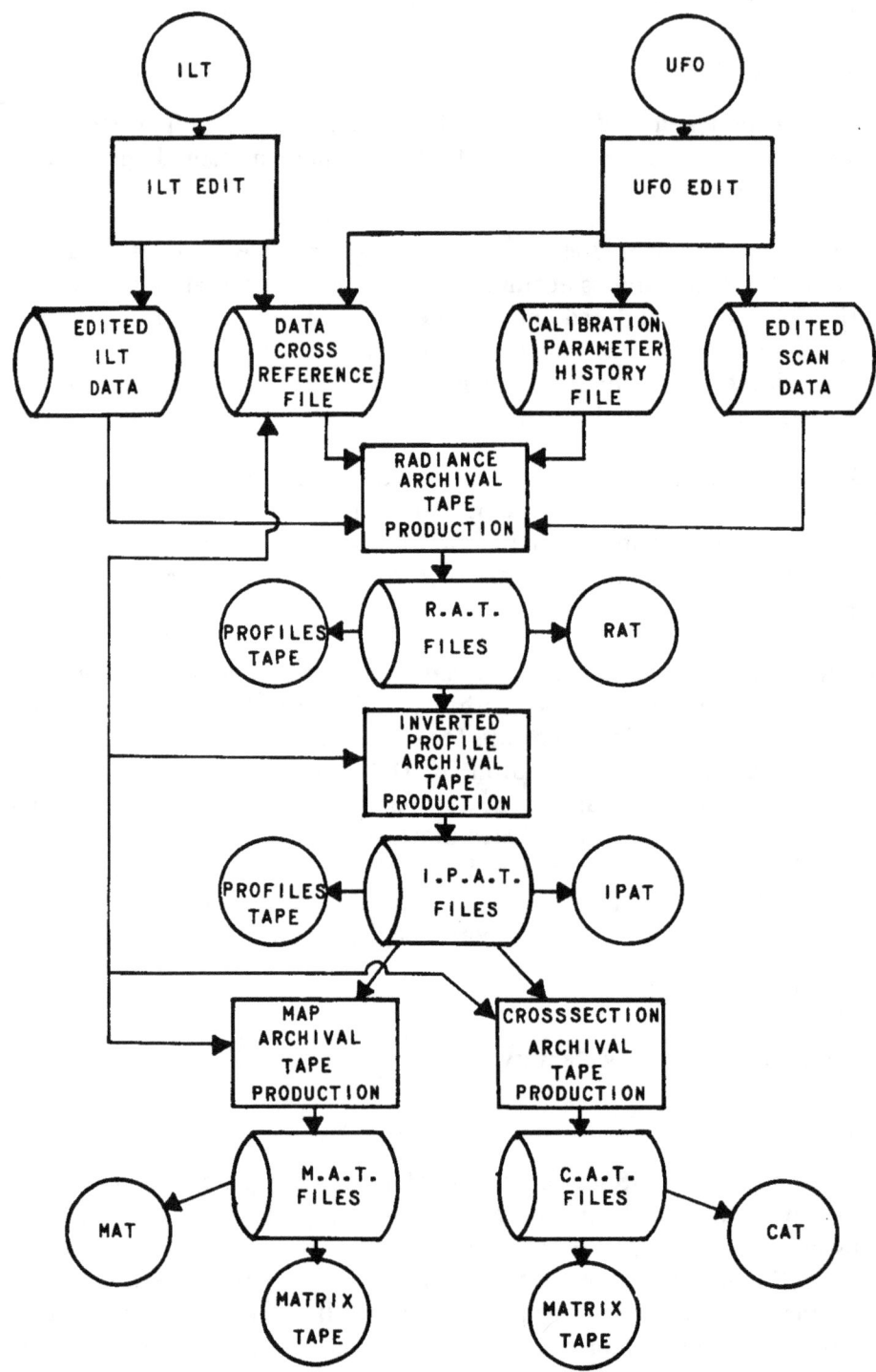

Figure 4-14. LIMS Data Processing Flow at NCAR

- RAT (Radiance Archival Tape)

    These contain the most elementary and complete form of the experimental observations consisting of all useful radiance scans plus time and scan angle references, spacecraft location and attitude, solar location, and sensor calibration.

- IPAT (Inverted Profile Archival Tape)

    These contain the most useful form of the basic data on the RAT's. These consist of parameter profiles of ozone, water vapor, nitric acid, and nitrogen dioxide at selected standard latitudes. The radiance profiles used to make parameter profiles are also given. All parameter profiles are given as a function of pressure at a vertical resolution of 1.5 km.

- MAT (Map Archival Tape)

    These contain daily world map grids (64°S to 84°N) of harmonic coefficients for each parameter at standard pressure levels. Coefficients are derived from the data on the IPAT's. Data are organized and averaged into 38 four-degree latitude bands symmetric about the equator. Data for the daily maps are interpolated to synoptic times (0000 GMT and 1200 GMT).

- SMAT (Stacked Map Archival Tape)

    These contain one month and three month world map grid averages of the daily maps on the MAT's. The data format is almost identical to the MAT.

- CAT (Cross-section Archival Tape)

    These contain north-to-south grid-point cross-section depictions of all parameters compiled in a cross section format. Data are latitudinally averaged over selected longitudinal sections (several orbits in a row) and for each day. The tapes also contain orbital values of cloudtop pressure values referenced by latitude. All data values are derived from processing the IPAT'S.

- SCAT (Summary Cross-section Archival Tape)

    These contain the same parameters as on the CAT's, but averaged over periods of one month and three months. Data format is almost identical to the CAT.

- MATRIX-M (Map Data Matrix Tape)

    These contain one northern and one southern hemisphere polar stereographic map (0° − 90°) and one Mercator map (±32°) record of countour values for each parameter for each time period to be on microfilm map displays.

- MATRIX-C (Cross-section Data Matrix Tape)

    These contain cross-section records of contour values (64°S − 84°N) for each parameter to be on microfilm as a cross-section display. Each record also contains all necessary reference and title information needed for annotation of each of these displays. It also contains all necessary reference and title information needed for annotation of each display.

- PROFILE-R (Radiance Profile Tape)

   These contain arrays of radiance versus scan angle selected from the RAT's for various time periods throughout the lifetime of the experiment. These arrays are used to make radiance versus scan angle profile displays on microfilm. Each tape record also contains all necessary reference and title information for complete annotation of each of these displays.

- PROFILE-I (Inverted Profiles and Radiance Tape)

   These contain arrays of matching parameter and corrected radiance profiles as a function of pressure selected from the IPAT's. Each record also contains all necessary reference and title information for complete annotation of each parameter and radiance profile display.

The form and content of each of these tapes is specified in a tape specification document for each tape type. The appropriate documents will accompany a tape shipment to a user. See Section 1.5 of this document for details.

4.6.3   Display Products

The LIMS data are displayed on 84 different map sets, 21 different cross section sets, 2 profile sets, and 1 plot. Six parameters are mapped: temperature, ozone, nitrogen dioxide, water vapor, nitric acid, and geopotential height. Each parameter is mapped at 14 pressure levels for each day, month, and three month time period. (See Figures 4-15 and 4-16 for map display examples.) The same six parameters are displayed on 64°S to 84°N cross sections for 60 degree longitude zones each day (Figure 4-17) plus the 64°S to 84°N latitudinal averages for each day, each month, and every three months (Figure 4-18). Periodically, selected LIMS values are displayed as profiles of radiances (Figure 4-19). Each profile is for a 24-second scan period. Routinely (approximately 64 displays per orbit), selected radiance profiles are inverted and displayed as parameter profiles of temperature, ozone, nitrogen dioxide, nitric acid, and water vapor (Figure 4-20). Plots of cloudtop heights (along the tangent point tracks) are generated for each orbit, and seven orbital plots are then produced on one display (Figure 4-21).

Table 4-2 lists the titles of all LIMS microfilm displays and the corresponding film specification number for each. Most parameters are produced at more than one time scale as Table 4-2 shows. Interpreting the fourth digit from the left in each specification number gives the frequency of production of the parameter listed in the title:

   XXX0XX = produced orbitally, or more often,

   XXX1XX = produced every day,

   XXX7XX = produced every month, and

   XXX8XX = produced every three months.

All map displays contain one north and one south polar stereographic projection (pole to equator for each) and one Mercator projection (to ±32°). Each map contains contoured data as specified in the display title. Immediately beneath each Mercator map is contouring information giving the contour units, interval between contour lines, and the maximum and minimum value contoured. All map displays contain an indicator of the quantity of data within a display. On the one day displays there is a "missing orbits per day" code specifying how many orbits of data are missing from that

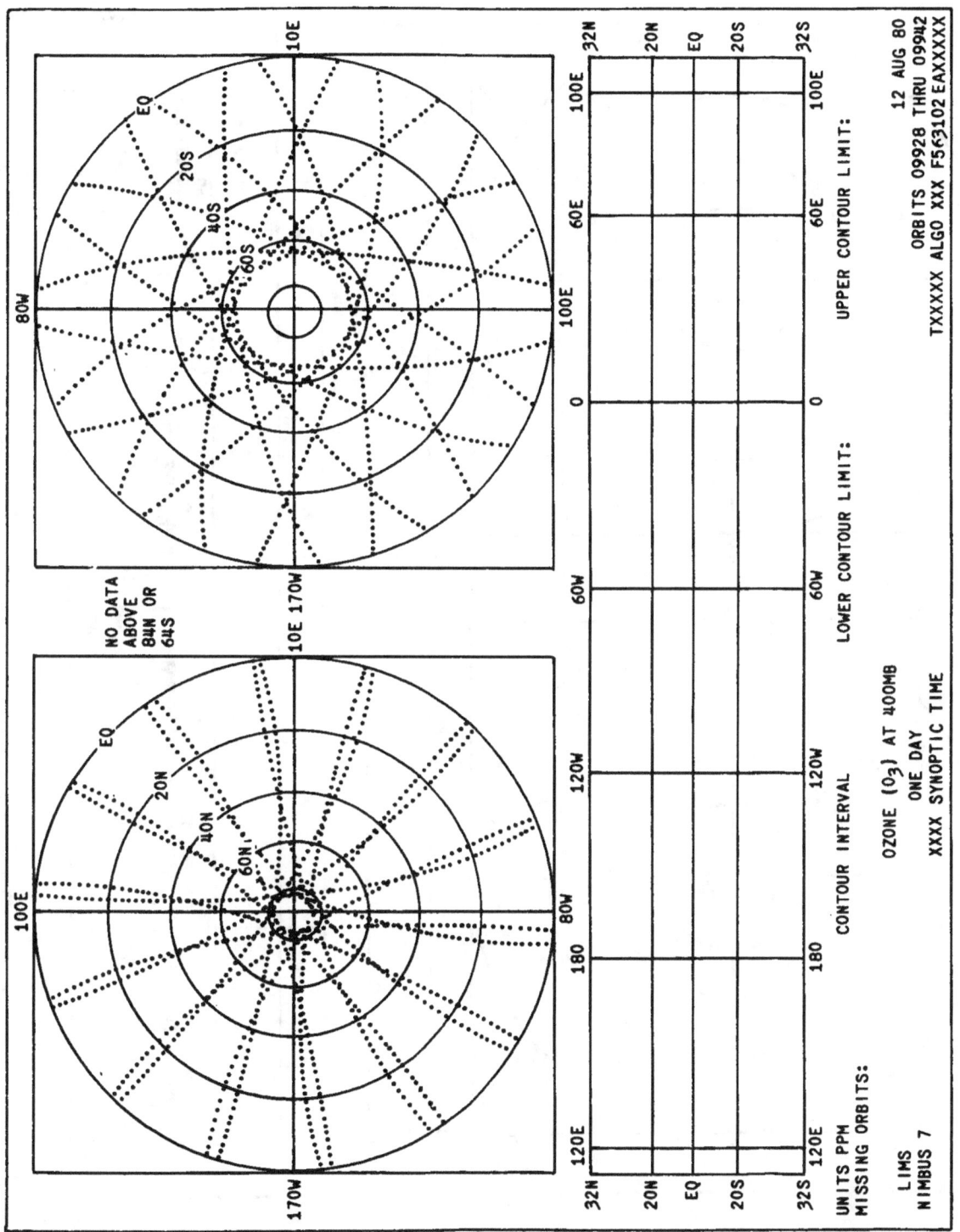

Figure 4-15. LIMS Microfilm Map Display Format for One Day Products

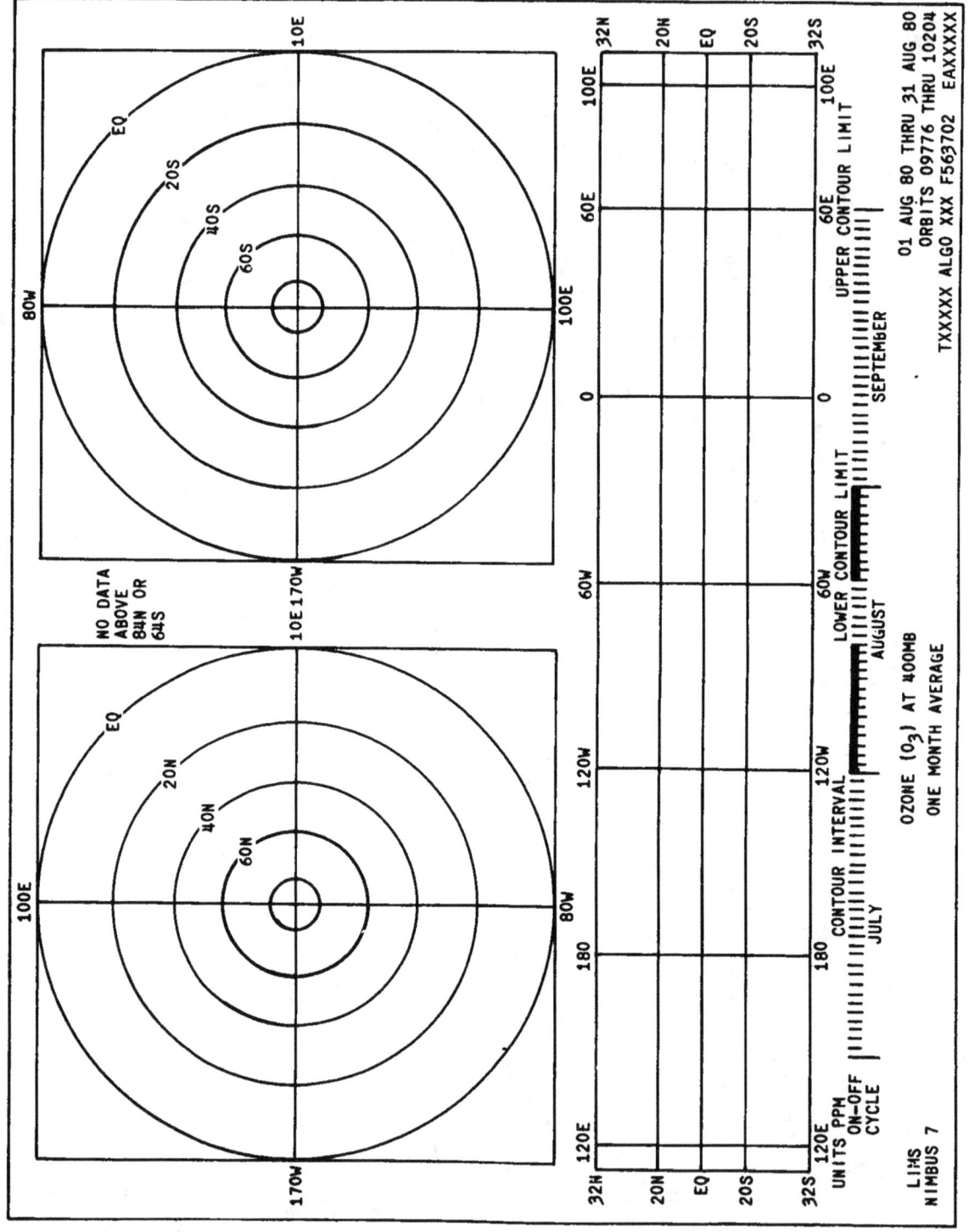

Figure 4-16. LIMS Microfilm Map Display Format for One Month and Three Month Products

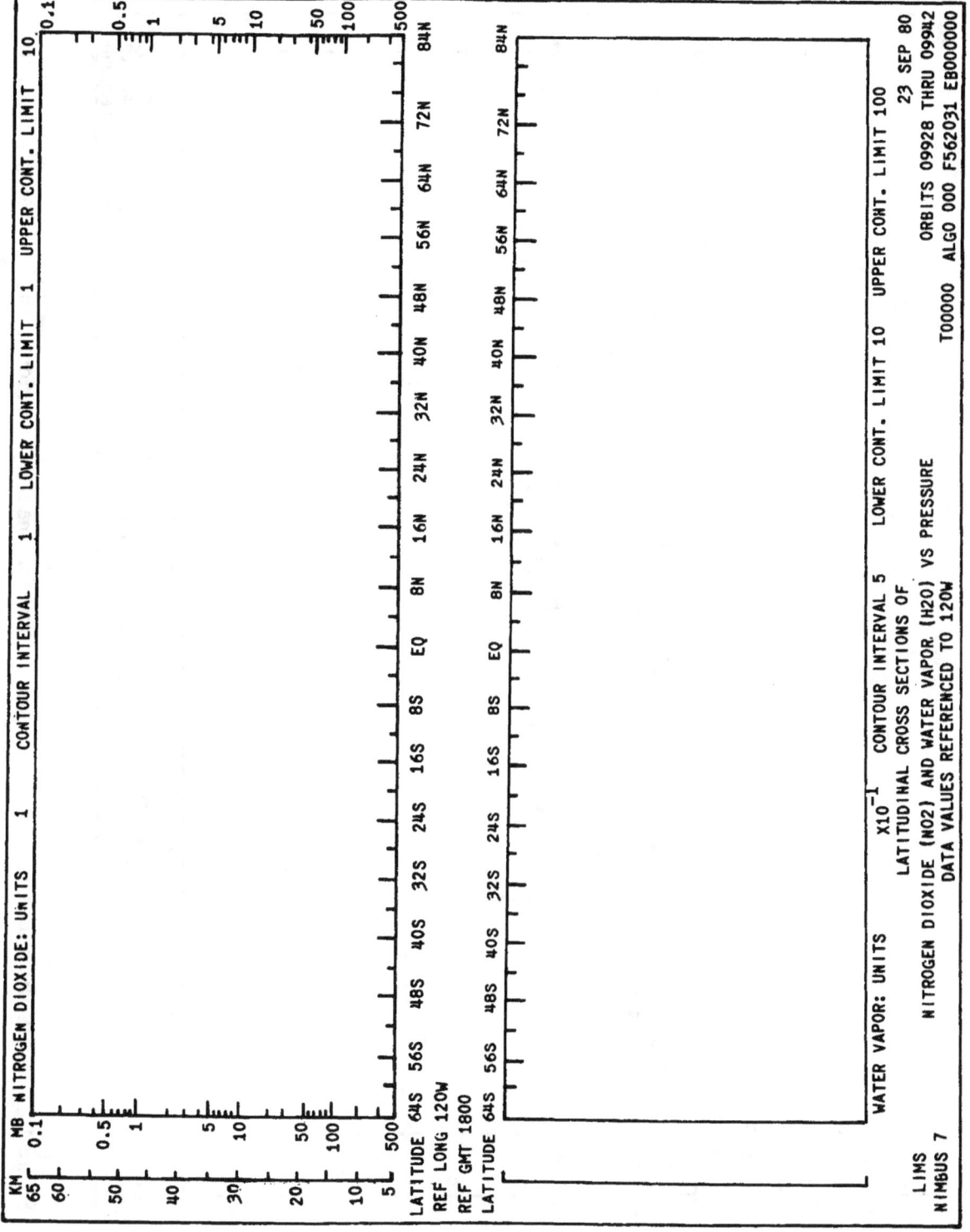

Figure 4-17. LIMS Microfilm Cross Section Display Format for 60 Degree Longitude Zones

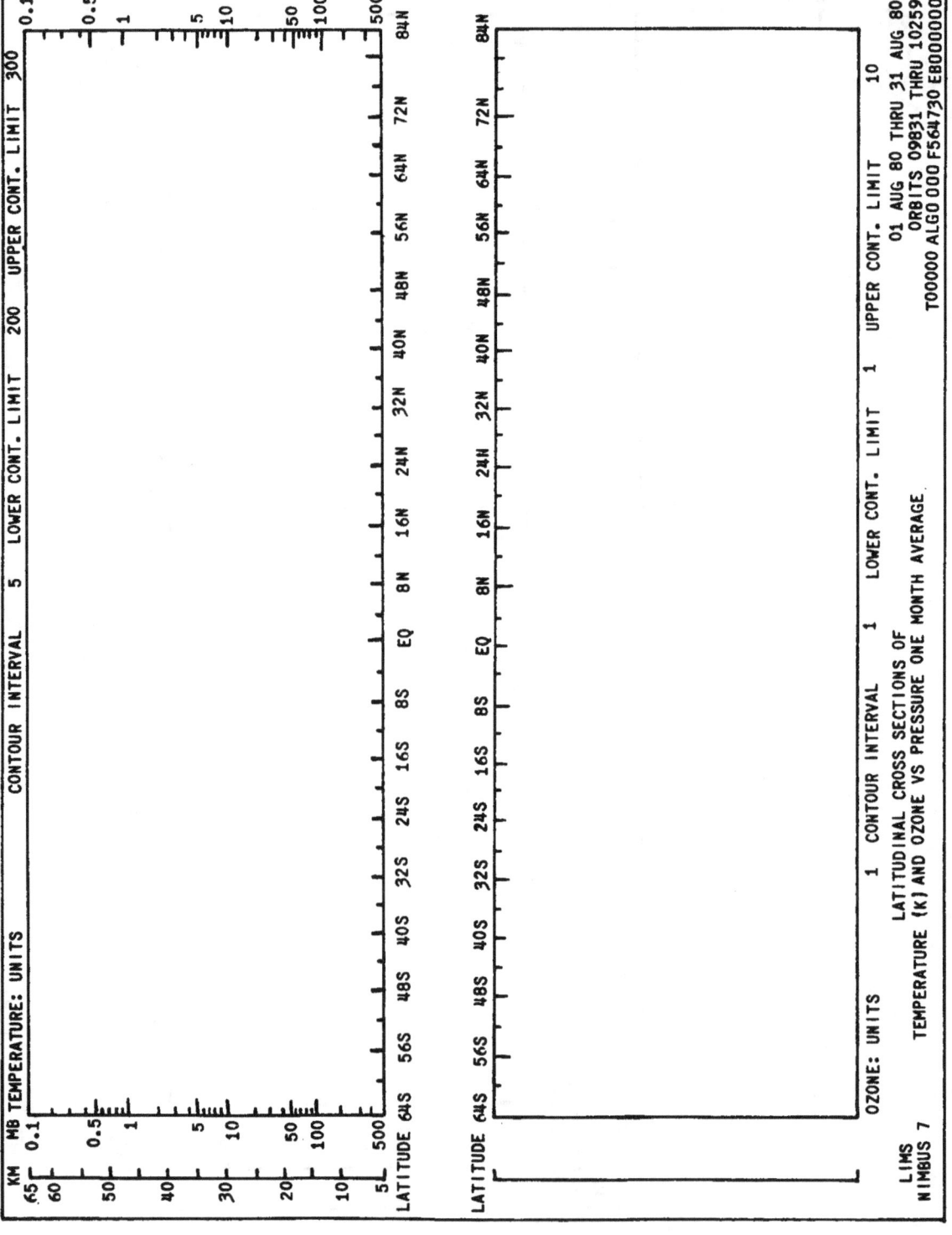

Figure 4-18. LIMS Microfilm Cross Section Display Format for One Day, One Month, and Three Month Products

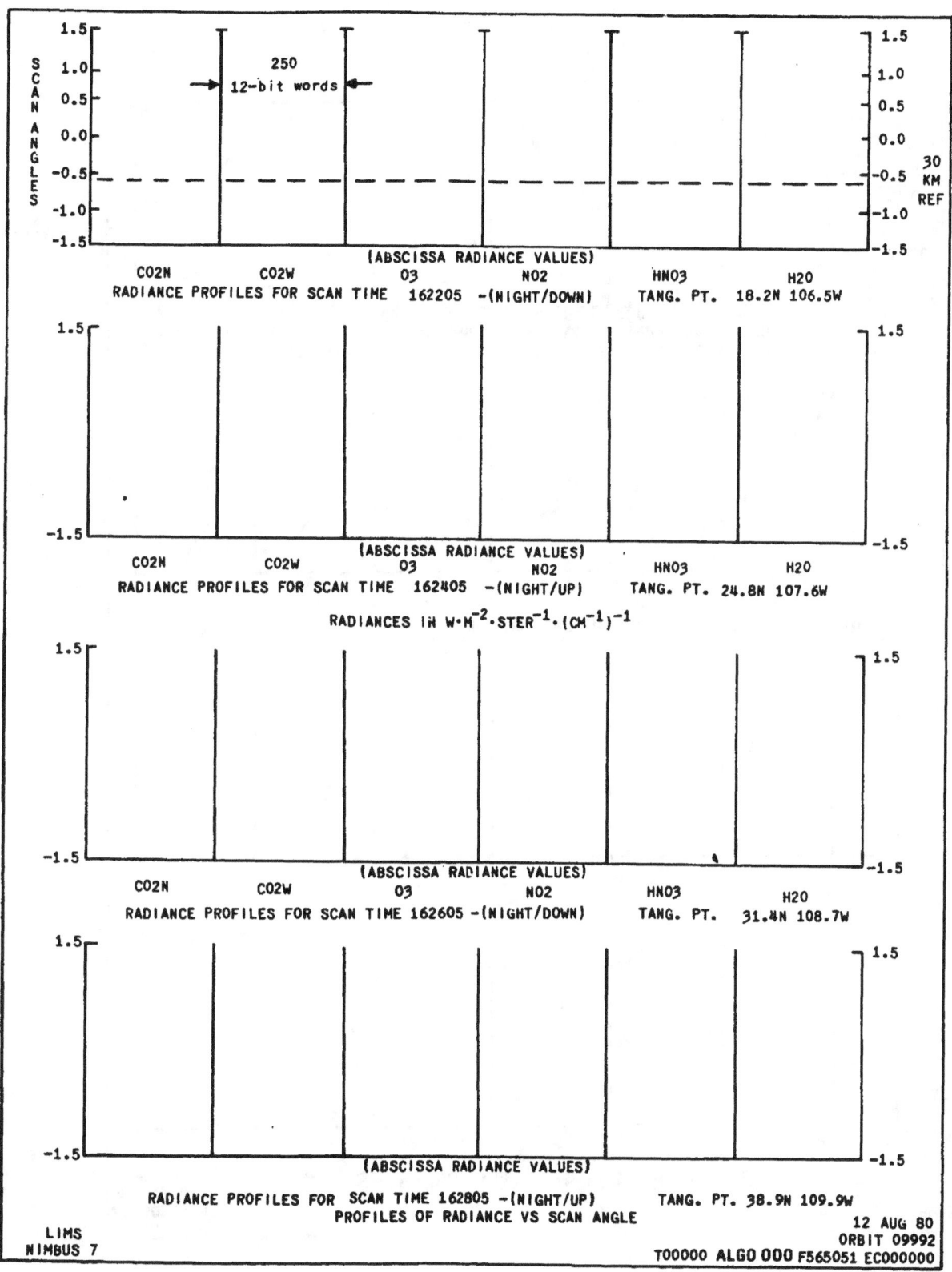

Figure 4-19. LIMS Microfilm Profile Display Format of Radiances versus Scan Angle

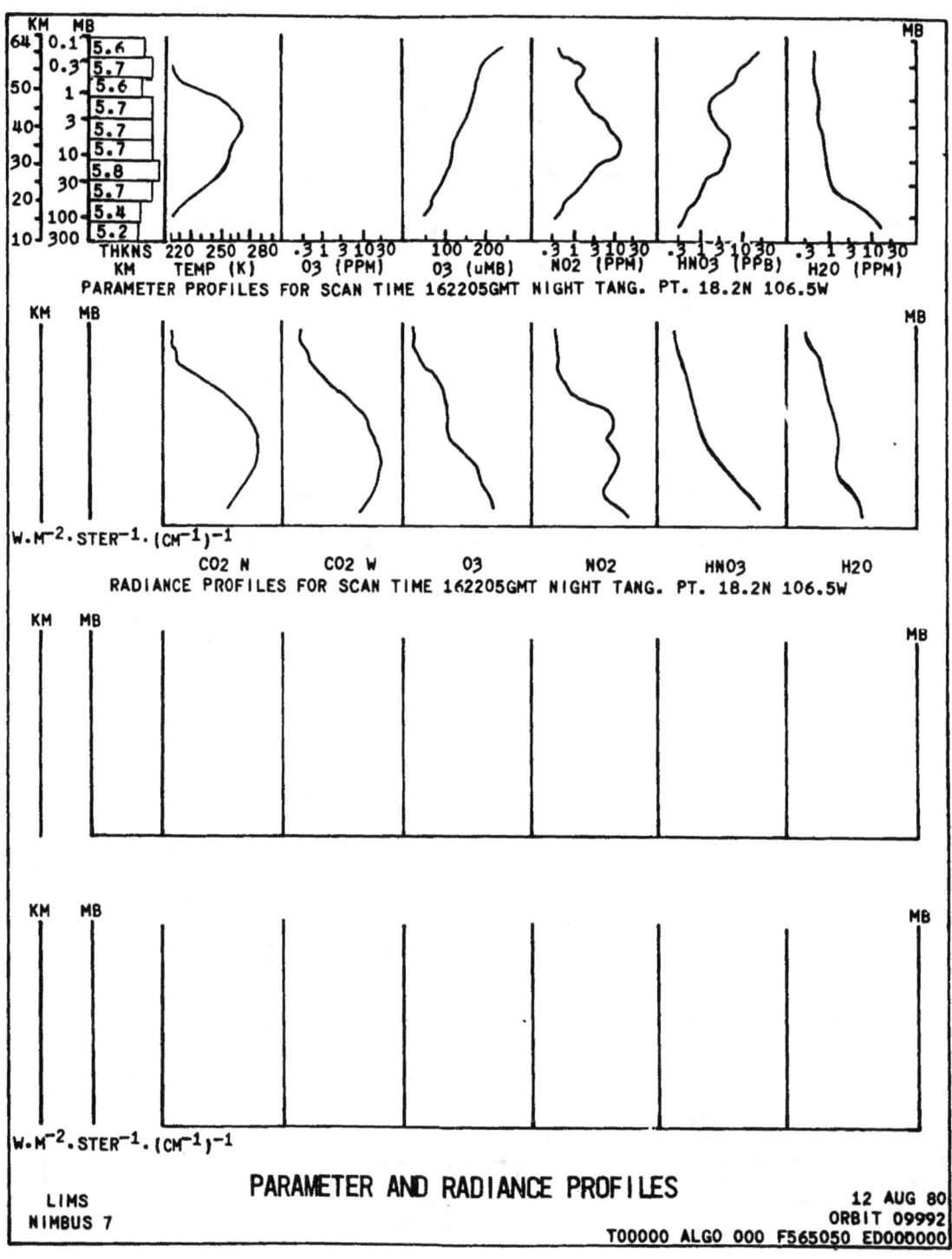

Figure 4-20. LIMS Microfilm Display Format of Parameter and Radiance Profiles

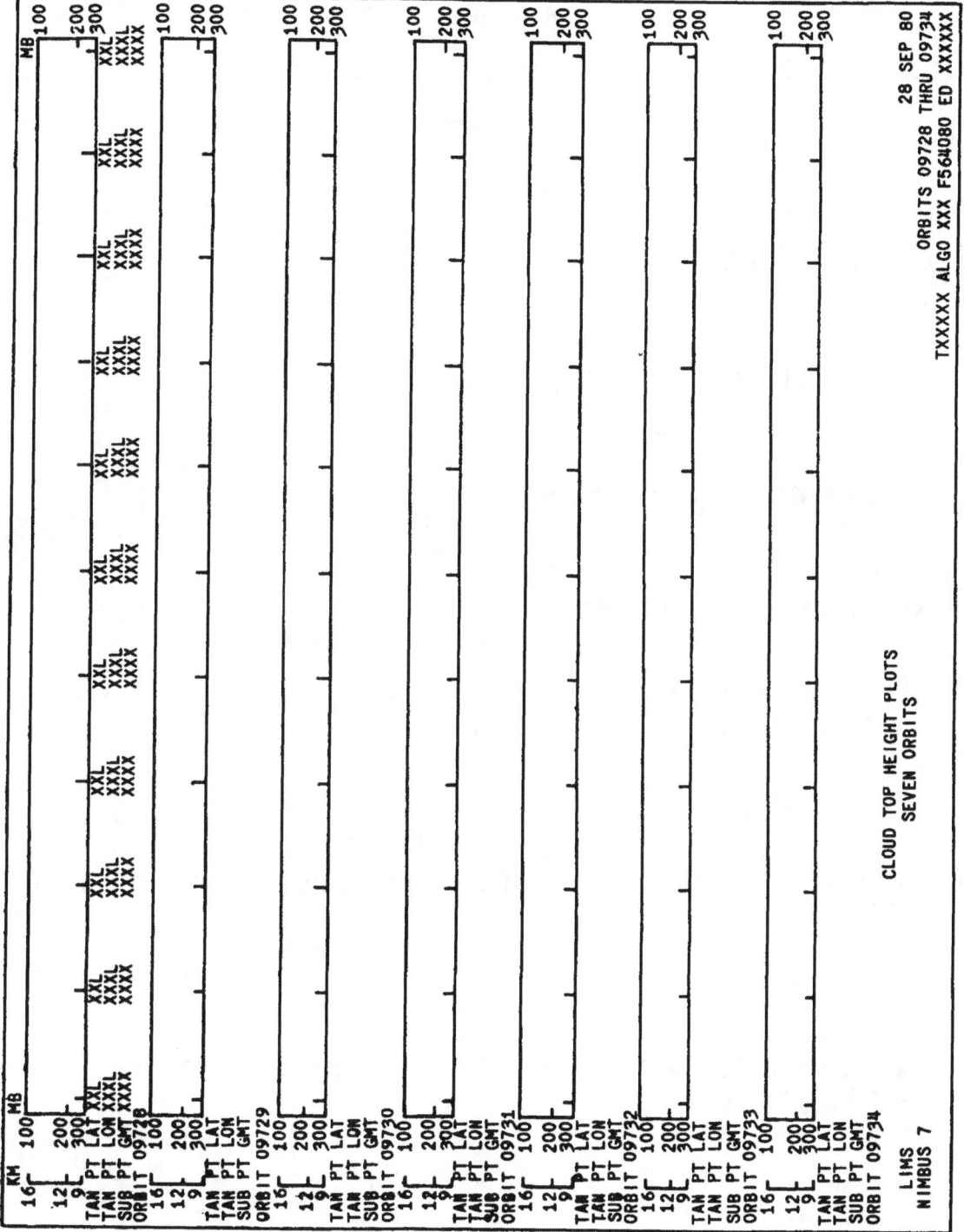

Figure 4-21. LIMS Microfilm Display Format of Cloudtop Height Plots

Table 4-2
Titles and Film Specification Numbers for LIMS
Microfilm Products

Film Spec Number	Film Product Title
	MAPS
563101	TEMPERATURE (K) AT XXX*MB
563701	
563801	
563102	OZONE (O_3) AT XXX MB
563702	
563802	
563103	NITROGEN DIOXIDE (NO_2) AT XXX MB
563703	
563803	
563104	WATER VAPOR (H_2O) AT XXX MB
563704	
563804	
563105	NITRIC ACID (HNO_3) AT XXX MB
563705	
563805	
563106	GEOPOTENTIAL HEIGHT (h) AT XXX MB
563706	
563806	
	CROSS SECTIONS
564030	LATITUDINAL CROSS SECTIONS OF TEMPERATURE (T) AND OZONE (O_3) VS PRESSURE
564130	
564730	
564830	
564031	LATITUDINAL CROSS SECTIONS OF NITROGEN DIOXIDE (NO_2) AND WATER VAPOR (H_2O) VS PRESSURE
564131	
564731	
564831	
564032	LATITUDINAL CROSS SECTIONS OF NITRIC ACID (HNO_3) AND GEOPOTENTIAL HEIGHT (h) VS PRESSURE

Table 4-2 (Continued)

Film Spec Number	Film Product Title
564132	
564732	
564832	
	<u>PROFILES</u>
565050	PARAMETER AND RADIANCE PROFILES
565051	PROFILES OF RADIANCES VS SCAN ANGLE
	<u>PLOTS</u>
567080	CLOUD TOP HEIGHT PLOTS

*Each mapped parameter is contoured and displayed at 14 pressure levels. These are: 400, 300, 200, 100, 70, 50, 30, 10, 5, 2, 1, 0.4, 0.2, and 0.1 millibars.

day's input to the map (See Figure 4-15). On the one month and three month maps (Figure 4-16) there is an "on-off cycle" scale specifying the days during the display period when the instrument was on and off. If a day is "filled in" the instrument was on, the data was collected, interpreted, and used in the contouring. If a day is not filled in, the instrument was either off or the data was unusable for some reason.

The two polar stereographic maps on the one day displays (Figure 4-15 is an example) contain "dot tracks" with each dot representing a tangent point data value location used in the construction of the contours on the maps.

Figure 4-17 is an example of a longitudinal zone (60°) cross section display constructed with data from two to four consecutive orbits. Two parameters are on each display; one cross section for each parameter. As the display indicates, the orbits used in constructing the cross section are interpolated to a reference GMT and a reference longitude. Data are referenced vertically by pressure and standard atmosphere altitudes, and horizontally by latitudes. Figure 4-18 is an example of the one day, one month, and three month cross section displays. The format is identical to Figure 4-17 except that these longer-period displays have no reference GMT and reference longitude.

Figure 4-19 illustrates the format for the radiance versus scan angle displays. On each set of profiles the data are referenced vertically by the LIMS instrument scan angle and the location of the 30 kilometer altitude location, and are located horizontally by radiances. Four profile sets are presented on each display.

Figure 4-20 illustrates the format for the parameter and radiance profile sets. Data values are referenced vertically by pressure and standard atmosphere altitudes, and horizontally by the parameter values as shown. (The "THKNS" parameter, at the left, provides the calculated distances in the atmosphere, in tenths of a kilometer, between two predefined pressure limits — as shown by the horizontal lines over to the pressure scale.)

Figure 4-21 illustrates the format for the cloudtop height plots. Each plot shows the pressure and corresponding height of the tops of the clouds above 300 mb — as determined from processing LIMS data. Where the tops are below 300 mb the plotted line drops below the lower limit of the pressure scale and is not shown. Data are referenced vertically by pressure and standard atmosphere altitudes, and horizontally by eleven tangent point latitude-longitude locations, plus eleven satellite subpoint GMT's corresponding to the tangent point locations.

Title and reference information at the bottom of all displays is mostly self-explanatory. The right half of the last line, however, requires explanation. These items are: the physical tape number the data is stored on (TXXXXX), the algorithm reference number used in processing the data (ALGO XXX), the film specification number (F56XXXX), the project data format code (EA, EB, EC, or ED), and the film frame number (XXXXXX).

4.6.4 Data Availability

The LIMS experimental data, consisting of the magnetic tapes described in Section 4.6.2 and the 16 mm microfilm displays listed and illustrated in Section 4.6.3, are archived at the NSSDC. The tape data are also archived at NCAR for easy access to universities having computer ties to NCAR.

All tape products except the RAT are in a form for direct use in scientific investigations of the stratosphere and mesosphere. The RAT product is available for algorithm research and other uses but such studies can only be carried out at large computing centers due to the large number of tapes involved. The RAT products will be archived first followed by the other data. It is anticipated that the first month of RAT data will be archived about six months after launch.

Users requesting LIMS data from NSSDC should read Section 1.5 of this document for general tape and film ordering information.

4.7 Planned Investigations and Data Applications

The LIMS NET has conceived a number of investigations for both validation of the data set prior to public archiving and for demonstration of initial data use in studying the stratosphere.

The validation investigations include evaluation of the accuracy and precision of the LIMS orbital measurements, study of the consistency of the LIMS measured temperature, O_3, HNO_3, NO_2, and H_2O with alternate techniques, and comparison of temperature and geopotential maps with NMC data. Alternative inversion methods will also be tested against the standard LIMS approach to ensure that the best possible data are achieved. Cloud information algorithms will be validated and tuned to allow for proper removal or consideration of cloud effects in all channels. Comparisons will also be made with other experiments on Nimbus 7 and other satellites. For example, LIMS and SAMS temperature and H_2O maps and cross sections will be compared, LIMS ozone will be compared with SBUV/TOMS and SAGE results, and all LIMS results will be correlated with SAM II and SAGE aerosol results to ensure that effects due to thin cirrus clouds and aerosols are not overlooked in the archived data. These investigations should provide the required knowledge and confidence in the LIMS data and a proper assessment of data quality prior to release of the results to the general science community.

A number of atmospheric research investigations are planned by LIMS NET members to show the initial utility of the data. These include the following:

- Study of the stratospheric ozone – nitrogen chemistry using SAMS and LIMS data

- Study of atmospheric dynamics, including propagation of planetary waves, detection and determination of properties for equatorial waves, five-day waves, stratopause instabilities and high frequency waves

- Correlation of temperature and ozone

- Study of LIMS data correlations with solar variability and other processes

- Study of transports and budgets of chemical constituents, heat, momentum, energy and potential vorticity

- Preparation of atmosphere climatology

- Study of diurnal variations in trace constituents and temperature

- Use of constituent profiles to study the radiative budget of the middle atmosphere

- Use of LIMS data in the lower stratosphere to study troposphere-stratosphere exchange

- Ozone secondary maxima studies

- Interpret and map global distribution of background cloud radiance

- Study to improve transmittances

- Study of improved inversion algorithms

These investigations will demonstrate the scope and usefulness of the LIMS data set and should stimulate and guide further studies of the stratosphere and mesosphere.

4.8 References

1. Gille, J. C. and Russell, J. M., 1975: The Nimbus 6 User's Guide: LRIR, Section 7, pp. 141-161.

2. Gille, J. C. and House, F. B., 1971: On the Inversion of Limb Radiance Measurements I: Temperature and Thickness. J. Atmos. Sci., $\underline{28}$, pp. 1427-1442.

3. Russell, J. M., and Drayson, S. R., 1972: Inference of Atmospheric Ozone Using Satellite Horizon Measurements in the 1042 cm^{-1} Band. J. Atmos. Sci., $\underline{29}$, No. 2, pp. 376-390.

Acknowledgements

We express appreciation to Paul Bailey, Douglas Roewe, Gail Anderson, and William Kohri of the National Center for Atmospheric Research for their help in preparing this section.

SECTION 5

STRATOSPHERIC AEROSOL MEASUREMENT (SAM II) EXPERIMENT

by

M. P. McCormick, L. E. Mauldin, III, L. R. McMasters, and W. P. Chu
National Aeronautics and Space Administration
Langley Research Center, Hampton, Virginia 23665

and

T. Swissler
Systems and Applied Sciences
17 Research Drive
Hampton, Virginia 23666

5.1 Introduction

The Stratospheric Aerosol Measurement (SAM II) experiment provides the vertical distribution of stratospheric aerosols in the polar regions of both hemispheres. SAM II is a one-spectral channel sun photometer, centered at 1.0 μm, which views a small portion of the sun through the earth's atmosphere during spacecraft sunrise and sunset as shown in Figure 5-1. The time-dependent radiance thus measured each sunrise and sunset is combined with the spacecraft ephemeris data and local atmospheric density profile, and then numerically inverted to yield a one-kilometer height-resolved vertical profile of aerosol extinction above the earth tangent point. The Nimbus 7 orbit yields tangent points in the 64 degree to 80 degree latitude band in both hemispheres as shown in Figures 5-1 and 5-2.

The data SAM II provides are from the inaccessible polar regions where little aerosol data exist. Since aerosols have the potential to modify climate significantly and their microphysical and chemical interactions enter into a number of important environmental processes, these SAM II data provide valuable and needed information to aid in these studies. In addition, the effect of aerosols on some remote sensor techniques must be understood for their unique interpretation.

5.2 Scientific Objectives

The scientific objective of the SAM II experiment is to produce a baseline data set on stratospheric aerosols in high-latitude regions with one-kilometer vertical resolution. This set of aerosol data can be used for investigations in the following areas:

- Radiative transfer and climatic studies

- Aerosol transport, sources and sinks in the stratosphere

- Seasonal variations and sudden warning phenomena

- Volcanic injection phenomena

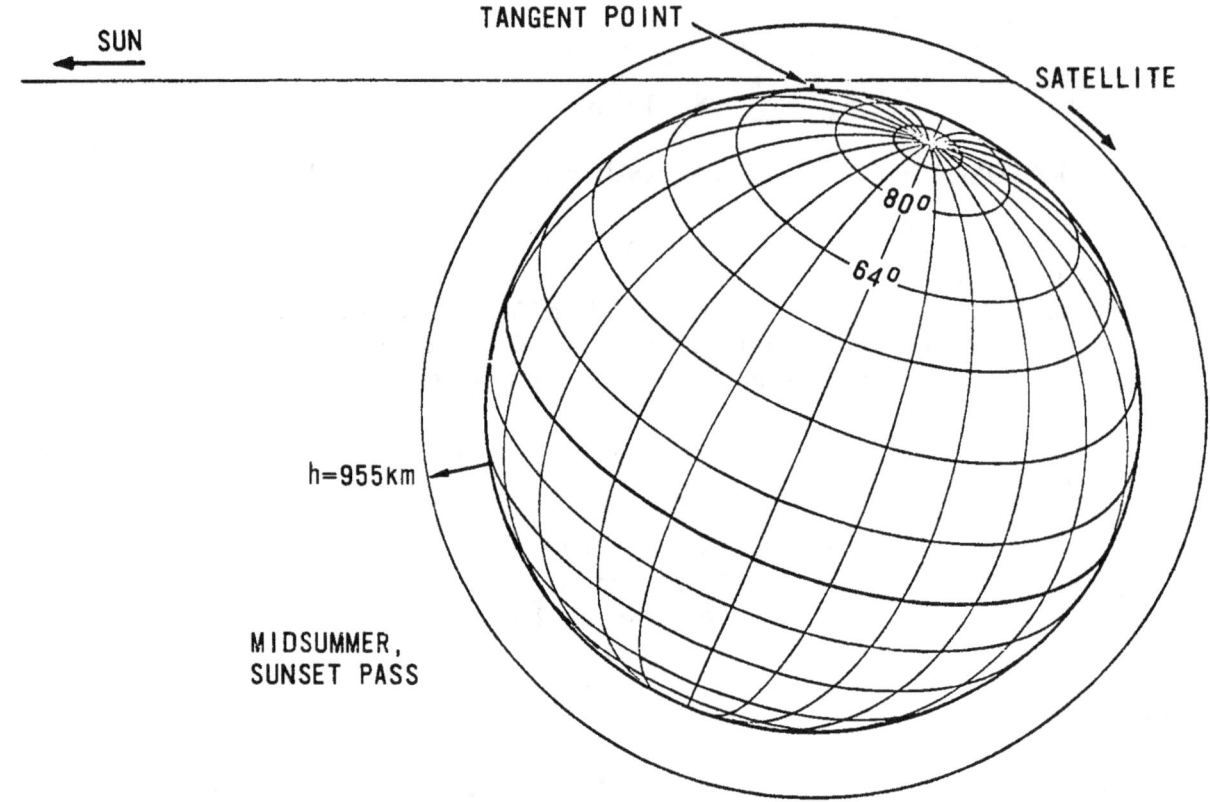

Figure 5-1. SAM II Orbital and Viewing Geometry

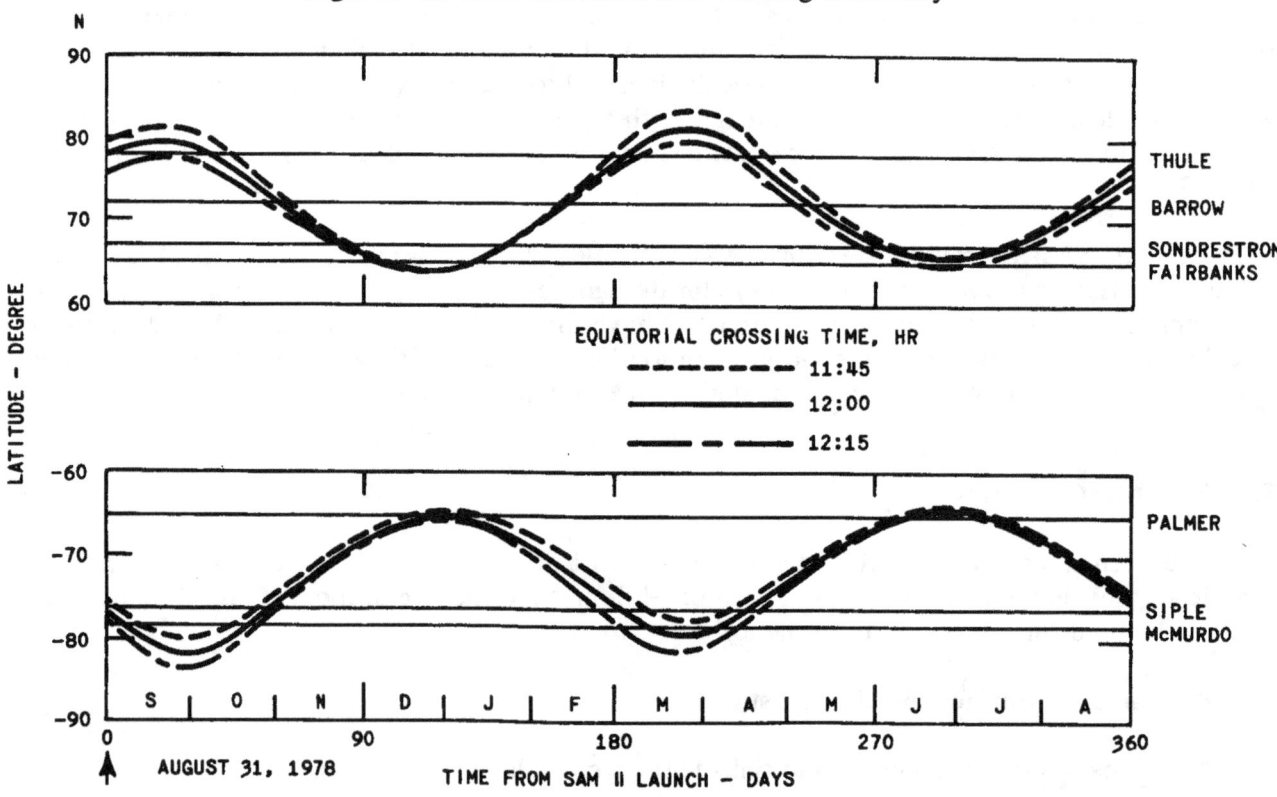

Figure 5-2. Latitude Coverage of SAM II Tangent Points For Sun-Synchronous High-Noon Orbit
(Effect of changing equatorial crossing time on tangent location dates is shown by dashed curves)

Mesospheric aerosols, noctilucent and nacreous clouds, and thin cirrus clouds near the tropopause will also be detected when their optical thickness is sufficiently large. These data are of use in all four areas listed above, as well as in water vapor studies.

5.3 Experiment Concept

The experimental approach for SAM II is solar occultation by the earth's atmosphere. The radiometer will track the solar disk during each spacecraft sunrise or sunset event in order to produce an atmospheric extinction profile at the 1.0 μm wavelength down to an atmospheric vertical altitude of below 10 km. The data acquisition mode for the SAM II radiometer is illustrated in Figure 5-3. The two solid lines denote respectively the image positions of the top and bottom of the solar disk during a sunset event as viewed from the spacecraft platform. The gradual shrinking of the vertical sun shape image is due to atmospheric refraction effects. The left ordinate denotes relative angle measured from the spacecraft coordinate system in units of arc minute, while the right ordinate denotes corresponding vertical tangent altitude for sun ray in kilometers. The horizontal abscissa denotes event time in seconds. The zig-zag dashed line in the figure represents a typical data-taking sequence. The radiometer vertically scans the solar disk up and down with respect to the earth's horizon. The nominal scan rate is 15 arc minutes per second. For a sunrise event, the data-taking mode is similar, with the time sequence in reverse. The radiometric data is sampled at a rate of 50 samples per second and is digitized to ten-bit accuracy. The instrument field of view of approximately 0.6 arc minute produces a vertical resolution better than one kilometer.

The measured irradiance from the SAM II instrument is related to atmospheric optical properties through the following equation:

$$H(t) = \int F(\theta,\phi) \, S(\theta,\phi,t) \exp[-\tau(\theta)] \, d\Omega \quad (1)$$

where $H(t)$ is the measured irradiance at time t, $S(\theta,\phi,t)$ is the extraterrestrial solar radiance profile at time t corrected for atmospheric refraction effects, $F(\theta,\phi)$ is the instrumental field-of-view function, and $\tau(\theta)$ is the optical thickness of the atmosphere for view angle θ. Since each view angle θ corresponds uniquely to an atmospheric tangent height h_T, the optical thickness $\tau(h_T)$ can be related to atmospheric extinction properties through the following equation:

$$\tau(h_T) = \int [\beta_a(h) + \beta_{ND}(h)] \, d\rho(h) \quad (2)$$

where $\beta_a(h)$ is the aerosol particulate extinction versus altitude profile, $\beta_{ND}(h)$ is neutral density versus altitude profile (Rayleigh scattering), and $\rho(h)$ is the path length through the atmosphere. The integral is evaluated from the spacecraft position to the sun.

The retrieval of aerosol extinction profiles from SAM II data is accomplished through the following two steps. First, the measured irradiance data are reduced together with spacecraft ephemeris data into a single profile of limb optical thickness, $\tau(h_T)$, as a function of tangent height, h_T, in the atmosphere. The high altitude solar scan profiles are used as a calibrated solar limb profile in this process. The second step is then to subtract the estimated neutral density contribution along each limb path to obtain the aerosol extinction profile. Dividing the atmosphere into N homogeneous layers, the integral equation can be reduced to a system of linear equations:

$$\tau_{ai} = \sum_{j=i}^{N} \rho_{ij} \beta_{aj} \quad (3)$$

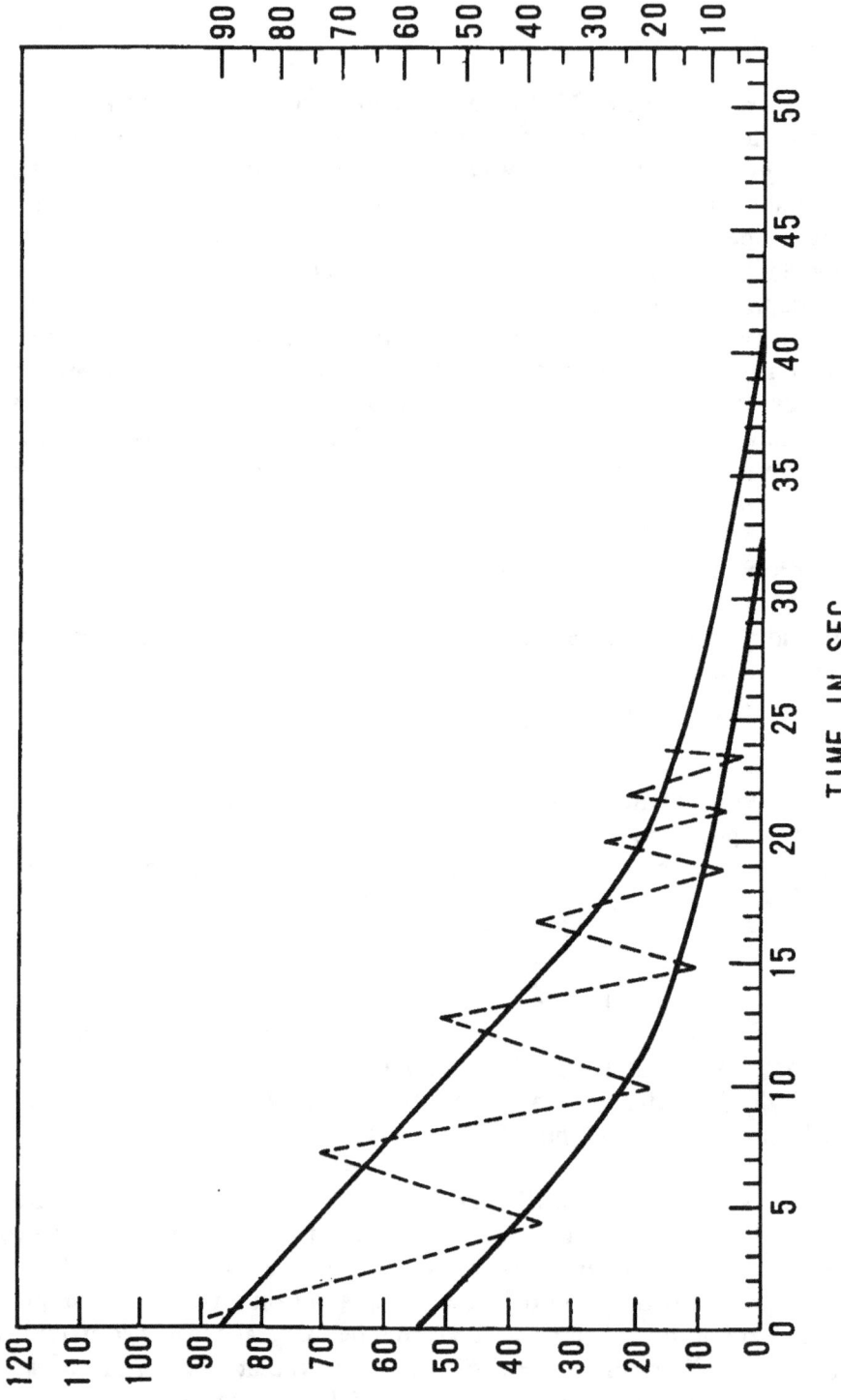

Figure 5-3. Sun Top and Bottom for Standard Atmosphere

where τ_a is the measured limb optical thickness profile for aerosols, ρ_{ij} is the sunray's path length in the jth layer with its tangent height at i^{th} layer, and β_{aj} is the averaged aerosol extinction coefficient for the jth layer. Equation (3) can be inverted to find the values of β_{aj} with different inversion methods such as the constrained smoothing method (Twomey) or an iterative approach (Chahine).

Figure 5-4 illustrates inversion results from synthetic data representative of background aerosol condition with typical instrument errors and a three percent uncertainty in neutral density. A total of ten computer simulations of the complete experimental event were performed and the subsequent inverted solutions analyzed. The maximum and minimum deviation of the ten inverted results are shown by error bars. The true profile is shown as a solid line. The inversion results indicate that aerosol layers between 10 to 20 km can be retrieved with an accuracy of about 10 percent for the assumed stratospheric aerosol conditions and uncertainty in neutral density profile.

5.4 SAM II Instrument

5.4.1 Physical Description

A diagram of the instrument is shown in Figure 5-5. The instrument weighs 17 kg, and the overall envelope is 36 cm by 20 cm by 51 cm. A cross section of the instrument is shown in Figure 5-6.

Solar input radiation is reflected from the scan mirror into the Cassegrainian telescope, and a solar image is formed at the slit plate. The slit plate contains two solar edge sensing detectors located on either side of the science detector aperture. Radiation passing through the science aperture is collected with a field lens, passes through an interference filter for wavelength discrimination, and finally is measured by a silicon photodiode detector.

The Instantaneous Field of View (IFOV) is controlled by the science aperture, and out-of-field rejection is accomplished by an optical baffling system (an image of the telescope secondary mirror is formed on the detector) and a mechanical baffle tube mounted to the telescope primary mirror. The scan mirror and telescope primary and secondary mirrors have silver coatings with an overcoat protective layer. The secondary mirror is mounted to a quartz entrance window which is coated to reject wavelengths shorter than 0.8 μm for thermal control and for providing wavelength discrimination for the solar edge sensing detectors.

The entire optical/detector system is contained in the azimuth gimbal. This gimbal is supported by ball bearings, a duplex pair at the bottom, and a single bearing at the top. Electrical power and data are transferred across the gimbal by two flexible ribbon cables mounted on each end of the gimbal shaft. Also mounted to the azimuth gimbal are two radiation balancing sun sensors to align the gimbal to the radiometric centroid of the sun, and a sun presence sensor.

The depression gimbal is contained within the azimuth gimbal, and consists of the scan mirror supported by flex pivots, a DC torque motor, a rotary variable differential transformer (RVDT), and the gimbal support structure.

The thermal control system is completely passive. Thermal insulation covers the instrument exterior surfaces to limit the radiative heat flow to outer space. The instrument is thermally tied to the observatory structure so that its base temperature will be at or near that of the observatory. The mechanical, optical, and electrical systems have been designed to be relatively insensitive to temperature within the -5°C to 45°C design range. The silicon science detector, however, is temperature

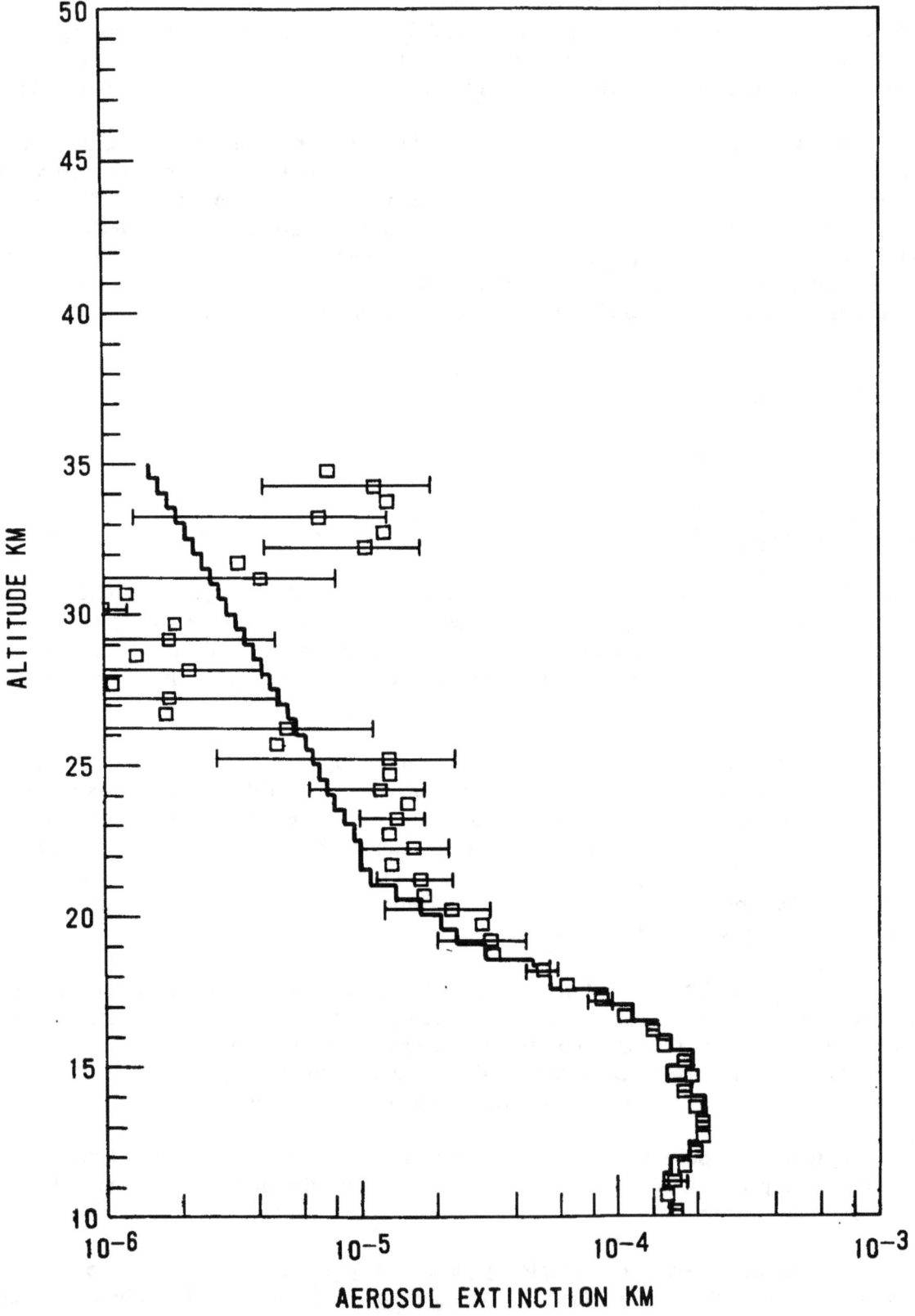

Figure 5-4. Inversion Results from Synthetic Data

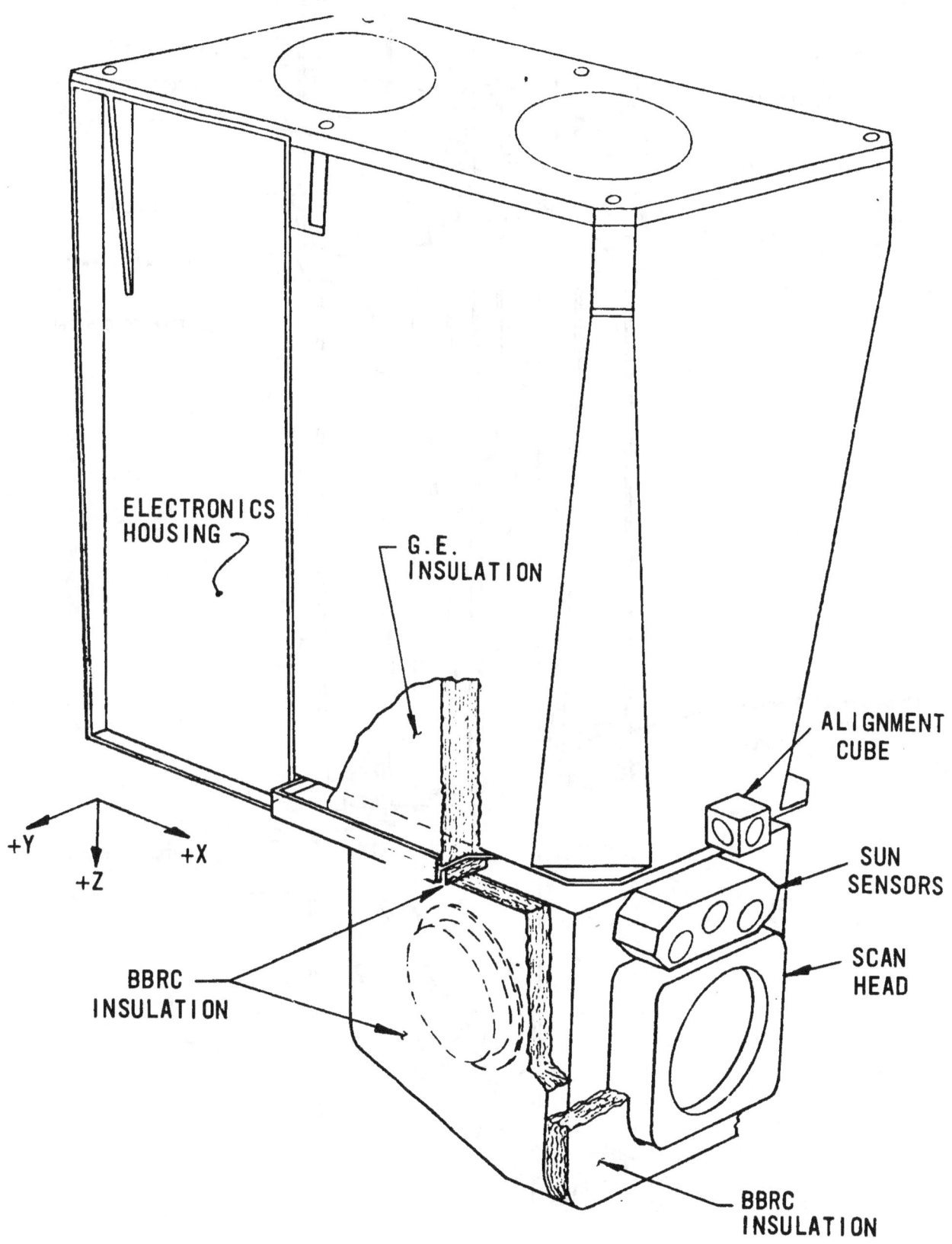

Figure 5-5. SAM II Instrument

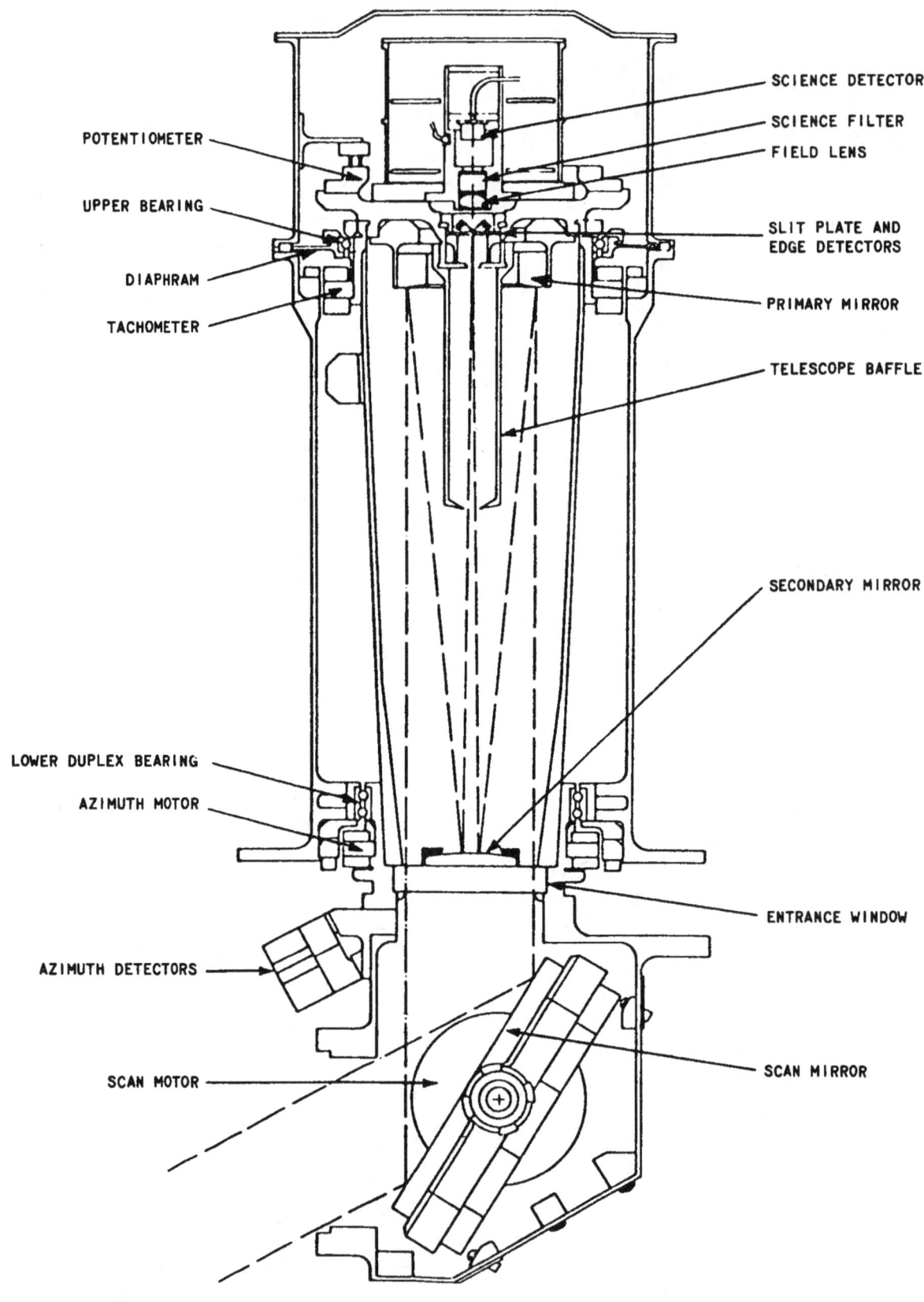

Figure 5-6. SAM II Instrument Cross Section

sensitive and its thermal design has concentrated on keeping it stable to the order of 0.1°C during a data-taking period of approximately 2.5 minutes.

The instrument contains 20 printed circuit (PC) boards mounted in a rectangular box, and these provide electronics for signal conditioning, pulse code modulation (PCM) encoding, timing and control, command, azimuth servo, depression servo, power conditioning, and housekeeping functions.

5.4.2 Functional Description

A block diagram of the instrument is shown in Figure 5-7. The SAM II instrument is designed to operate twice an orbit, once during observatory sunrise and once during sunset. Two data commands are stored in the spacecraft memory, one for sunrise and one for sunset. When these commands are activated the instrument automatically sequences through three modes. The three modes are described below.

5.4.2.1 Slew Mode

Execution of a data command initiates a timed slew mode where the azimuth gimbal slews at five degrees per second to the expected sun position (angle) stored in the azimuth register. The azimuth register contains a stored potentiometer position that is initially loaded from ground commands and is automatically updated to the proper position after the initial sun acquisition during a sunset event. This register is automatically updated during each sunset event to the proper position for the next sunrise and sunset event.

The azimuth position in the register is changed to an analog level in a D/A converter and compared to the actual position as indicated by the azimuth potentiometer. The difference in these signals is used as an error signal which is applied to the servo gain and compensation network and then to the motor driver which drives the motor until the error is nulled. A rate tachometer is used to limit the slew motion to 5 degrees per second. At the end of this timed mode, the azimuth gimbal has aligned itself to within 5 degrees of the radiometric centroid of the sun, and the instrument sequences to the pointing mode.

5.4.2.2 Pointing Mode

When the slew mode timer times out, the azimuth gimbal control is switched to the azimuth sun sensors. These two radiation balancing sensors generate a linear error signal about the radiometric centroid of the sun. When control is switched to these sensors, the azimuth gimbal is aligned to the radiometric centroid of the sun within 1 arc minute, and remains aligned to within 1 arc minute for the remainder of the data-taking event. When these sun sensors have reached null and the sun presence sensor output is above a set threshold, the instrument sequences to the scan mode.

5.4.2.3 Scan Mode

In the scan mode, which is also a timed mode, the depression servo drives the scan mirror in a linear scan cycle and science data are taken of solar intensity. The initial mirror scan rate is 180 arc minutes per second (line of sight). During this scan the solar disk image is driven toward the slit plate in the focal plane of the telescope. A diagram of the solar image scanning across the slit plate is shown in Figure 5-8. When both leading and trailing edges of the sun have passed the science aperture, the

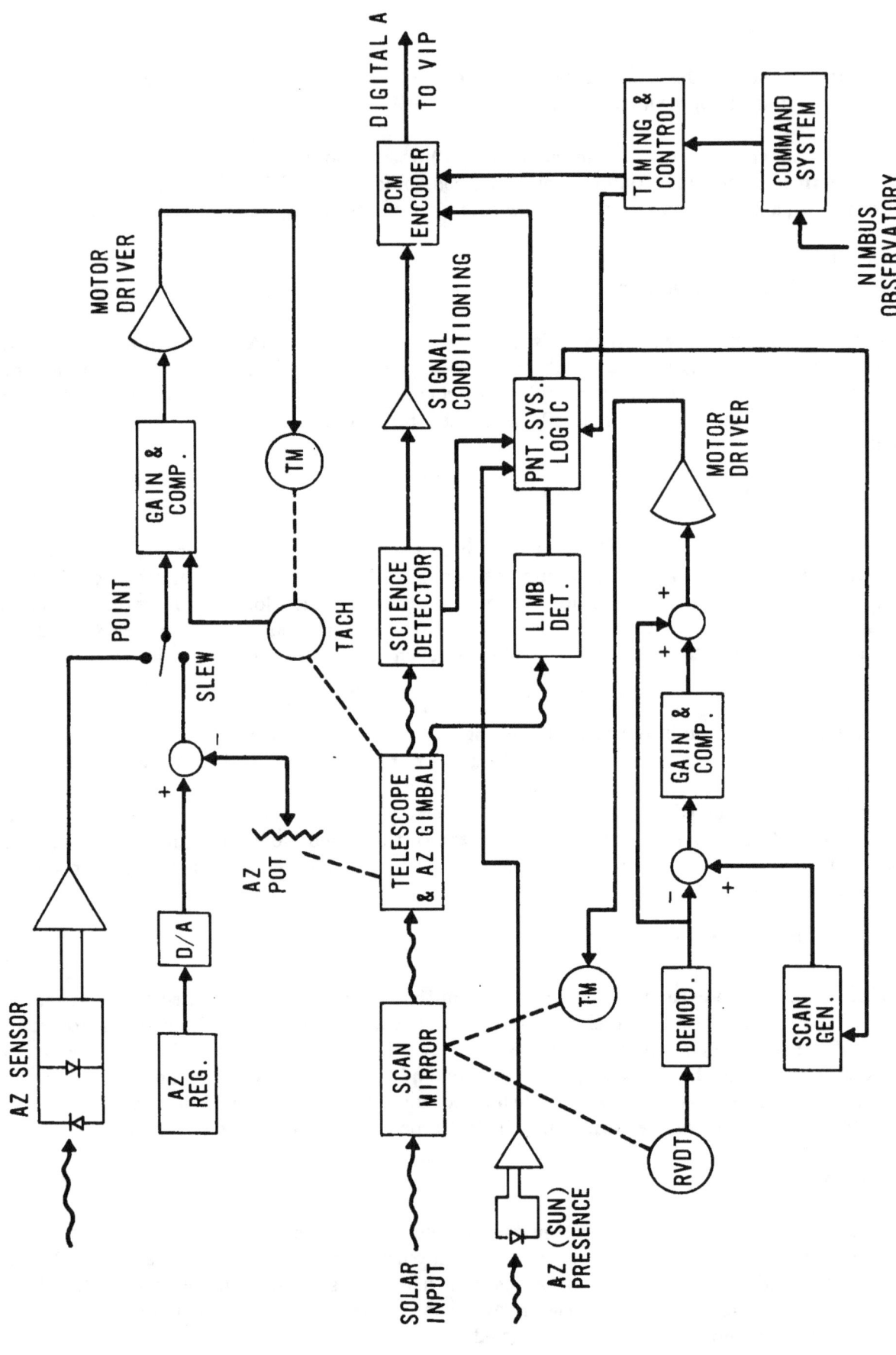

Figure 5-7. SAM II System Block Diagram

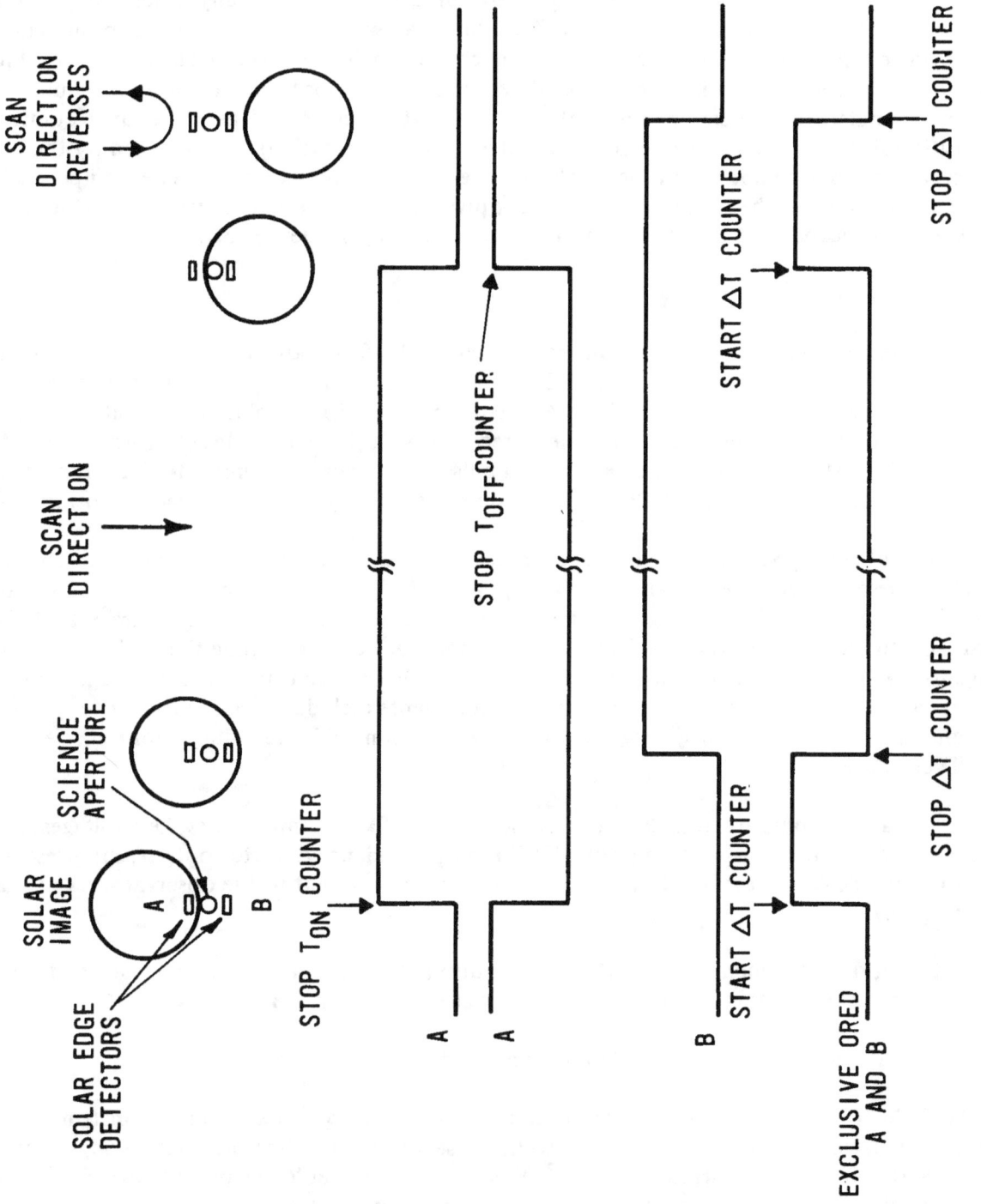

Figure 5-8. Timing Diagram For T_{on}, T_{off}, and ΔT

depression servo switches to a 15-arc minutes per second (line-of-sight) scan rate. In addition, when the trailing edge of the sun drops off the science aperture, the solar image scans an additional five arc minutes and then the mirror scan direction reverses. In the slow scan mode the mirror scan direction is reversed each time the trailing solar edge has scanned approximately five arc minutes past the center of the science aperture.

In the scan mode the scan mirror is controlled by comparing an electrically generated scan waveform to the mirror position indicated by the RVDT. The scan waveform is a linear ramp that changes direction as necessary to make the solar image scan back and forth across the science aperture. The turnaround signal is generated when the science detector output drops below a threshold value which is set at approximately one percent of full-scale solar intensity. The RVDT generates an accurate mirror position signal. The error between this signal and the scan waveform is processed by the depression servo gain and compensation network and the motor driver network. The limited angle brushless DC motor drives the scan mirror on its flex pivot mounts to null the error. The scan mode continues for 144 seconds and then the instrument automatically sequences off.

5.4.3 Data System

The instrument data is composed of one science channel, 10 supporting measurements, 25 analog housekeeping measurements, and 15 status checks. The science channel contains the science detector output and is multiplexed into a 10-bit A/D converter at a 50-samples per second rate. A submultiplexer combines the ten supporting measurements with the science data to form a serial data stream. The observatory provides clock and gating pulses to alternately dump this data from two storage buffers into the observatory VIP digital A data system storage matrix for later transmission.

Although all ten supporting measurements are used in the inversion of the science data, the most important of these are T_{ON}, T_{OFF}, and ΔT. T_{ON}, T_{OFF}, and ΔT are time measurements of leading and trailing solar edge crossings on edge sensing detectors A and B, as depicted in Figure 5-8. Using these data in conjunction with the observatory data system, one can determine the GMT crossing of each edge for each scan, the angular width of the sun for each scan, and the scan rate at the beginning and end of each scan. The remaining supporting measurements include the azimuth register position, azimuth gimbal position, mode, a sync word, and three precision calibrate voltages for the science detector A/D converter.

The 25 analog outputs are housekeeping measurements of key temperatures, key voltages, key currents, azimuth and depression servo errors, RVDT output, science detector output, sun presence detector output, and clock status. These analog measurements are sent to the observatory VIP analog data system.

In addition, the instrument contains 15 binary outputs that indicate operating mode status and command status. These outputs are sent to the observatory VIP digital B data system.

5.4.4 Viewing Geometry and Instantaneous Field of View (IFOV)

The SAM II IFOV is circular, 0.61 arc minutes in diameter, centered about the instrument optical axis. The viewing geometry, with respect to the observatory coordinate system, is shown in Figure 5-9 (a) and 5-9 (b). The instrument axes $\overline{X}$, $\overline{Y}$, and $\overline{Z}$ are parallel to the observatory X, Y, and Z axes, respectively. The azimuth gimbal axis of rotation is about the $\overline{Z}$ axis. The IFOV can be rotated 210 degrees, centered about the $+\overline{Y}$ axis. Sunrise events occur within ±15 degrees of the $+\overline{X}$ axis, and

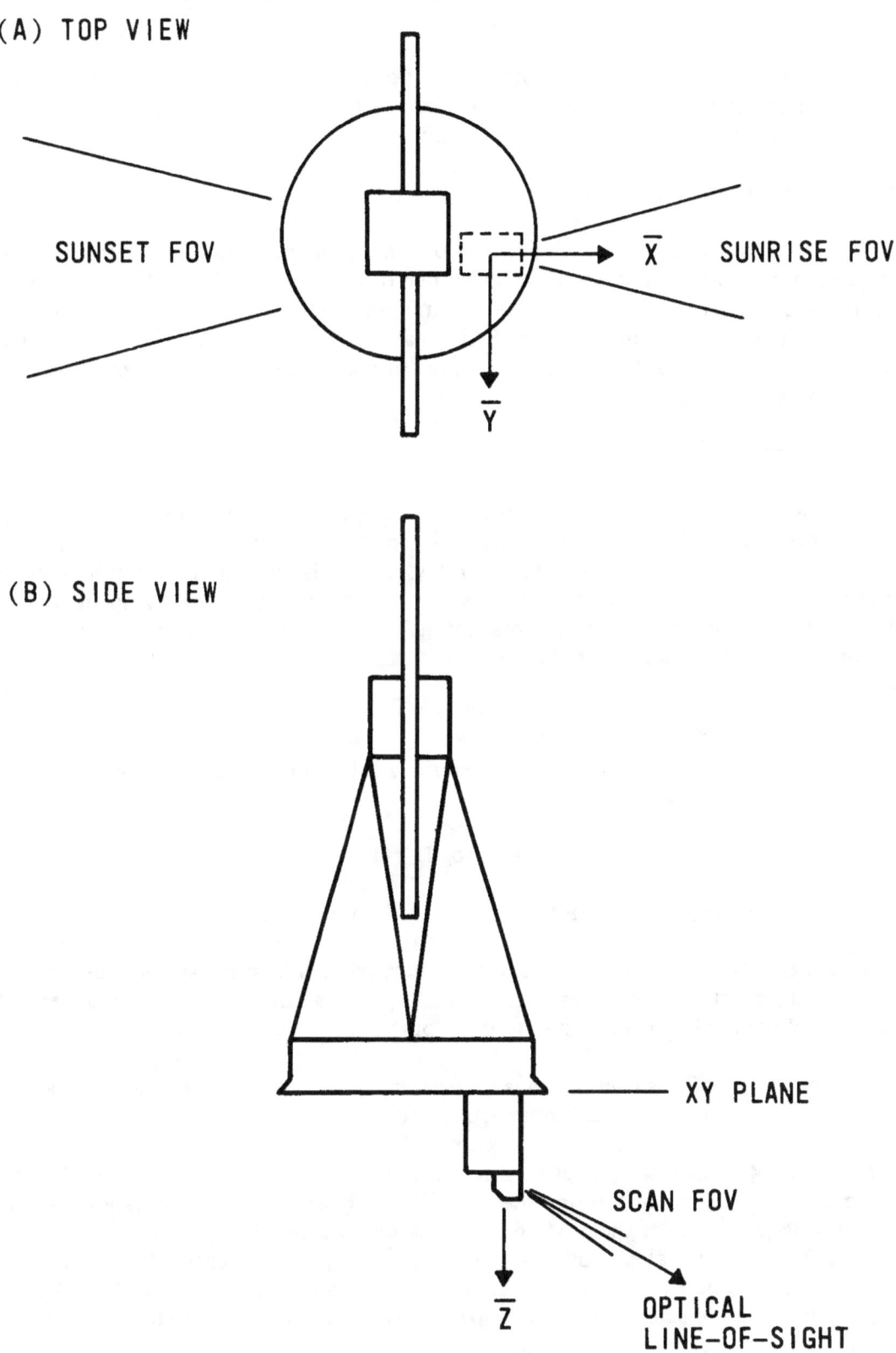

Figure 5-9. SAM II Field-of-View and Viewing Geometry

sunset events occur within ±15 degrees of the $\overline{X}$ axis. The exact location of sunrise and sunset depends on the insertion beta angle of the satellite, satellite drift during the mission, and the analemma function.

When the scan mirror is at the flex pivot null position, the instrument optical line of sight is 28.25 degrees in depression angle from the local horizontal (observatory XY plane). The line-of-sight depression scan is ±5 degrees about flex pivot null.

5.4.5 Performance Characteristics

The SAM II spectral bandpass, 0.98 μm to 1.02 μm, is determined by the combination of the bandpass of the front entrance window, the interference filter, and the silicon photodiode spectral sensitivity. The bandpass characteristics are shown in Figure 5-10. The instrument is calibrated such that a full-scale solar intensity (sun viewed by SAM II above earth's atmosphere) measurement is set at 900 counts on the 1,023-count scale. The accuracy of a measurement is 0.3 percent full-scale intensity (2.7 counts).

5.4.6 Calibration

The calibration technique used is often called the Langley Technique. In using this technique to calculate zero air mass solar intensity, one measures intensity of the sun over a several-hour period on a clear, optically stable day. By using the calendar day and the latitude and longitude of the measurement location, one can calculate a table of air mass versus time of day. Then by noting the time of day on each solar intensity reading, one can plot intensity (detector op-amp output voltage) versus air mass. From the Lambert-Bouguer relationship,

$$I = I_o e^{-kT} \tag{1}$$

where I is solar intensity at air mass T, I_o is unattenuated solar intensity (zero air mass) and k a constant (equal to 1 air mass). Then

$$\log I = \log I_o - kT \tag{2}$$

This equation defines a straight line with $\log I_o$ as the ordinate intercept.

Thus, intensity data were taken over a several-hour period, a least squares straight line was fit to the data, and the intercept (I_o) was determined. Finally, the op-amp gain resistor was set so that the op-amp output voltage for I_o corresponded to 900 counts.

The confidence level of this calibration is ±5 percent. No attempt is made to correlate this signal level to an absolute intensity since this is not required for science data inversion.

The data for SAM II are self-calibrating since, in the science inversion process, each data point is normalized to the full-scale reading during the same event. Therefore, slight variations of the full-scale reading during the course of the mission do not seriously affect inversion accuracy. Linearity of the data system, however, must be maintained. The detector/op-amp combination is inherently an extremely stable device, and its linearity is not expected to change during the mission. Three precision calibrate voltages are injected immediately after the op-amp once every 16 seconds to check the linearity of the remainder of the data system.

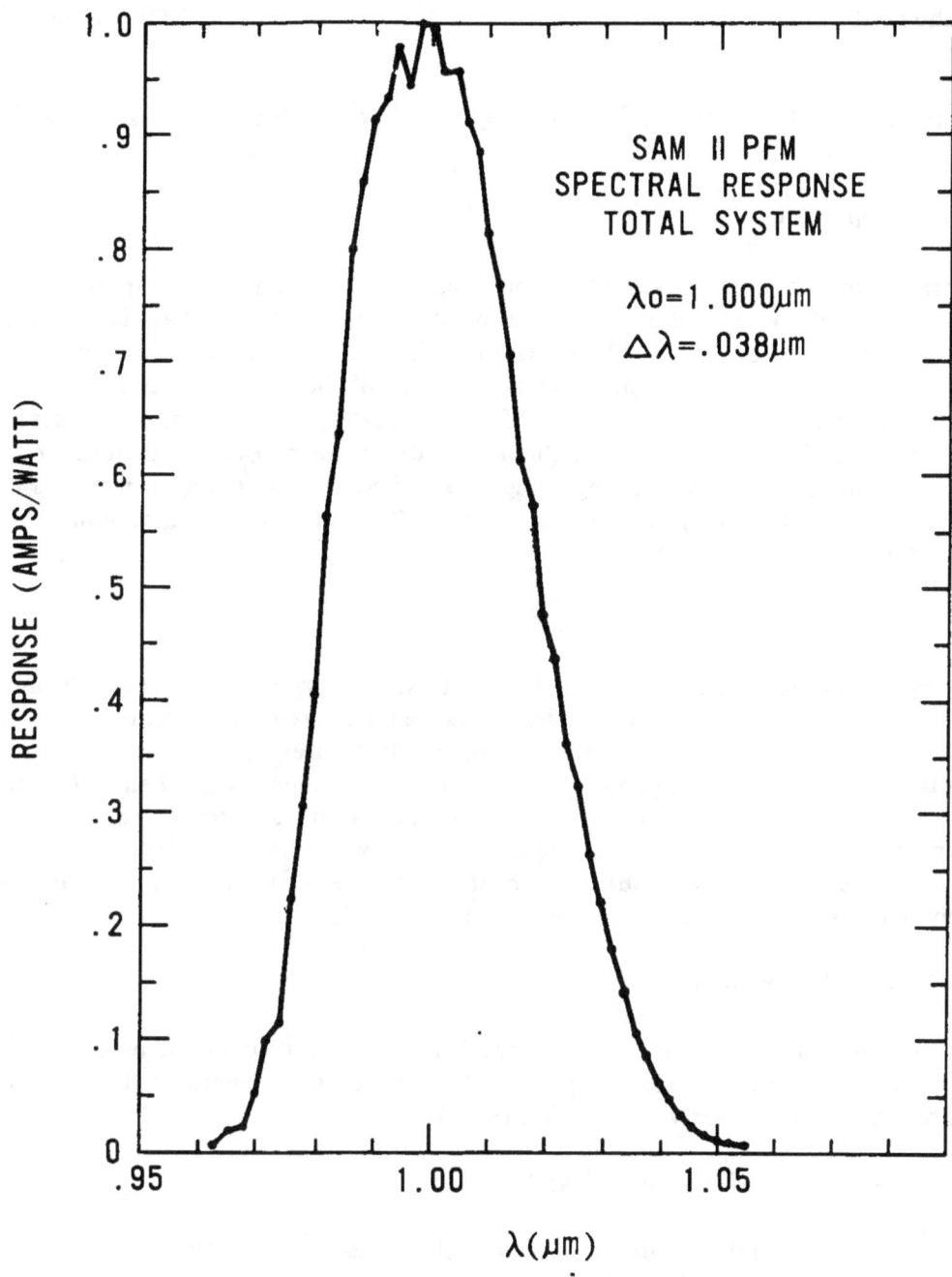

Figure 5-10. System Spectral Response

5.5 Data Validation

Before the SAM II data products (vertical profiles of aerosol extinction and number density as a function of latitude and longitude) are archived, the data will be validated by comparisons with other measurements made by sensors of appropriate accuracy, resolution, and reliability. Inversions using a typical SAM II inversion algorithm, simulated radiance data, and simulated errors of the magnitude expected for a typical measurement cycle indicate that ground truth measurements for SAM II data validation must provide a vertical profile of an aerosol parameter from cloud tops to heights of 40 km with a vertical resolution of 1 km or better and with a measurement accuracy of approximately ten percent over the height range where aerosol 1.0 μm extinction exceeds about 50 percent of molecular 1.0 μm extinction.

The sensors discussed in Sections 5.5.1 through 5.5.4 were selected for use in validating the SAM II data.

5.5.1   Airborne Lidar

An airborne lidar system was developed to provide exact ground point comparison with SAM II vertical profiles and to check aerosol horizontal homogeneity along the SAM II viewing path since this homogeneity is assumed in the SAM II inversion algorithms. The lidar system will provide profiles of aerosol scattering ratios at 0.69 μm with a vertical resolution of better than 1 km and an uncertainty in aerosol backscattering coefficients of approximately ten percent at the aerosol peak height. The uncertainty in the conversion of the lidar-measured aerosol backscattering coefficients to aerosol 1.0 μm extinction coefficients (by using an assumed refractive index and aerosol size distribution) is about 20 percent. (Other aerosol parameters will be measured simultaneously which may allow this uncertainty to be reduced.)

5.5.2   Dustsondes

Dustsondes (balloon-borne optical particle counters) were selected for comparison with the SAM II data because of their extensive use and acceptance by the scientific community. The dustsondes provide number density profiles of aerosols with radii greater than 0.15 and 0.25 μm. The vertical resolution is better than 1 km and the uncertainty in number density below 25 km is about 8 percent. The uncertainty in the conversion of aerosol number density to aerosol 1.0 μm extinction coefficient (by using an assumed refractive index and a two-parameter size distribution fitted to the two-channel dustsonde data) is about 25 percent. (Other aerosol parameters will be measured simultaneously which may allow this uncertainty to be reduced.)

5.5.3   Balloon-borne Sun Photometer

A balloon-borne sun photometer which views the total solar disc will be used for comparison with the SAM II data because it provides the only other direct measurement of the aerosol 1.0 μm extinction. The vertical resolution will be better than 1 km.

5.5.4   Airborne Polar Nephelometer and Impactors

Optical properties of the stratospheric aerosol, including size distribution and complex refractive index, must be known to convert the SAM II extinction profiles to profiles of aerosol number density and to convert the various ground truth measurements to 1.0 μm extinction profiles. The airborne polar nephelometer and impactor measurements are desired to supply information on these optical properties. But, since they do not very easily yield vertical profile information, these measurements will not be made unless such flights under other research programs can be coordinated with the SAM II program.

5.5.5   Ground Truth Experiments

Four ground truth experiments will be conducted using the selected sensors to validate the SAM II data. The first experiment, conducted at Laramie, Wyoming, in July 1978, provided a test for the newly developed sensors, and established multisensor and multiplatform coordination procedures. It also tested data reduction and comparison techniques. Following this practice comparative experiment, a ground truth experiment will be conducted during November 1978 at Sondre Stromfjord, Greenland. (This site was selected because of the SAM II high latitude coverage, favorable meteorological conditions, and available logistics support that will be required for each sensor.) This experiment will include the airborne lidar, the balloon sun photometer, dustsondes, rawinsondes, and possibly the airborne polar nephelometer and impactors.

The third ground truth experiment using only the airborne lidar will be conducted in December 1978 over the Antarctic Palmer Peninsula. This site was selected for validation of the southern hemisphere data since all northern hemisphere measurements are sunset measurements and all southern hemisphere measurements are sunrise measurements due to the nature of the spacecraft orbit and SAM II field of view.

The fourth ground truth experiment will be conducted during May 1979 at Sondre Stromfjord, Greenland. This site was selected for a repeat experiment to ensure the SAM II was still operating properly and because it provided a unique opportunity for comparison of the SAM II data with the Stratospheric Aerosol and Gas Experiment (SAGE) aerosol 1.0 μm extinction measurement. This experiment will include the airborne lidar, the balloon sun photometer, dustsondes, rawinsondes, and possibly the airborne polar nephelometer and impactors.

The data from each ground truth experiment will be processed, compared with the appropriate SAM II data, and archived with the same SAM II data.

5.6 Data Processing, Formats, and Availability

5.6.1 Data Processing

The SAM II data is collected onboard the Nimbus 7 observatory and stored in the Versatile Information Processor (VIP) data system for later transmission to the ground. After reception at the Spaceflight Tracking and Data Network (STDN) these data are forwarded to the Meteorological Operations Control Center (MetOCC) facility at GSFC for preprocessing operations including telemetry documentation and data formatting. MetOCC then combines the SAM II telemetry data, consisting of the science data, the 10 supporting instrument measurements, the 25 analog housekeeping measurements, and the 15 binary status checks, with spacecraft time corrections, orbital attitude and ephemeris data, as well as solar location information, on an Image Location Tape (ILT). The ILT is shipped to the NASA/LaRC for further science processing. GSFC also sends LaRC a National Meteorological Center (NMC) data tape containing temperature and pressure information needed for the SAM II data processing.

The block diagram in Figure 5-11 presents an overview of the SAM II data reduction. LaRC merges the SAM II instrument and selected housekeeping data with the spacecraft attitude and orbit data. These data are screened using quick-look analyses programs to ensure proper instrument operation and to identify anomalies in the radiometric data. This basic raw radiance data is saved on magnetic tapes for archival. The attenuation of solar radiation produced by the earth's atmosphere during a sunrise or sunset event, i.e., the ratio of the solar intensity measured at a specific location on the solar disk during a scan sequence to the intensity measured at the same solar location during a scan of the unattenuated sun, is computed as a function of the tangent altitude. Atmospheric effects due to

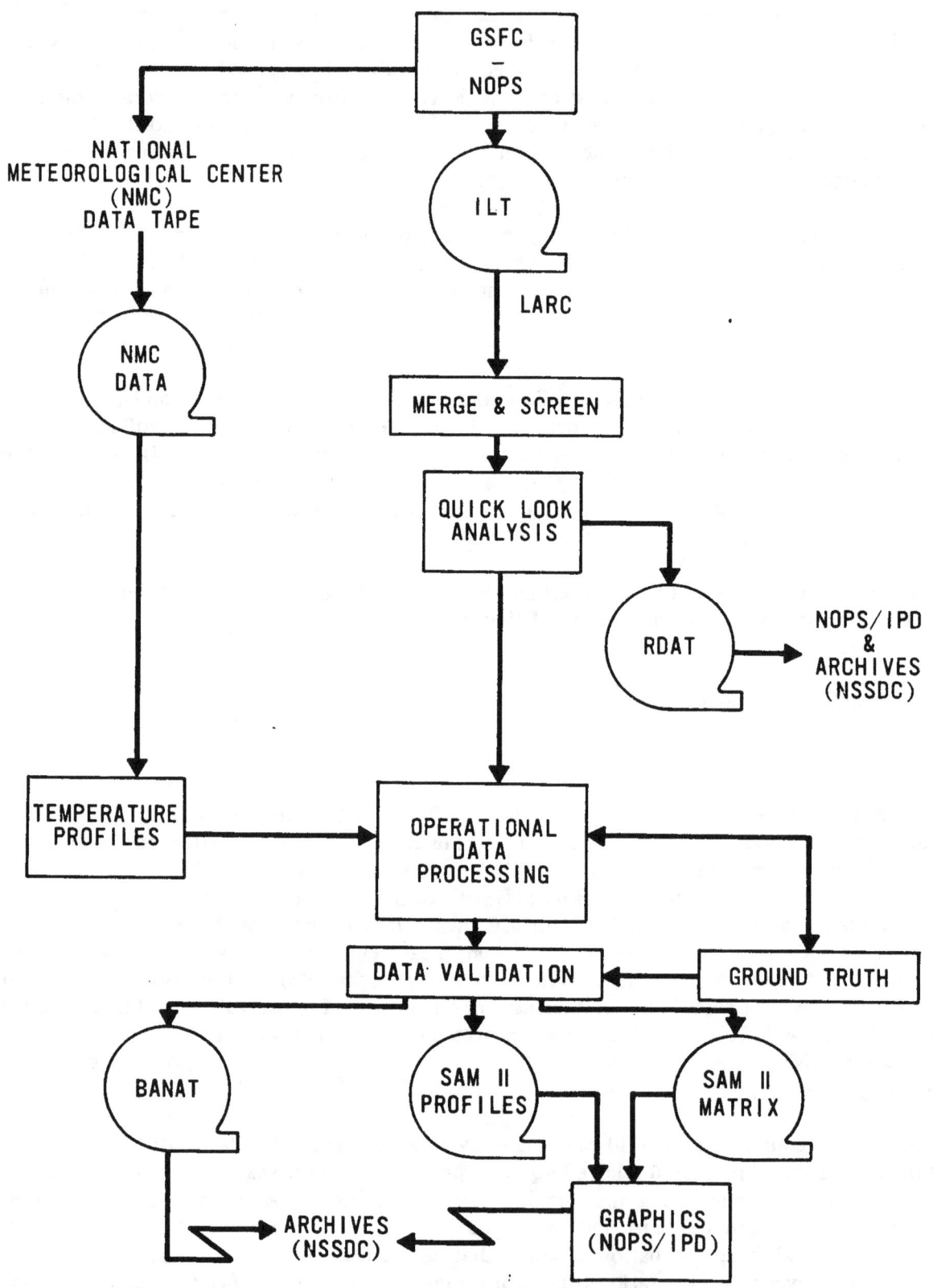

Figure 5-11. SAM II Data Processing

refraction are incorporated into the data reduction using temperature and pressure profile information for the event provided in the NMC tape. The extinction coefficient as a function of altitude is determined by calculating the attenuation of successive light rays as they pass through the atmosphere. The output profile data is compared with ground truth measurement data to ensure its accuracy. After validation, the output profiles (aerosol extinction coefficient and aerosol number density) are stored on tape and transferred to the National Space Science Data Center (NSSDC) for archiving. In addition, the data are assembled on SAM II Profiles and SAM II Matrix data tapes and sent to the Image Processing Division (IPD) of GSFC for display processing. The display products are stored on 16 mm film and archived at NSSDC.

A description of the science processing algorithms, inversion procedures, and specific methods employed in reducing the SAM II data, will be provided with the material archived at the NSSDC.

5.6.2 Tape Products

The following tapes are produced by Langley Research Center and used by IPD before being sent to the NSSDC for archiving. Brief descriptions of these tapes are as follows:

- RDAT (Raw Data Archive Tape)

  Contains the basic validation radiance data from each SAM II sunrise or sunset event as a function of time and tangent viewing location.

- BANAT (Beta-aerosol Number Density Archive Tape)

  This tape contains the derived aerosol extinction coefficients, modeled aerosol number densities, atmospheric molecular extinction coefficients, and total extinction ratios as a function of altitude for each SAM II sunrise and sunset observation. The tape also contains a summary description of the aerosol models used to generate number densities from extinction coefficients.

- MATRIX (Mapped Data Matrix Tape)

  Contains the projected map matrices of derived SAM II parameters for plots of stereographic polar maps and cross sections of latitude (or longitude) versus altitude. This tape is utilized for the display product processing.

- PROFILE (SAM II Output Product Profiles)

  This tape contains the profiles of derived SAM II products. It is also utilized for the display product processing.

The form and content of each of these tapes are specified in a tape specification document for each tape type. The appropriate document will accompany a tape shipment to a user. See Section 1.5 of this document for details.

5.6.3 Display Products

There are 18 different types of SAM II output products produced on 16 mm film and consisting of profiles, cross sections, maps, and time histories. Table 5-1 lists the titles and reference information for all displays. Figures 5-12 through 5-21 are examples. Brief descriptions of these displays are as follows:

Table 5-1
SAM II Display Products

No.	Title of Display Products	Frequency	No./Year	Film Spec #	Applicable Tape Spec
	Profiles				
1	Solar Irradiance vs Time (SR)	Daily	365	F455150	T454011
	Solar Irradiance vs Time (SS)	"	"	"	"
2	Solar Irradiance vs Altitude	Daily	365	F455151	T454011
	• (SR) 0–40 km	"	"	"	"
	• (SR) 40–160 km	"	"	"	"
	• (SS) 0–40 km	"	"	"	"
	• (SS) 40–160 km	"	"	"	"
3	Aerosol Coef. of Extinction vs Altitude				
	• (SR) 0–40 km	Daily	365	F455152	T454011
	• (SR) 40–160 km	"	"	"	"
	• (SS) 0–40 km	"	"	"	"
	• (SS) 40–140 km	"	"	"	"
4	Total Extinction Ratio				
	• (SR) 0–40 km	Daily	365	F455153	T454011
	• (SR) 40–160 km	"	"	"	"
	• (SS) 0–40 km	"	"	"	"
	• (SS) 40–160 km	"	"	"	"
5	6 Day Avg. – Aerosol Coef. of Ext. vs Altitude	1/6 Days	61	F455452	T454011
	Contours				
6	Long. Cross Sections of Aerosol Coef. of Ext. vs Altitude	1/6 Days	61	F454430	T454021
7	Aerosol No. Density vs Altitude	1/6 Days	61	F454431	T454021
8	Total Ext. Ratio vs Altitude	1/6 Days	61	F454432	T454021
9	Lat. Cross Sections Aerosol Coef. of Ext. vs Altitude	36/Quarterly	144	F454833	T454021
10	Aerosol No. Density vs Altitude	36/Quarterly	144	F454834	T454021
11	Total Ext. Ratio vs Altitude	36/Quarterly	144	F454835	T454021
	Maps				
12	Aerosol Coef. of Ext.	14/Quarterly	56	F453801	T454021
13	Aerosol No. Density	14/Quarterly	56	F453802	T454021
14	Total Ext. Ratio	14/Quarterly	56	F453803	T454021
15	Integrated No. Density	3/Quarterly	12	F453804	T454021
16	Optical Depth	3/Quarterly	12	F453805	T454021
	Time Histories				
17	Optical Depth From –	15/Quarterly	60	F457880	T454011
18	Alt. and Peak Ext. Ratio – 1	2/Quarterly	8	F457881	T454011

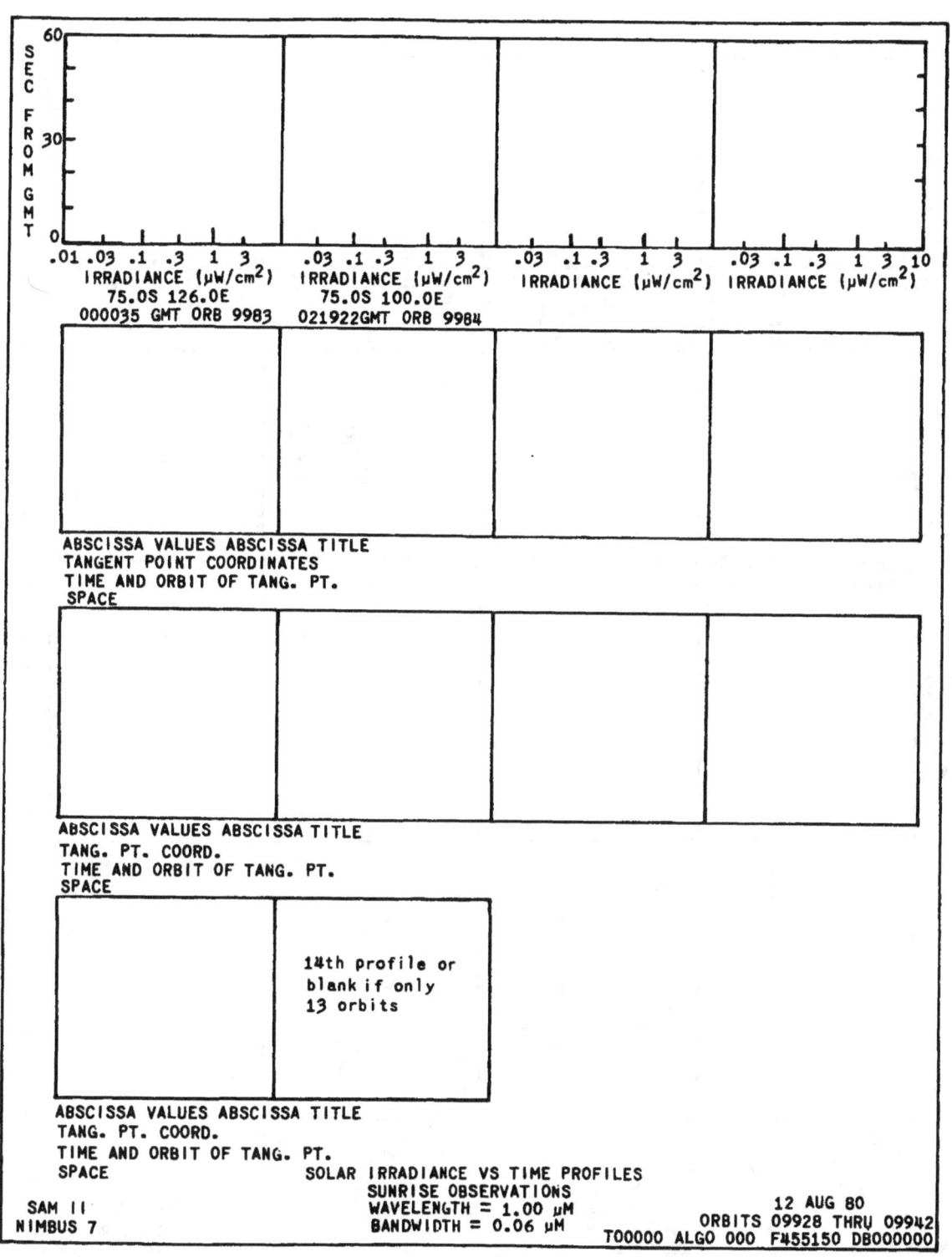

Figure 5-12. SAM II Microfilm Display Format of Solar Irradiance versus Time Profiles

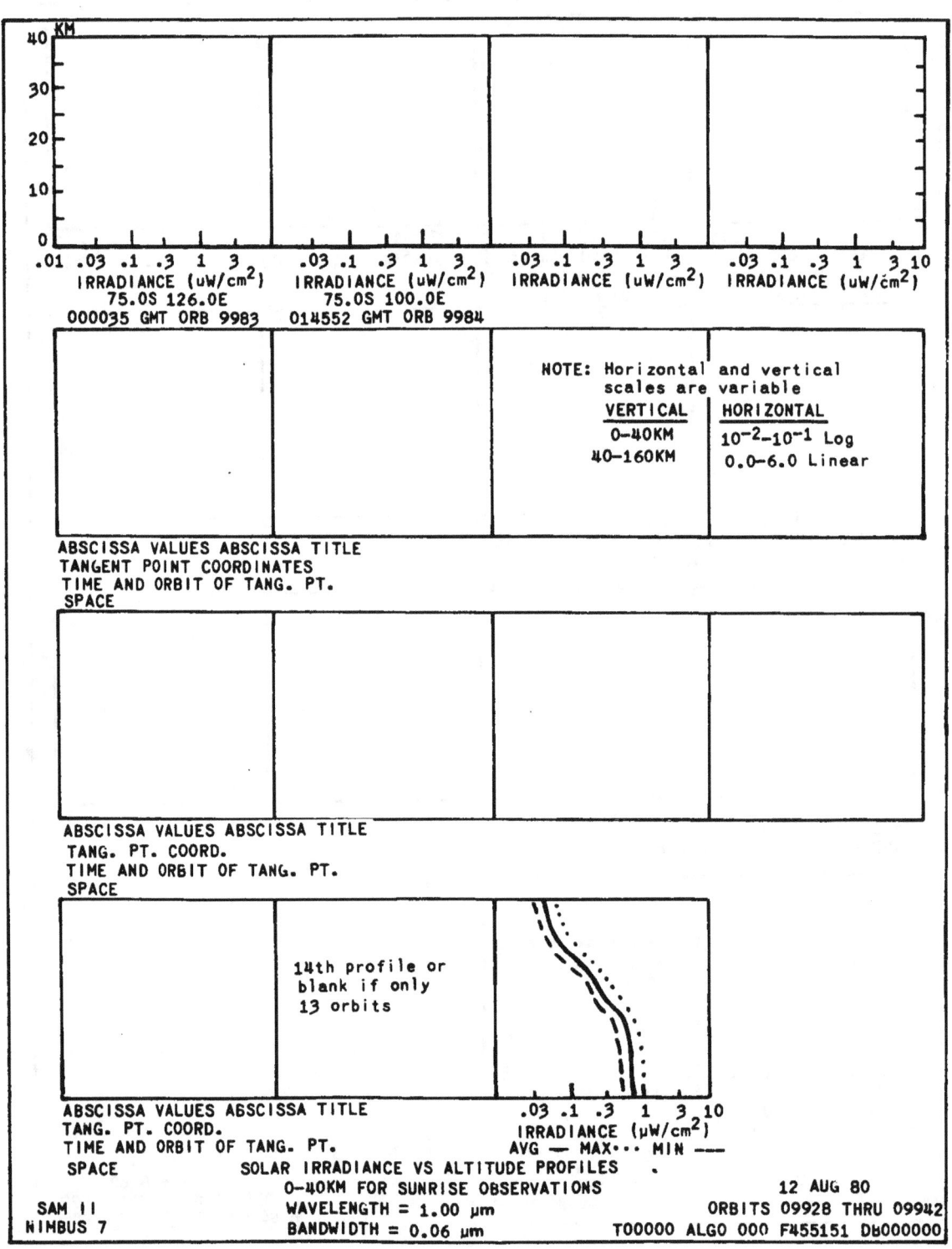

Figure 5-13. SAM II Microfilm Display Format of Solar Irradiance versus Altitude Profiles

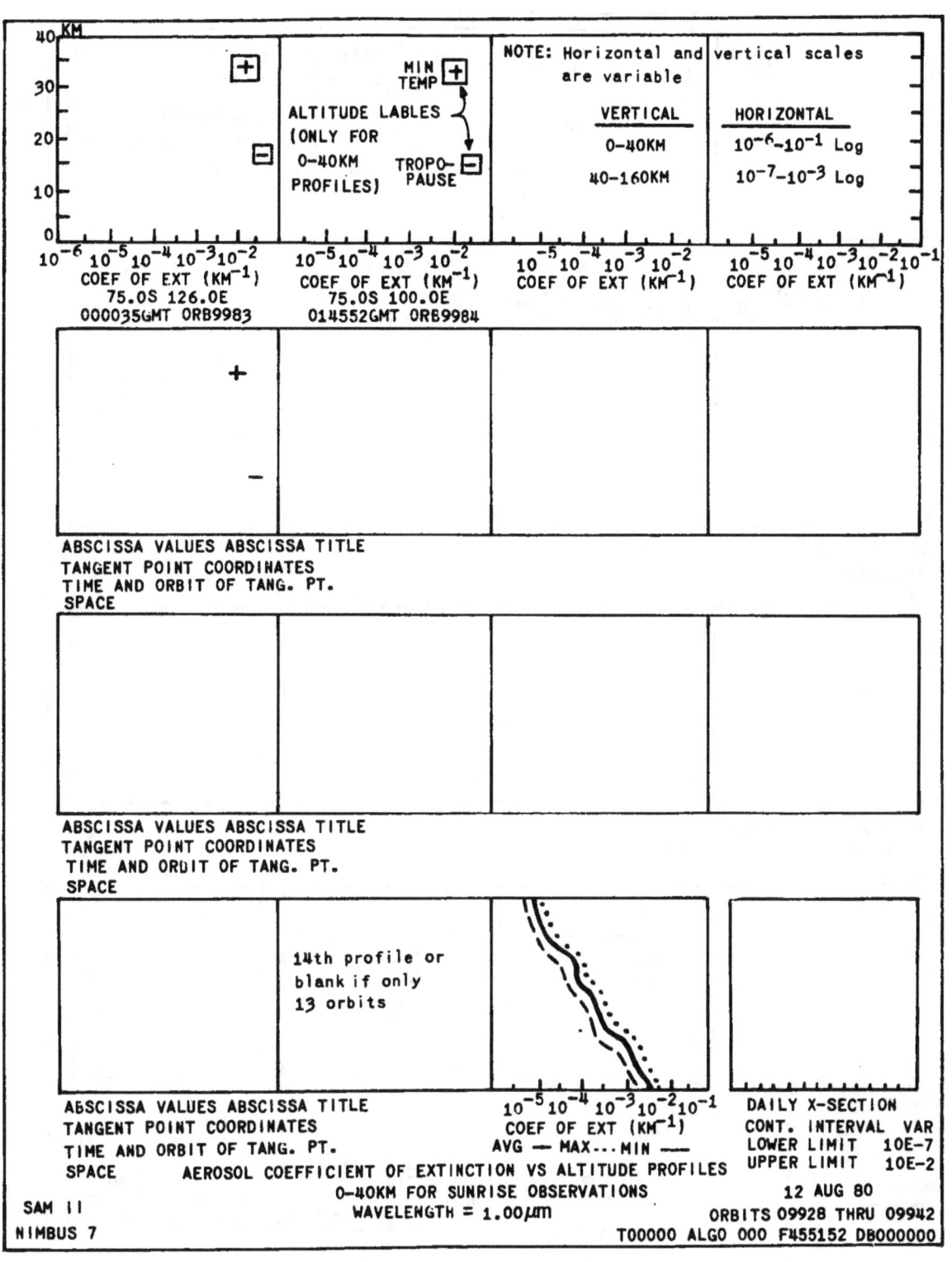

Figure 5-14. SAM II Microfilm Display Format of Aerosol Coefficient of Extinction versus Altitude Profiles

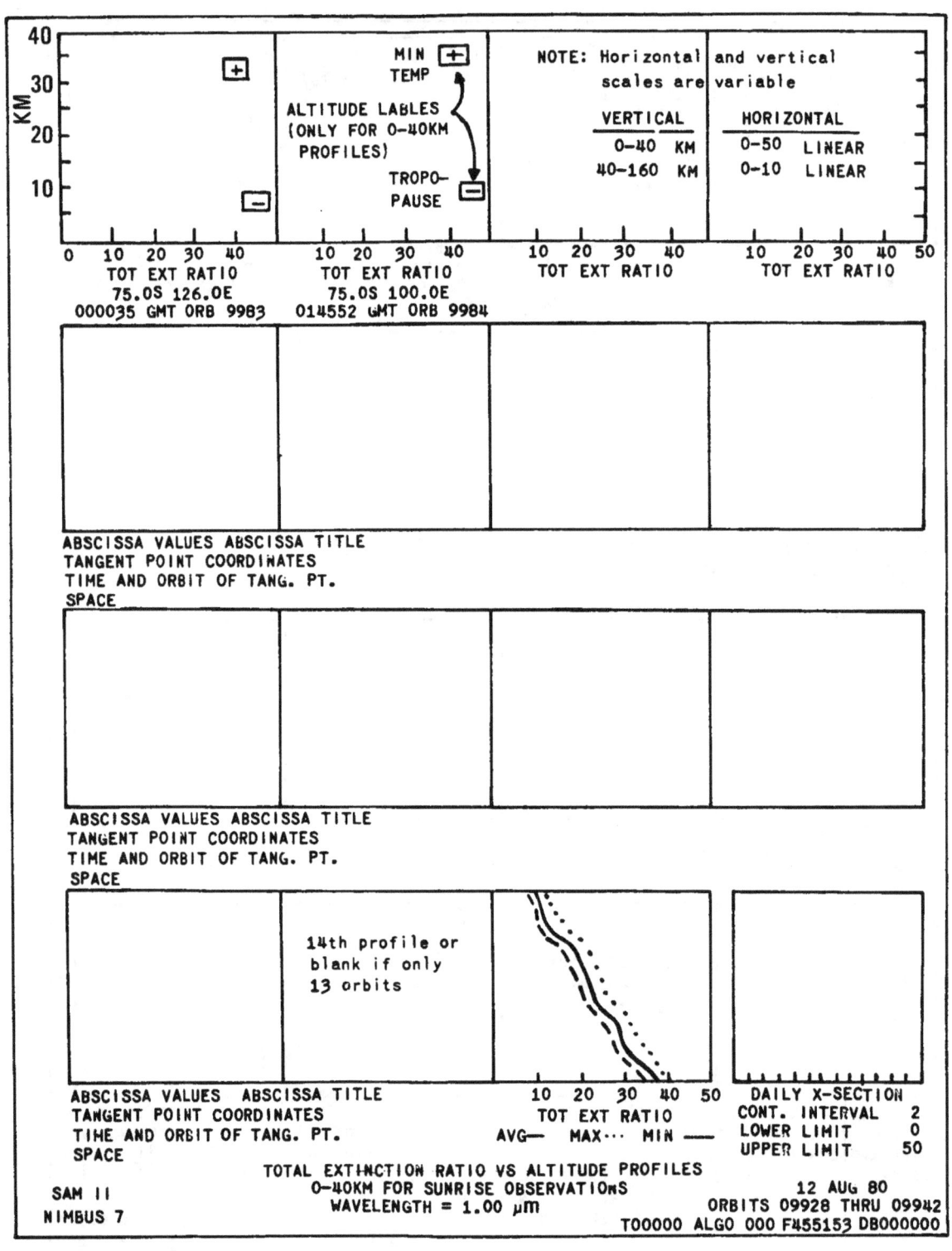

Figure 5-15. SAM II Microfilm Display Format of Total Extinction Ratio versus Altitude Profiles

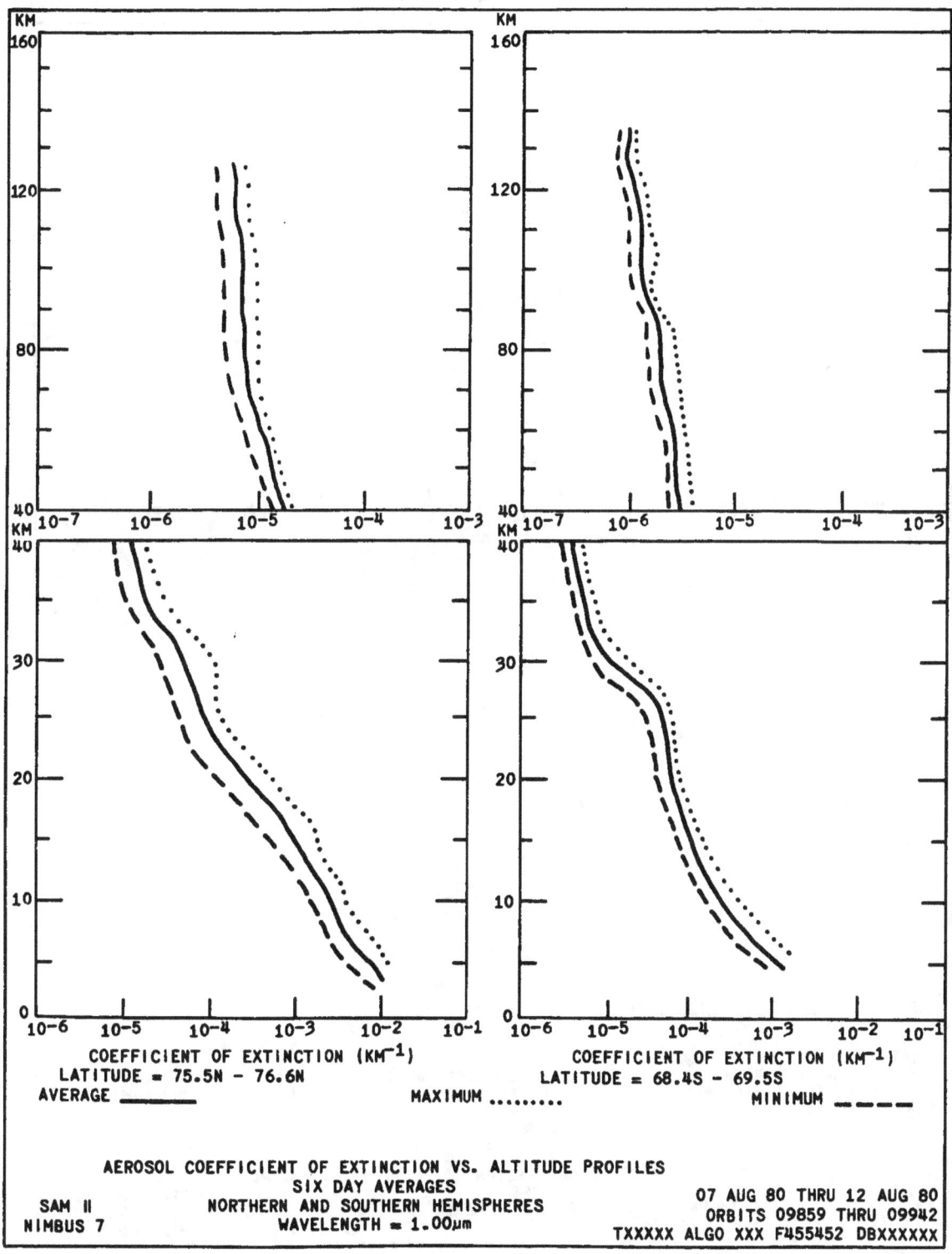

Figure 5-16. SAM II Microfilm Display Format of Aerosol Coefficient versus Altitude Profiles

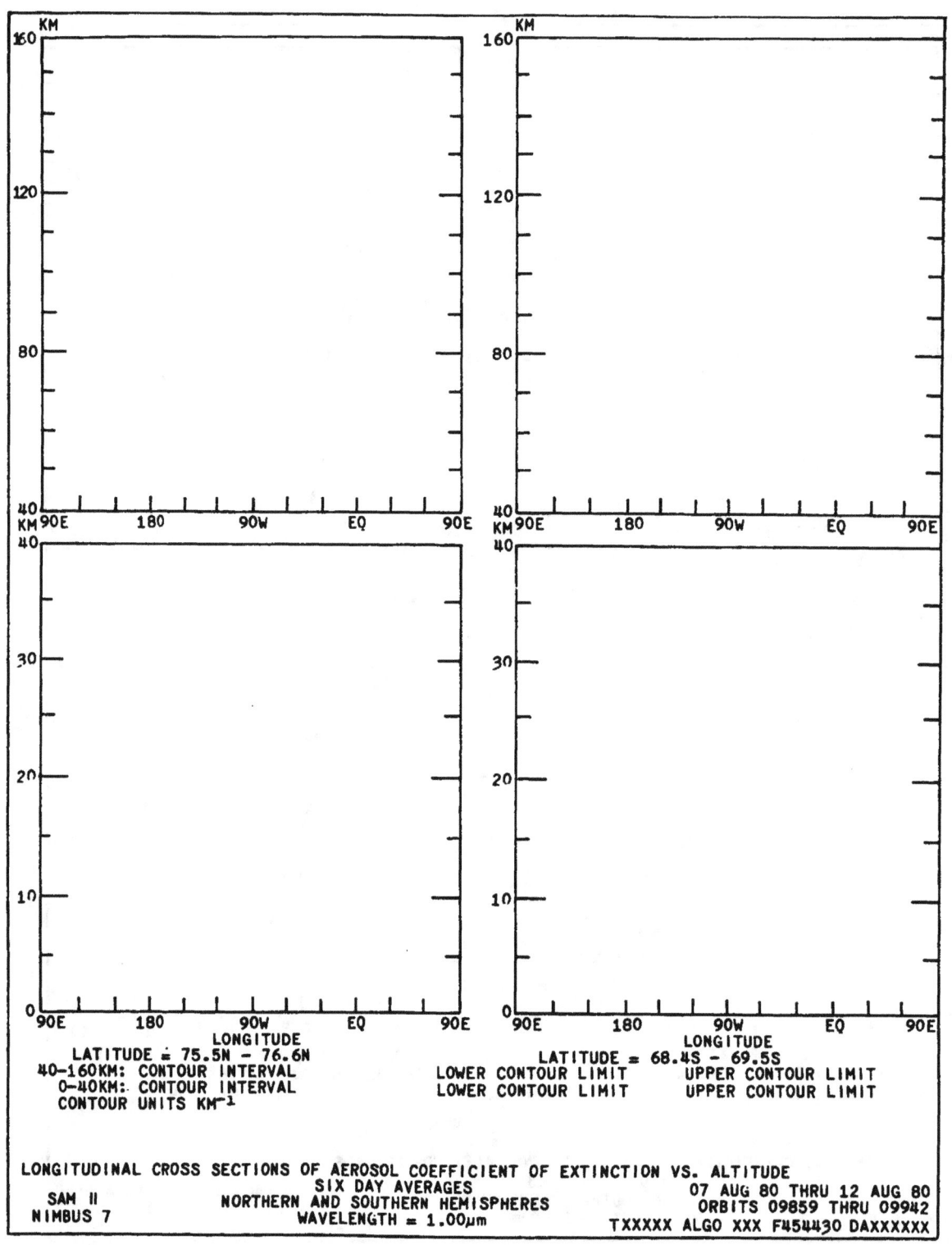

Figure 5-17. SAM II Microfilm Display Format of Longitude Cross Section of Aerosol Coefficient of Extinction versus Altitude

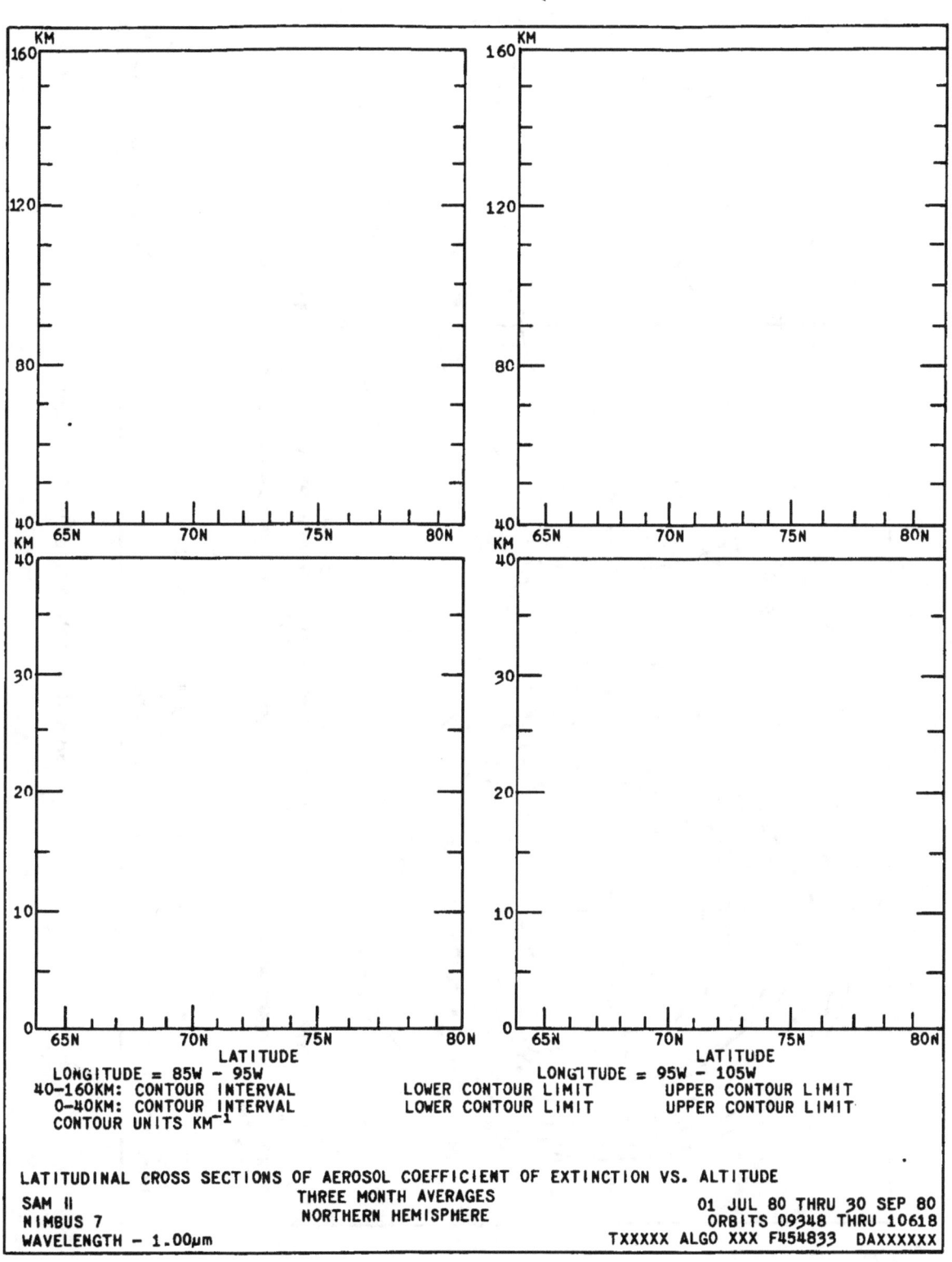

Figure 5-18. SAM II Microfilm Display Format of Latitude Cross Section of Aerosol Coefficient of Extinction versus Altitude

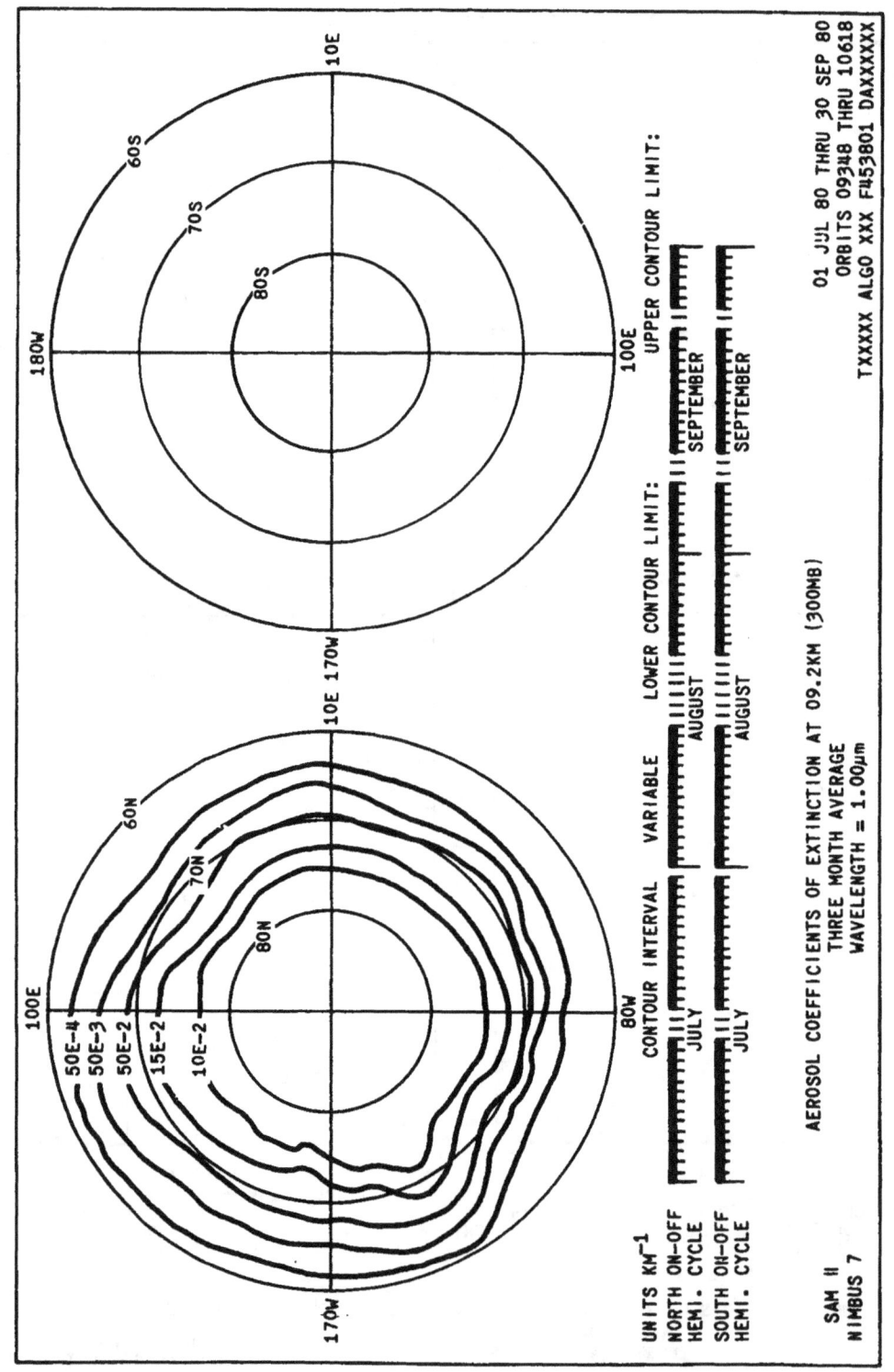

Figure 5-19. SAM II Microfilm Map Display Format

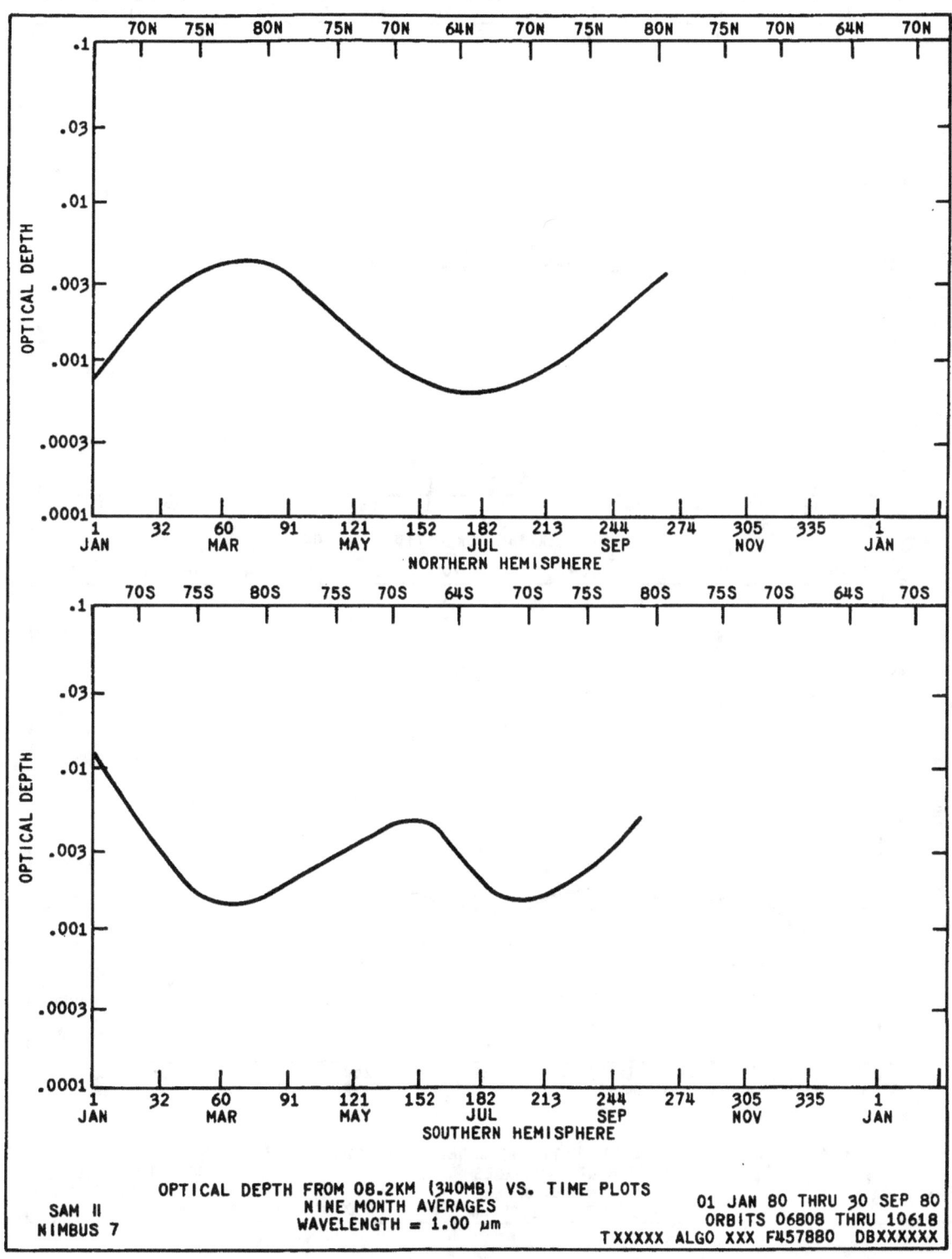

Figure 5-20. SAM II Microfilm Display Format of Optical Depth versus Time Plots

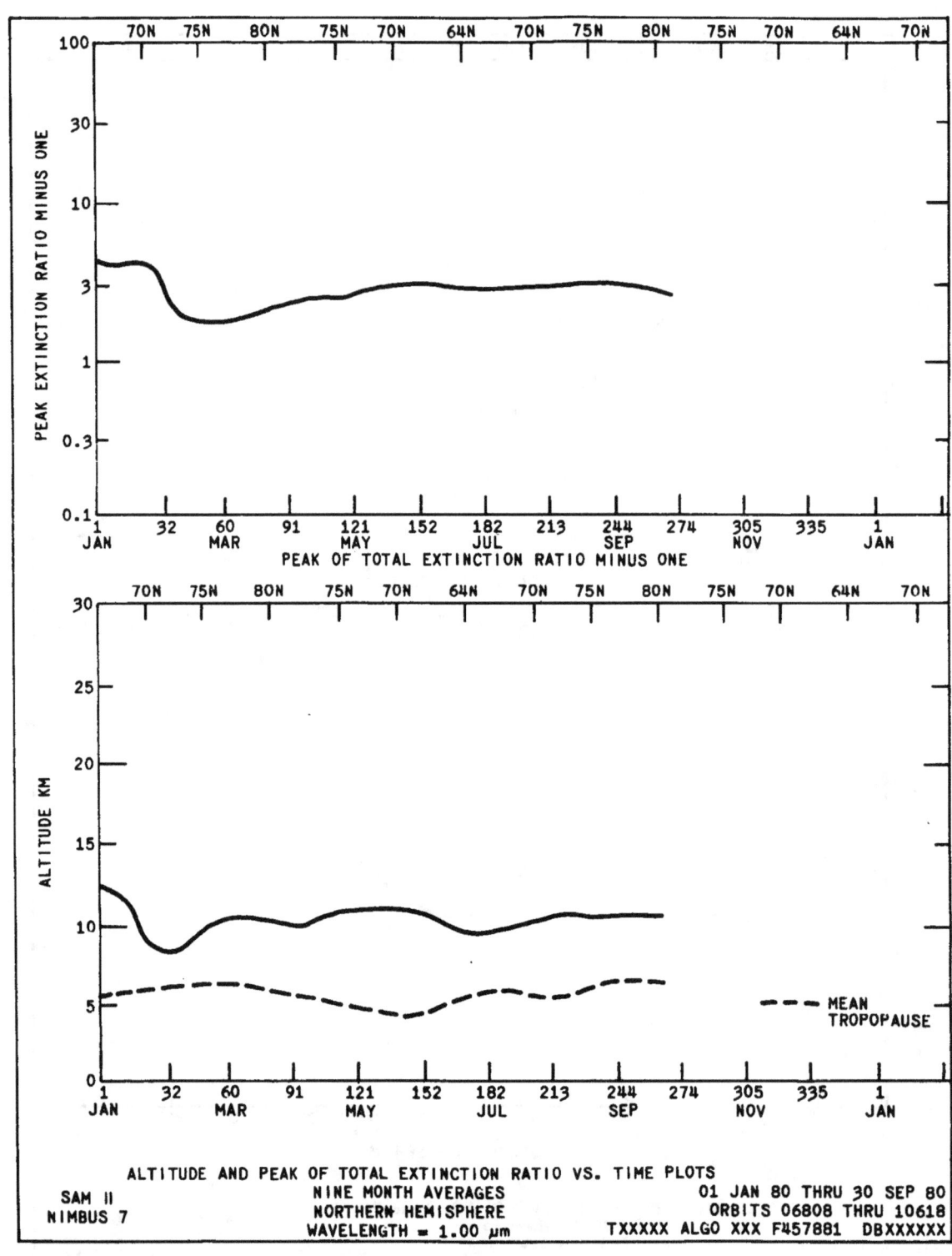

Figure 5-21. SAM II Microfilm Display Format of Altitude and Peak of Total Extinction Ratio versus Time Plots

5.6.3.1 Profiles

These products are produced on a daily basis at the rate of 14 profiles per hemisphere. A composite is formed by these 14 as illustrated in Figures 5-12 through 5-16. Each observation is referenced to single tangent viewing time and location on the earth's surface. The SAM II parameters profiled in this way are:

- Solar Irradiance vs Time (Figure 5-12)

- Solar Irradiance vs Altitude (Figure 5-13)

  A fifteenth profile has been added representing the average sunrise/sunset observation for that day with variability of the individual products indicated by the maximum and minimum limits. The profiles are shown in two altitude ranges. One for 0 to 40 km with one scaling factor and another for an altitude range of 40 to 160 km with a more sensitive scaling factor.

- Aerosol Coefficient of Extinction vs Altitude (Figure 5-14)

  Same display format as above. The sixteenth chart on each daily profile frame is a cross section of sunrise/sunset observations of that day.

- Total Extinction Ratio vs Altitude (Figure 5-15)

  This display format is similar to the above. The SAM II parameter plotted is the sum of the aerosol and molecular extinction coefficient ratioed to the molecular extinction coefficient. The molecular extinction coefficient is derived from the atmospheric temperature/pressure information for the event location.

- Six-day Average Aerosol Extinction Coefficient vs Altitude (Figure 5-16)

  The dotted lines indicate the variability of the individual profiles. Separate graphs are produced for the low and high altitude range and for the sunrise/sunset observations. The latitude interval for each 6-day period is indicated.

5.6.3.2 Cross Sections

Figure 5-17 is an example of a longitude cross section of the parameters listed in Table 5-1. Each display is for a six-day interval, for two altitude ranges, 0 to 40 km and 40 to 160 km, and for two one-degree latitude bands, one in both hemispheres.

Figure 5-18 is an example of a latitude cross section of the parameters listed in Table 5-1. Each display is a three-month average for two altitude ranges for each of two ten-degree longitude bands. There are 18 separate displays for the northern hemisphere and 18 for the southern.

5.6.3.3 Contour Maps

These products represent average values of aerosol extinction coefficient, aerosol number density, total extinction ratio, integrated aerosol number density and optical depth over a three-month period and plotted as contours on polar maps at various atmospheric levels.

- Aerosol Coefficient of Extinction maps (Figure 5-19)

- Aerosol Number Density maps

- Total Extinction Ratio maps

    North and south polar maps are produced for 14 different atmospheric levels consisting of 8 km (340 mb), 9.2 km (300 mb), 10 km (255 mb), 11.8 km (200 mb), 14 km (140 mb), 16.2 km (100 mb), 17 km (90 mb), 18 km (76 mb), 18.5 km (70 mb), 20 km (56 mb), 23.9 km (30 mb), 26.6 km (20 mb), 28 km (16 mb), and 31.2 km (10 mb).

- Integrated Number Density maps

    North and south polar maps of the aerosol number density are integrated over three atmospheric layers: 8 to 12 km, 12 to 20 km, and 20 to 28 km. Particle model parameters used to derive this parameter are indicated on the display frame.

- Optical Depth Map

    As above except contours are for optical depth for each of three altitude layers.

5.6.3.4 Time History Plots

These products are time histories of the SAM II parameters: optical depth, peak extinction ratio minus one, and the altitude of the peak extinction ratio plotted up to the current time and updated seasonally.

- Optical Depth vs Time (Figure 5-20)

    The total optical depth for each hemisphere is integrated upward through the stratosphere from the tropopause level as well as from the same atmospheric levels utilized in the aerosol coefficient of extinction maps for the averaged profiles in one-degree latitude bands between 64 degrees and 80 degrees of latitude in both hemispheres.

- Altitude and Peak of Total Extinction Ratio vs Time (Figure 5-21)

    The peak extinction ratio minus one as well as the altitude of the peak extinction ratio are plotted as a function of time for the averaged profiles in one-degree latitude bands between 64 degrees and 80 degrees of latitude in both hemispheres. The mean tropopause level as a function of time is also indicated on this display.

5.6.4 Data Availability

The SAM II experimental data consisting of the RDAT and BANAT magnetic tapes, as well as 16 mm microfilm of the display products listed in Table 5-1, are archived at the National Space Science Data Center. These data will be available to the archival center six months after the Nimbus 7 launch. Users requesting SAM II data should read Section 1.5 of this document for general tape and film ordering information.

For further information regarding these data contact the SAM II sensor scientist:

Dr. M. P. McCormick
Mail Stop 475
NASA Langley Research Center
Hampton, VA 23665

5.7 Planned NET Experiment Investigations and Data Applications

In addition to the primary scientific objective of mapping the extinction coefficient and number density of stratospheric aerosols as a function of altitude, latitude and longitude, the SAM II experiment team is planned to use the satellite data in the following scientific investigations:

- Effect of stratospheric aerosols on radiative transfer and climate

    Using the SAM II aerosol measurements and radiative climatological models, investigations will be conducted to determine the effect of stratospheric aerosols on the earth's radiological balance and climate in the polar regions.

- Investigation of source and sinks

    The SAM II aerosol data will be searched for stratospheric aerosol injections and removal mechanisms such as volcanic events and tropospheric/stratospheric exchange phenomena, and, if such mechanisms are observed, they will be investigated.

- **Investigation of sudden warming effects**

    Climatological data will be searched for sudden warming phenomena; and, if observed, the coincident SAM II aerosol data will be investigated to determine the effect of sudden warmings on the stratospheric aerosol vertical profile.

- Study of clouds in polar regions

    Because of the sensitivity of SAM II to clouds, nacreous and cirrus clouds in the arctic and antarctic will be investigated.

- Atmospheric motion studies

    The SAM II aerosol data, with the optical model determined from the ground truth program, will be used to investigate both vertical diffusion and horizontal transport phenomena. In addition, the SAM II data will be used to study hemispherical differences and provide an indication of interhemispheric transport.

- Investigation of high altitude aerosols

    The SAM II data will be searched for observations of mesospheric aerosols and such observations investigated.

- Synergistic studies

    The SAM II aerosol data will be used with data from the ground truth program, as well as data from the other Nimbus 7 experiments, to investigate the effect of aerosols on stratospheric chemistry. For example, LIMS nitric acid data will be compared with SAM II aerosol data to look for correlation between such aerosol-forming materials and observed aerosol number density.

Acknowledgements

The SAM II science team has been involved in all aspects of the SAM II experiment development including instrument calibration, choice of data product and formatting, validation program, and in the initial data use investigations of the data. They are: M. P. McCormick (leader), NASA Langley Research Center; G. W. Grams, Georgia Institute of Technology; B. M. Herman, University of Arizona; T. J. Pepin, University of Wyoming; and P. B. Russell, SRI International. The science team has been supported by L. R. McMasters, W. P. Chu of NASA Langley Research Center, and T. Swissler, SAS Corporation on all aspects of the experiment. The engineering team of NASA Langley Research Center is headed by L. E. Mauldin, III (Instrument Manager) and is comprised of K. D. Headgepeth (Technical Project Engineer), R. M. Holloway (Instrument Project Engineer), and R. L. Baker (Servo Engineer). All of the above have contributed to this section.

SECTION 6

THE STRATOSPHERIC AND MESOSPHERIC SOUNDER (SAMS) EXPERIMENT

by

J. R. Drummond, J. T. Houghton, G. D. Peskett, C. D. Rodgers,
M. J. Wale, J. Whitney, E. J. Williamson
Dept. of Atmospheric Physics, Clarendon Laboratory
Oxford, England

6.1 Introduction and Objectives

6.1.1 Introduction

The Stratospheric and Mesospheric Sounder (SAMS) instrument is the fourth in a series of multi-channel infrared radiometers designed to measure emission from the upper atmosphere, for which conventional spectral filtering techniques do not give adequate performance.

The technique used in these radiometers is known as gas correlation spectroscopy and is based on the use of gas cells to select emission from chosen spectral lines or from particular parts of spectral lines.

In the Selective Chopper Radiometer (SCR) on Nimbus 4, a beam-chopping technique was employed to switch the scene (at 10 Hz) between the atmosphere and space view in a differential manner between two gas cells containing different amounts of CO_2. The "difference" signal was then detected by a thermistor bolometer. The chief limitation in performance of the Nimbus 4 SCR was the difficulty in maintaining a balance condition due to stray thermal emission from within the instrument.

The Nimbus 5 SCR employed a variation of this technique, in which cells containing different amounts of CO_2 were switched in sequence (one each second) into the optical path to a pyroelectric detector. The difference signals were then extracted on the ground. Performance in this case was limited by gas leakage with time, by uncertainties in the effects of degradation, and by contamination of the cell windows (which give spurious difference signals).

The Pressure Modulator Radiometer (PMR) on Nimbus 6 overcame earlier difficulties by employing a single gas cell and no moving parts in front of the detector. The gas (CO_2) amount in the cell is modulated at approximately 35 Hz by an oscillating piston, and the oscillatory component of signal arriving at the detector is related directly to the radiance of the scene, but only at the frequencies corresponding to the variation in absorption of the spectral lines of the gas in the modulator cell.

The SAMS instrument extends this technique to gases other than CO_2, in addition to viewing the limb of the atmosphere rather than employing vertical sounding as in the earlier radiometers.

6.1.2 Scientific Objectives

The SAMS is a 12-channel infrared radiometer observing thermal emission and solar resonance fluorescence from the atmospheric limb. Global measurements are made of radiation from the molecular species as noted in Table 6-1. These measurements, when interpreted, together with results from the LIMS and SBUV/TOMS instruments, provide extensive data for chemical and dynamic models of the stratosphere and mesosphere.

Specific objectives of the SAMS experiment are to derive:

- Temperature from emission in the 15 μm CO_2 band from 15 km to 80 km altitude

- Vibrational temperature of CO_2 bands where they depart from local thermodynamic equilibrium (LTE) between 50 km and 140 km

- Distribution of CO, NO, CH_4, N_2O, and H_2O from 15 km to 60 km

- Distributions of CO_2 (4.3 μm) and CO (4.7 μm) from 100 km to 140 km and H_2O from 60 km to 100 km to study dissociation in the lower thermosphere

In addition an attempt will be made to measure zonal wind speed from 60 km by using a Doppler shift technique. These measurements will be compared with calculated thermal winds.

The study of planetary waves at present being carried out using data from the Nimbus 5 SCR and Nimbus 6 PMR will be continued and extended by making use of these new data. In addition, the measurements allow calculation of the transfers of momentum, energy, and trace gases by mean motions and eddies. These calculated results will be compared with the predictions of dynamic models.

6.1.3 Technical Objectives

The SAMS instrument is designed to exploit the selectivity, energy grasp, and tuning capability of the pressure modulation technique proved earlier for CO_2 emission measurements in the Nimbus 6 PMR described in Reference 1.

Table 6-1
Molecular Species and Spectral Bands

Constituent	Spectral Band
Carbon dioxide	4.3 μm and 15 μm
Water vapor	2.7 μm and 25 μm to 100 μm
Carbon monoxide	4.7 μm
Nitrous oxide	7.7 μm
Methane	7.7 μm
Nitric oxide	5.3 μm

The main technical innovation in SAMS are:

- The extension of the pressure modulation technique to other gases

- The simultaneous use of conventional chopping and pressure modulation to:

    (1) extend the range of heights that can be sounded

    (2) determine the pressure at the viewing level

    (3) enable some interfering radiance signals to be eliminated

    (4) provide additional calibration information and confidence checks

- The use of a programmable stop scan system with two independent axes to enable the best usage of the observing time and to accommodate uncertainties in spacecraft attitude.

6.2 Principles of Operation

6.2.1 Basic Theory

The radiated power W given off by an atmospheric limb path and incident on one of the detectors is:

$$W = A\Omega \int_0^\infty \int_0^\infty \frac{d\tau_\nu(x)}{dx} J_\nu(x) f_{i\nu} \tau_{i\nu} dx\, d\nu \tag{1}$$

Where A is the collecting area of the telescope,

Ω is the field of view, and

$\tau_\nu(x)$ is the transmission at frequency ν between the satellite and the position along the path described by the co-ordinate x. $\tau_\nu(x)$ exhibits large and rapid variations with frequency ("lines") whose shapes in the part of the atmosphere being viewed by SAMS are determined by both collision-broadening and Doppler-broadening mechanisms.

$J_\nu(x)$ is the source function, equal to the Planck function $R_\nu(T)$ at the temperature T of the path at x, if LTE obtains. It is a smooth function of frequency.

$f_{i\nu}$ is the static component of the optical transmission in the ith channel (i.e., the filter profile).

$\tau_{i\nu}$ is the oscillatory component of the optical transmission in the ith channel (i.e., the modulation).

A small part of the atmospheric emission spectrum lying within a group of lines (band) of a particular gas may appear as in Figure 6.1 (A) where the widths of the lines have been exaggerated compared with their typical spacing.

In the SAMS radiometer, the beam passes through a high frequency mechanical modulator (chopper) operating at a frequency f_B, and an absorbing path of gas whose pressure is varied at a low

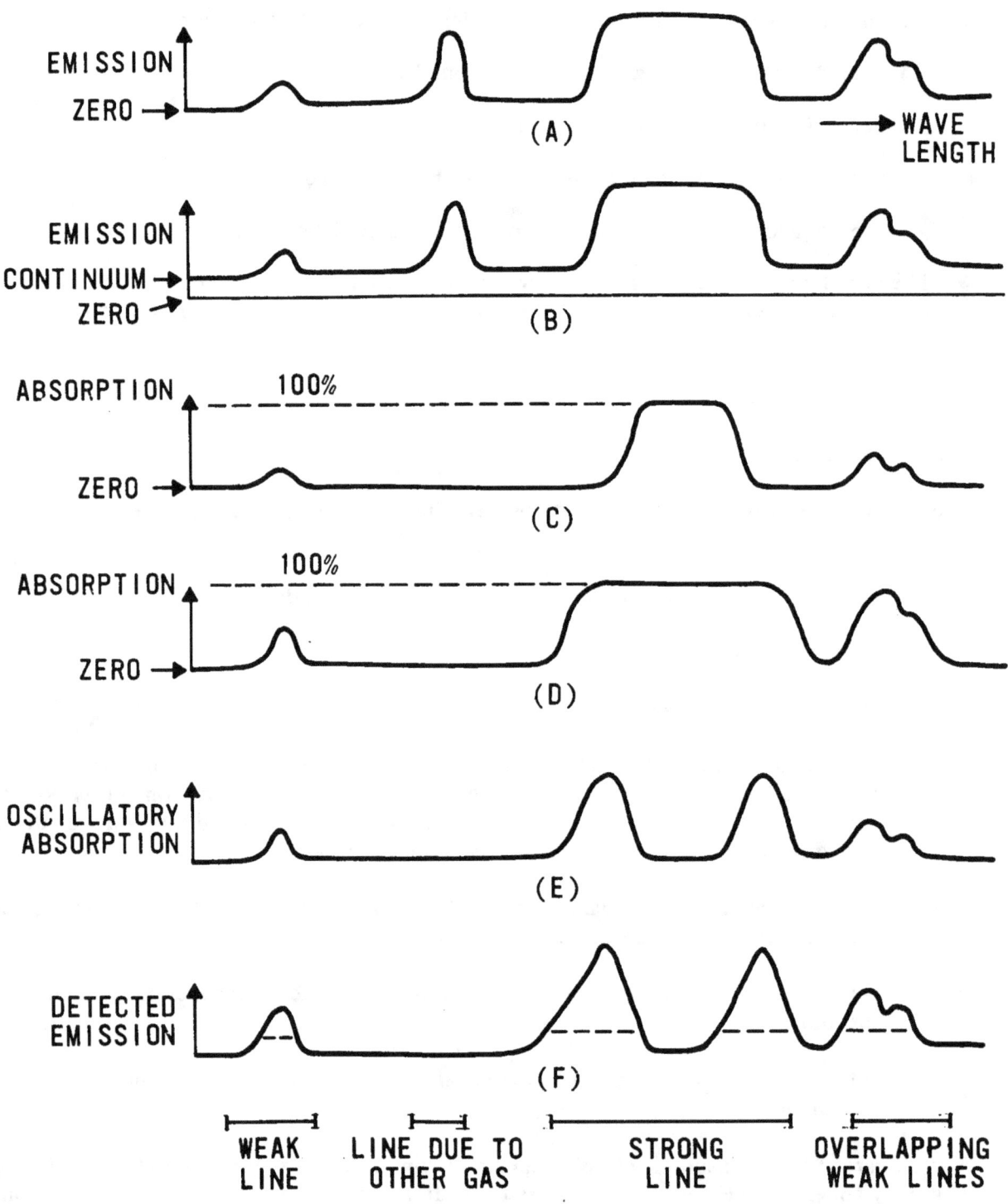

Figure 6-1. Typical Line Modulation Spectra

frequency f_L. Two different kinds of channels are obtained by separating the signals present at frequencies f_B and f_L in the detector output:

- Mechanically modulated (conventional) channels where $\tau_i(f_B)$ has no spectral dependence, and the time average of the absorption of the gas cell is incorporated into $f_{i\nu}$

- Pressure modulated channels where $\tau_{i\nu}(f_L)$ has a complicated spectral behavior and the mechanical modulator is treated as a simple attenuating factor in $f_{i\nu}$

The two kinds of channels, henceforth referred to as black and line modulated respectively, have identical fields of view. They also occupy the same spectral interval but differ in detailed spectral response within the interval.

6.2.2 Conventional Radiometry

Six of the 12 channels in SAMS are of the conventional type. Ignoring the effect of the cell and any weighting due to the optional components over the very small spectral interval shown in Figure 6-1 (A), the spectral distribution of the energy measured with these channels is simply the sum of the incoming emission and the continuous emission generated by the optical components in front of the chopper as pictured in Figure 6-1 (B).

These channels are adequate for observing radiation from spectral regions where the emission lines of a given constituent are strong, wide, numerous, and not intermingled with foreign lines. Such is the case over significant height ranges for CO_2 and H_2O, but it is not the case for most other minor atmospheric constituents. Also, radiometry of this kind fails at high altitude because atmospheric emission becomes so small the signal is lost in the fluctuations of the instrument continuous emission.

6.2.3 Pressure Modulation Radiometry

The six remaining channels are of the pressure modulated type. An absorption cell of the gas under consideration in which the pressure may be modulated, is placed in the optical path of the radiometer (Figure 6-2). The gas in the cell possesses an absorption spectrum which matches line for line the emission spectrum of the same gas in the atmosphere. The amount of absorption and the shapes of the absorption lines in the cell are determined by the pressure and temperature in the cell and may be varied. Shown in Figure 6-1 (C) is the absorption at particular values of pressure and temperature in the same spectral interval as Figure 6-1(A) while Figure 6-1 (D) shows the absorption at higher values of pressure (and temperature).

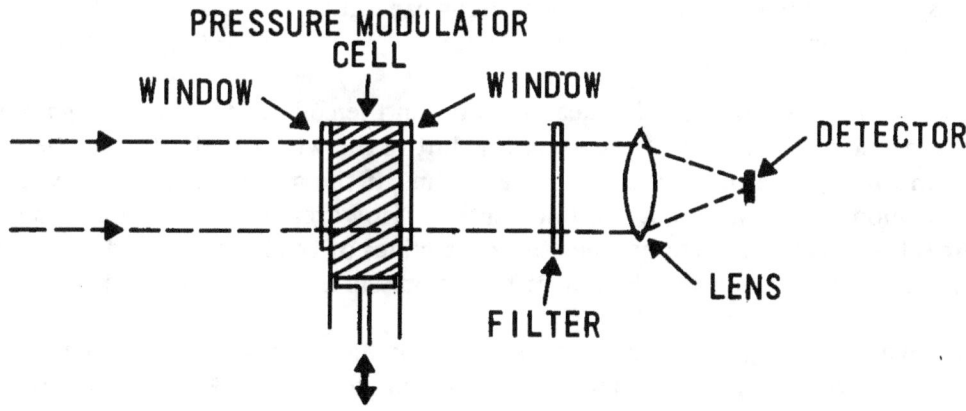

Figure 6-2. Pressure Modulator Optical Arrangement

If the pressure is cycled between these two values at a frequency f_L, there is an oscillatory component of absorption which has a spectral distribution as shown in Figure 6-1 (E). (The pressure cycling is not isothermal for typical values of f_L (25 Hz to 50 Hz) and some temperature cycling is therefore also present.)

There is no component between the lines or at the center of strong lines where the transmission is close to zero throughout the pressure cycle. The modulated energy falling on the detector and giving rise to an output at frequency f_L has a spectral distribution given by the convolution of curves in Figure 6-1 (A) and Figure 6-1 (E) and the overall filter profile of the radiometer curve in Figure 6-1 (F).

Comparing pressure modulator radiometry (curve A) with conventional radiometry (curve B), it is clear that there is no contribution from the foreign line and the contribution from continuum emission is greatly reduced. The optical bandwidth of the system is restricted by the modulation process to just those very narrow intervals occupied by emission lines of the chosen gas. A typical pressure modulator arrangement will select 10^{-4} to 10^{-2} of the spectral region covered by the filter, an approximate expression for the effective spectral bandwidth being

$$\int_0^\infty f_{i\nu} \left[\tau_{i\nu}(1) - \tau_{i\nu}(2) \right] d\nu \qquad (2)$$

where $\tau_{i\nu}(1)$ and $\tau_{i\nu}(2)$ are the transmissions at the maximum and minimum values of pressure in the modulation cycle.

When two or more gases having intermingled lines are to be observed, the radiation may be passed through a series of modulator cells to a common detector. In SAMS, the CH_4 and N_2O cells are in series, sharing a spectral interval around 7.7 μm as are the CO, NO and one of the CO_2 cells sharing a spectral interval from 4.3 μm to 5.3 μm. In normal operation, one cell in each chain is modulated, while the mean pressures in the unmodulated cells are set to high values so the effect of any lines overlapping those in the operating cell is reduced.

The side effect of pressure modulation as previously mentioned is that the pressure cycling of the gas in the cell accompanied by a temperature excursion and a radiation (emission) signal is generated, coherent with the modulation of the incoming radiation. For stable conditions in the radiometer the effect of this cell emission signal is to add a constant offset at the output of the signal processing electronics. It is readily subtracted by the normal radiance calibration procedure.

6.2.4 Tuning of Pressure Modulated Channels

The pressure modulated channels possess a tuning capability which is limited but very significant. Tuning may be achieved in two ways.

The first of these is by altering the mean pressure in the cell. In the example shown in Figure 6-1, modulation occurs at the center of the weak line but in the wings of the strong line since there is no transmission through the cell at the line center. Thus, if the mean cell pressure is chosen so most lines are weak, modulation occurs at the line centers. However, if the cell pressure is increased so most lines are saturated at their centers, then only the wings of the lines are modulated. Each pressure modulator cell includes a means of selecting different pressures by ground command.

A second way of tuning a pressure modulator channel is to shift the atmospheric emission lines with respect to the cell absorption lines by use of the Doppler effect. A systematically varying shift can be obtained by scanning about the vertical (Z) axis as shown in Figure 6-4, thereby introducing a small component of the spacecraft velocity into the line of sight. Figure 6-3 illustrates that a shift in either direction and of greater than a linewidth can be readily obtained by this method if both sets of lines are in the narrow Doppler-broadened regime.

6.2.5 Temperature and Concentration Measurements

For the purpose of illustration, it is useful to approximate radiance from a real non-uniform atmospheric path as the product of the emissivity ϵ of the path (i.e. one minus the transmission) and the Planck function appropriate to an effective temperature of the path, i.e.

$$\text{Radiance} = \epsilon \, B(\overline{T}). \tag{3}$$

With a given filter profile, the emissivity of the path depends on whether black or line modulation is considered and in the latter case, the mean cell pressure and amount of Doppler shift. In most of what follows, a viewing direction with no Doppler shift is assumed. For instance, if the wings of the strong lines are selected by a high cell pressure, then in general a lower emissivity (more transparent) path results than if the centers were selected by a low cell pressure. For temperature measurement an atmospheric constituent is required which has a known distribution, and which is sufficiently plentiful that opaque paths can be obtained. CO_2 is the obvious choice and temperature may be sounded up to about 80 km where, for the 15 μm band, deviations from LTE begin to occur. Information on the breakdown of LTE is obtained above this level. At lower levels other gases give rise to black paths and redundant information is obtained and confidence checking between channels is possible. This comparison is particularly useful for channels at shorter wavelengths because of the much stronger temperature dependence of the Planck function there.

With the temperature field known from measurements of CO_2 emission, the distribution of other constituents may be deduced from the appropriate radiance measurements by inferring their emissivity (Equation 3). When sounding a particular altitude range, conditions in the cell are chosen so the path is semi-transparent with ϵ lying in the range 0.2 to 0.8. The most accurate composition data may then be derived. As an example, the emissivity of CO in a limb path as a function of tangent height for various cell pressures and atmospheric mixing ratios is shown in Figure 6-5.

6.2.6 Zonal Wind Measurement

If for a fixed angle of limb view the azimuth view is directed forward of the YZ plane and then stepped back to behind the ZY plane at the uniform rate which provides image motion compensation at the tangent point (Figure 6-4), the signal will vary due to the accompanying increments of added

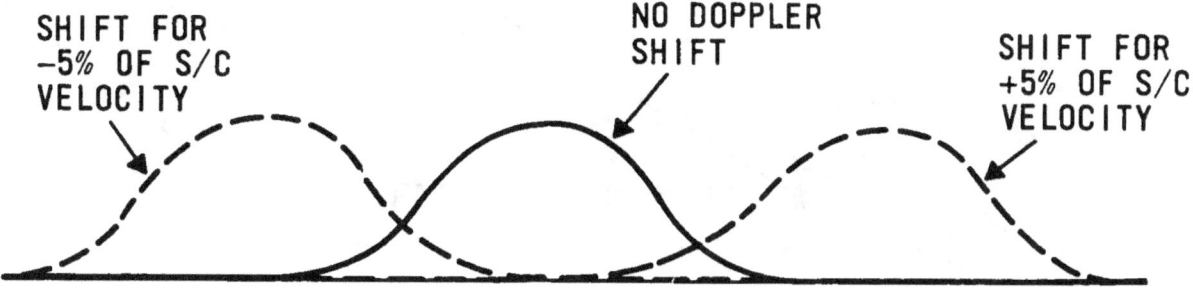

Figure 6-3. Effect of Doppler Shift on Line Spectra

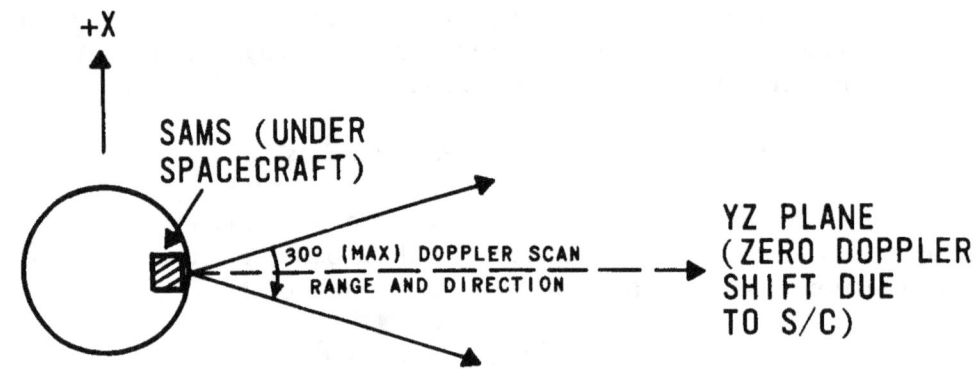

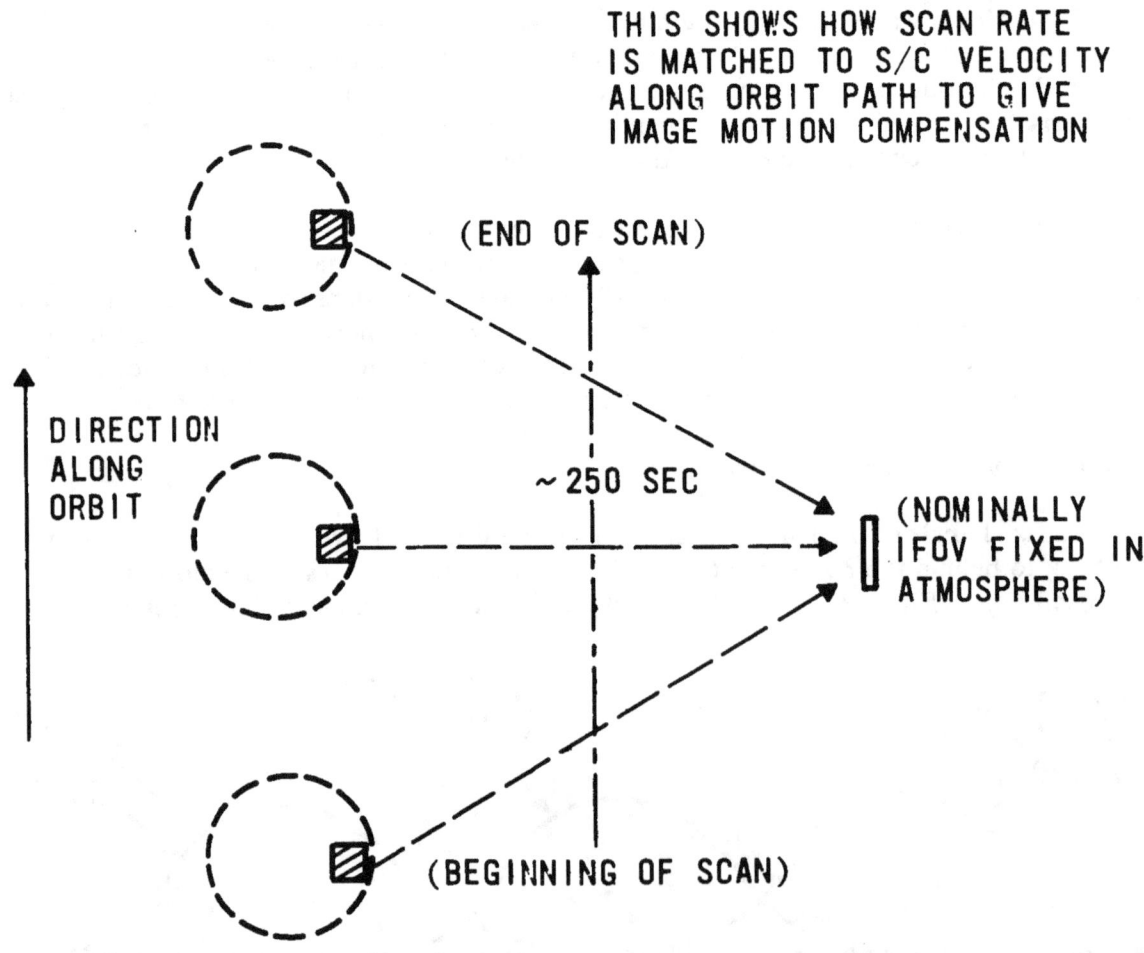

Figure 6-4. Azimuth Scan Angles

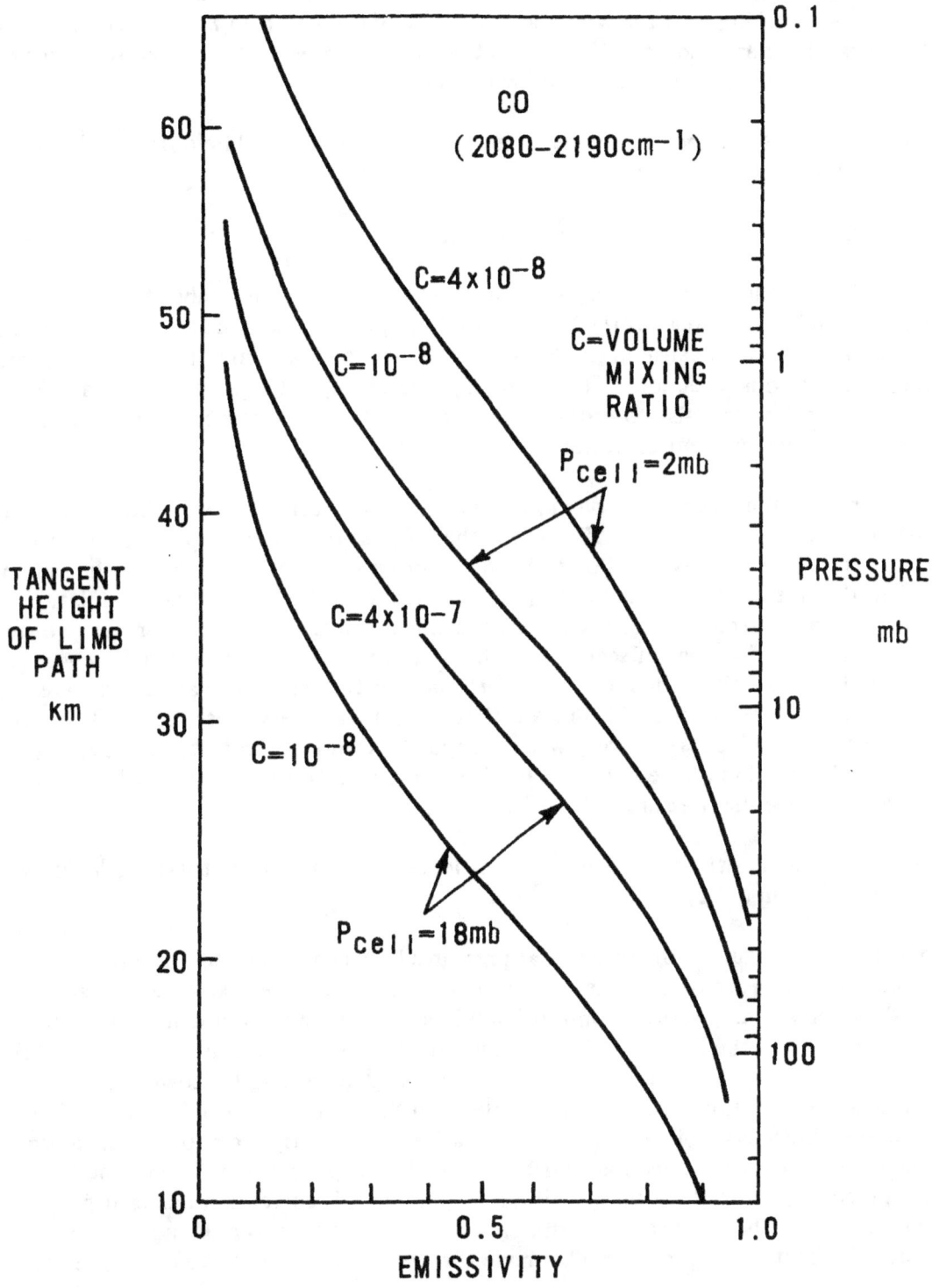

Figure 6-5. Emissivity of CO versus Tangent Height in the Atmosphere

Doppler shift. If the lines in the atmosphere and in the cell are narrow, the signal will have an appreciable dependence on the Doppler shift and will pass through a maximum at some azimuth angle near the normal to the flight direction. At this angle, the combined effects of the components of spacecraft velocity, the earth's rotation, and the wind along the line of sight sum to zero. If the velocity and attitude of the spacecraft are known with sufficient accuracy throughout this scan sequence of approximately 250 seconds, the zonal wind component may be deduced.

For this measurement it is expected the best signal-to-noise-ratios will be obtained from the shortwave (resonance fluorescence) channels.

6.2.7 Reference Pressure Determination

Where the emissivity of a limb path seen by a pressure modulated channel is about 0.5, the change in signal which corresponds to a change of 0.01 in emissivity (equivalent to about six percent in mixing ratio) could also result from a change in the level being viewed of 0.25 km (three percent in pressure or 0.004 degree in viewing direction). It is necessary, therefore, to keep a very accurate record of changes in viewing direction and also to provide the means to derive the atmospheric pressure at the level being viewed to three percent or better.

In SAMS, a channel in the 15 μm CO_2 band having a modulator cell filled with CO_2 at higher pressure than the other cells is included specifically for the purpose of reference pressure determination. The way in which this is done is as follows. In the approximation used previously the radiance from a given path in the case of black modulation is $\epsilon_B B(\overline{T_B})$, and in the case of line modulation it is $\epsilon_L B(\overline{T_L})$. CO_2 in the atmosphere is uniformly mixed and it is possible to choose the mean pressure in the cell so the relative contributions of segments of the path to the radiance detected are very similar for the black and line modualted channels over a large part of the scan range and for a wide variety of atmospheric temperature profiles. This means that $\overline{T_L}$ is always very close to $\overline{T_B}$. The ratio of the two signals is then ϵ_B/ϵ_L which is only very weakly dependent on temperature. However, ϵ_B and ϵ_L vary differently with elevation of the line of sight. Thus, the ratio gives a measure of the amount of CO_2 in the path and hence the pressure at the tangent height.

This pressure is also used as the reference for the other components of the total IFOV whose positions are known (see Section 6.3.1).

Calculations show (see Figure 6-6) for levels at pressures between 2 mb and 1 mb the ratio ϵ_B/ϵ_L is substantially independent of atmospheric temperature, even for the rather extreme atmospheric profiles used for the calculation. Over the 1 mb to 2 mb height range, measurement of the ratio ϵ_B/ϵ_L to 0.01 is adequate to infer the pressure with a maximum error of five percent even without taking into account the very different temperature structures which may be involved. However, information about the vertical temperature structure is available from the SAMS data (and from the Tiros Stratospheric Sounding Unit) making it possible to obtain a first order correction to the measurements of ϵ_B/ϵ_L. The pressure measurement accuracy with the SAMS data probably will be of the order of one percent which is equivalent to 0.08 km in altitude. Applying this correction makes it possible for pressure measurements to be inferred when this channel is observing lower altitudes. It is likely that measurements of acceptable accuracy can be made throughout the one mb to ten mb region.

6.2.8 Elimination of Contaminating Signals

If, within the passband of a particular channel, spectral lines from three molecular species are intermingled but uncorrelated, then the transmissions of the atmospheric paths for black and line

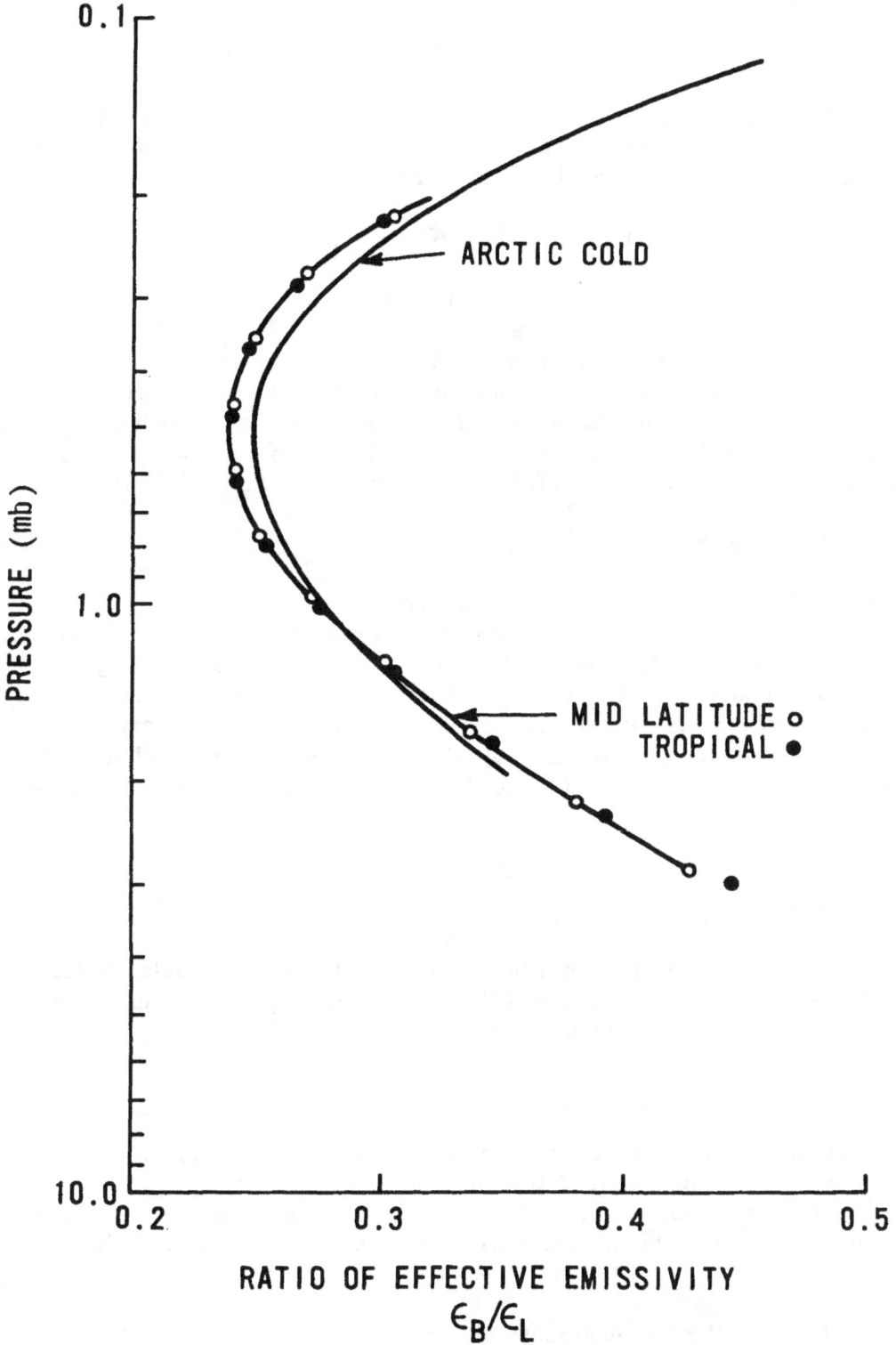

Figure 6-6. Variation of E_B/E_L with Height

modulated channels may be written,

$$\tau_B = \tau_B(W) \cdot \tau_B(C1) \cdot \tau_B(C2)$$

and

$$\tau_L = \tau_L(W) \cdot \tau_L(C1) \cdot \tau_L(C2),$$

where (W) refers to the wanted emission and (C1), (C2) to the contaminating emission. If the line modulation density is constant over the filter passband (similar to the black modulation), the amount by which the overall transmission is reduced by C1 and C2 is the same in both cases,

$$\tau_B(C1) \cdot \tau_B(C2) = \tau_L(C1) \cdot \tau_L(C2)$$

so that,

$$\tau_B/\tau_L = \tau_B(W)/\tau_L(W)$$

and the ratio of the measured transmissions is equal to the ratio of the uncontaminated transmission which is chiefly a function of the amount of (W) but with some slight temperature dependence. This technique for eliminating signals from gases other than the one of interest which emit within any spectral interval has been applied to measurements in the NO band at 5.3 μm, and the NO_2 band at 6.0 μm and is described by Chaloner et al. (1978) and Drummond and Jarnot (1978).

6.2.9 Basic Measuring Procedure

The line of sight of the radiometer can be stepped in elevation to view any altitude in the limb. A measurement sequence includes clear views to space and returns to levels where good altitude data can be derived. The purpose of the space view is to provide the instrumental offset signal (including the optics emission) which must be subtracted from all other signals; the differences are small fractions of the available dynamic range for most atmospheric views. (An important exception is the 2.7 μm water vapor channel where large fluorescence signals are received when the limb is in sunlight.) Calibration points at the other end of the range are obtained occasionally by moving a blackbody of known temperature into the beam.

6.3 The SAMS Radiometer

The SAMS radiometer consists of a sensor housing and an electronics module, the former being mounted under the sensory ring on the +Y side of the spacecraft and viewing the limb of the atmosphere on that side, i.e. normal to the direction of flight.

6.3.1 Optics

A central section through the optical system in the YZ plane is shown in Figure 6-7. The paraboloid (M2), ellipsoid (M3), folding mirror (M4), and composite field stop (M5) form a telescope with a collecting aperture of 177 cm^2 and a field of view with three separate components (A, B, C) each of solid angle 7.8 x 10^{-5} steradians. A fourth (plane) facet on M5 is used for alignment purposes.

The viewing direction of the radiometer may be stepped independently in tangent height and in azimuth by tilting the plane scanning mirror (M1) about axes parallel to the X and Z (roll and yaw) axes of the spacecraft.

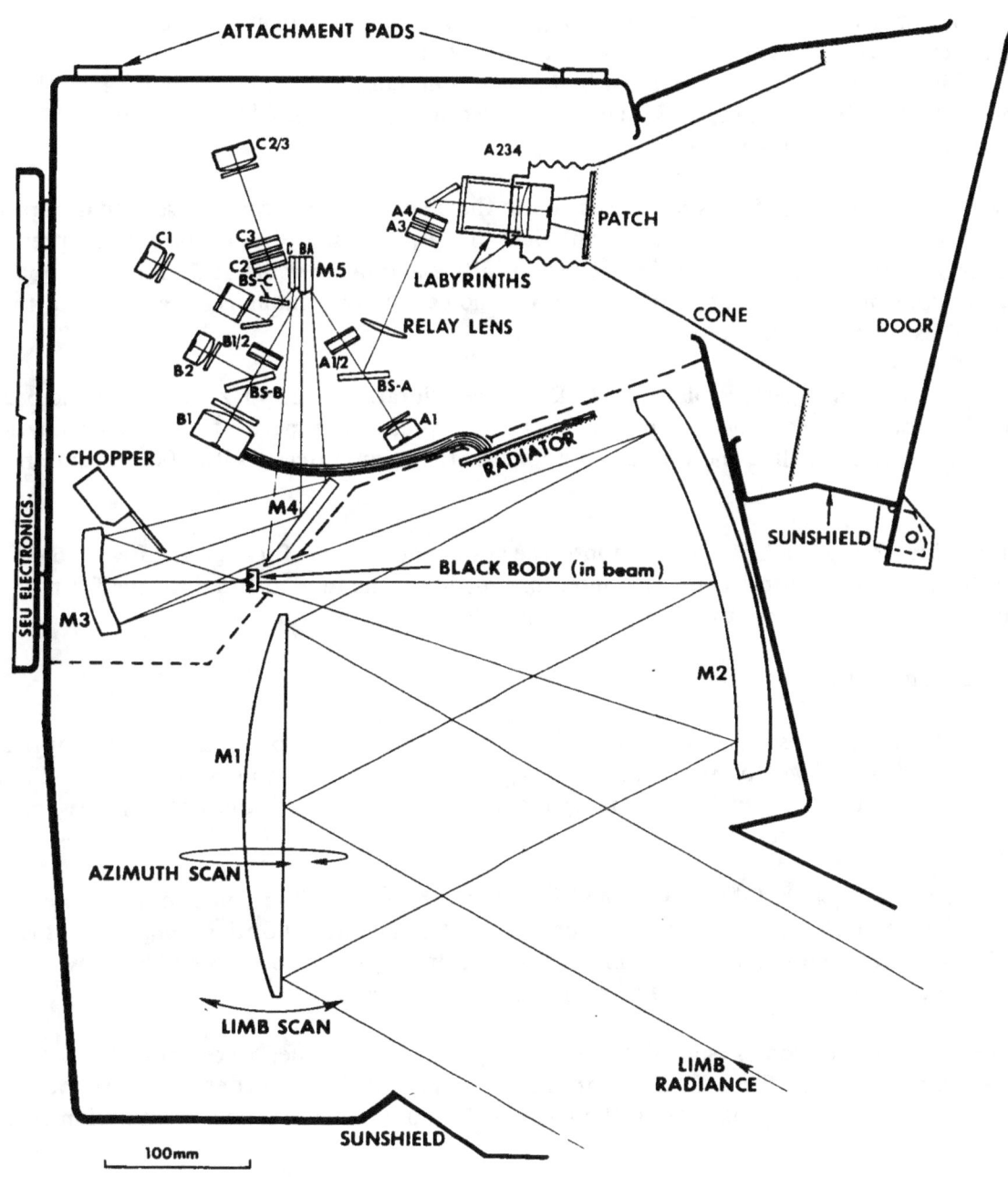

Figure 6-7. Central Section through Sensor

The view in azimuth is either set normal to the direction of flight or stepped to perform an image-motion-compensated scan, taking about 250 seconds, from 15 degrees forward to 15 degrees behind the normal. The limb scanning covers a range from a clear view of space down to the earth's surface. Each step moves the line of sight (LOS) by half a FOV component (i.e. 5 km at the tangent height). The range may be extended down to an angle approximately 70 degrees intersection of the LOS with the local vertical (Figure 6-8). Allowance must be made for spacecraft orbit and attitude uncertainties. The actual limb scan range is set by ground command to the desired portion of the 6 degree range available. The FOV and LOS angles are shown in Figure 6-8. Note the "smearing" due to the basic 1.8 second measuring period.

As shown in Figure 6-7, the facets of M5 distribute the beam to three trains of secondary optics, each associated with one of the components of the field of view. These optics trains contain pressure modulator cells operating in the 25 Hz to 50 Hz region, filters and beam splitters defining the optical passbands, relay components, and detectors. Table 6-2 shows how the channels are assigned to the three FOV's and lists their functions.

Positioned near M3, an aperture plane, a black chopper vibrating at 254 Hz modulates the beam to a depth of five percent peak to peak. Shallow chopping avoids the 50 percent average attenuation of the beam, and hence of the line-modulated components that would occur with 100 percent chopping.

Calibration blackbody can be introduced into the system at the first focal point (Figure 6-7). It consists of a thick aluminum disc 1 cm in diameter, machined on the inward face into the form of a re-entrant cone and painted black; the outward face is gold plated.

6.3.2   Thermal Design

Thermally, the SAMS sensor housing can be considered to be in four parts, the primary optics, the secondary optics, the A234 (InSb) detector cooler, and the sensor electronics unit (SEU). These may be identified in Figure 6-7 which shows the position of a thermal partition (dotted) between the primary and secondary optics.

The scan mirror and paraboloid are intended to cool to about $-5°C$ to reduce continuum emission, and the B1 (PbS) detector radiator to about $-10°C$. The secondary optics compartment is coupled to the spacecraft and maintains a nominal temperature of about $18°C$. The SEU is weakly coupled to the optics housing but thermally strapped to the spacecraft.

The components of the cooler (sunshield, cone and patch) are designed to operate at $-15°C$, $200°K$ and $145°K$ respectively. The InSb detector assembly is provided with heaters to enable the lens temperatures to be raised to approximately the spacecraft temperature for decontamination if required.

The dynamic response to simulated orbital variations of spacecraft mount temperature is shown in Figure 6-9.

6.3.3   Detectors

Three different kinds of detectors are used, all having active areas of 3.2 by 0.32 mm and doublet optics designed to image the active area on the relevant facet (A, B or C) of M5. Triglycine sulfate

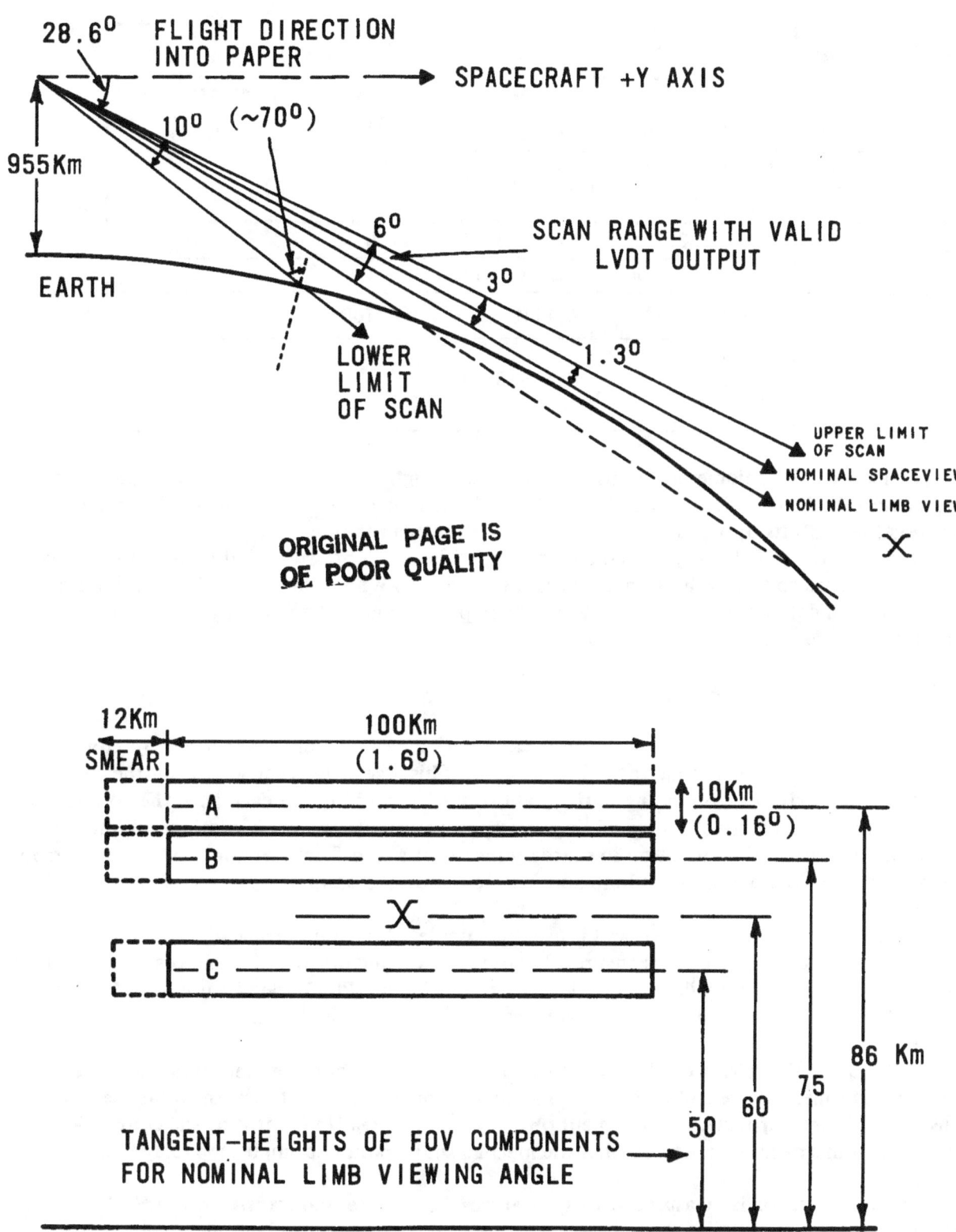

Figure 6-8. Fields of View and Line-of-Sight Angles

Table 6-2
Signal Channel Functions

FOV	CHAN	GAS	BAND	DATA PRODUCT
A	A1	CO_2	15μm	TEMPERATURE; VIBRATIONAL TEMPERATURE; ATTITUDE; CO_2 DISTRIBUTION; WIND.
A	A2	CO_2	4-5μm	TEMPERATURE; VIBRATIONAL TEMPERATURE; ATTITUDE; CO_2 DISTRIBUTION; WIND.
A	A3	CO	4-5μm	DISTRIBUTION
A	A4	NO	4-5μm	DISTRIBUTION
B	B1	H_2O	2.7μm	DISTRIBUTION, WIND.
B	B2	H_2O	25-100μm	DISTRIBUTION
C	C1	CO_2	15μm	ATTITUDE, TEMPERATURE
C	C2	N_2O	7.7μm	DISTRIBUTION
C	C3	CH_4	7.7μm	DISTRIBUTION

(TGS) pyroelectric flake bolometers are used for the wavelengths longer than 7 μm, an InSb photovoltaic diode for the 5 μm region, and a PbS photoconductive plate for the 2.7 μm H_2O channel. Each detector together with its optics and preamplifier is built into a module. A listing of some key optical parameters is given in Table 6-3 which indicates how the seven pressure modulator cells and six detectors are assigned to the twelve signal processing channels. Where there are cells in series in the path to a particular detector only one will normally be pressure modulated at any time. The desired combinations are selected by ground command.

6.3.4 Scan Mirror Assembly

The scanning mirror is mounted on a frame carried on a pair of cross-leaf springs (flexipivots) which allow vertical (limb) scanning of the line of sight. This assembly is carried on another frame, also supported on flexipivots, providing horizontal (azimuth) scanning. In each axis, the position is set by a jack screw, driven directly by a 45 degree stepping motor, and a recirculating-ball nut. The jack screws are lubricated for operation in space by means of a lead film. The smallest increments in the line of sight are 4.8 arc minutes in limb and 7.3 arc minutes in azimuth.

The mirror movement is initiated and is usually completed during the signal channel integrator run-down plus read-out period. This ensures that data lost during mirror movement is minimized. The mirror mechanism is controlled by the program control logic (PCL). See Figure 6-10 for a block schematic of the SAMS electronics and mechanisms. Overall timing is shown in Figure 6-11.

In the limb axis the mirror can be moved to any step position but in azimuth only a sequential scan can be performed. By reloading the PCL memory from the ground, the most suitable scanning patterns for achieving particular measurement objectives can be realized. Most of the control logic can be bypassed in a backup mode in which the mechanism responds to single step relay commands.

In each axis, position is measured with a linear variable differential transformer (LVDT) coupled directly to the mirror cradle. The limb LVDT is connected to a 14 bit triple-slope integration-type ADC which gives the system a resolution of 0.045 arc minutes in sight line within the 6 degree angular range.

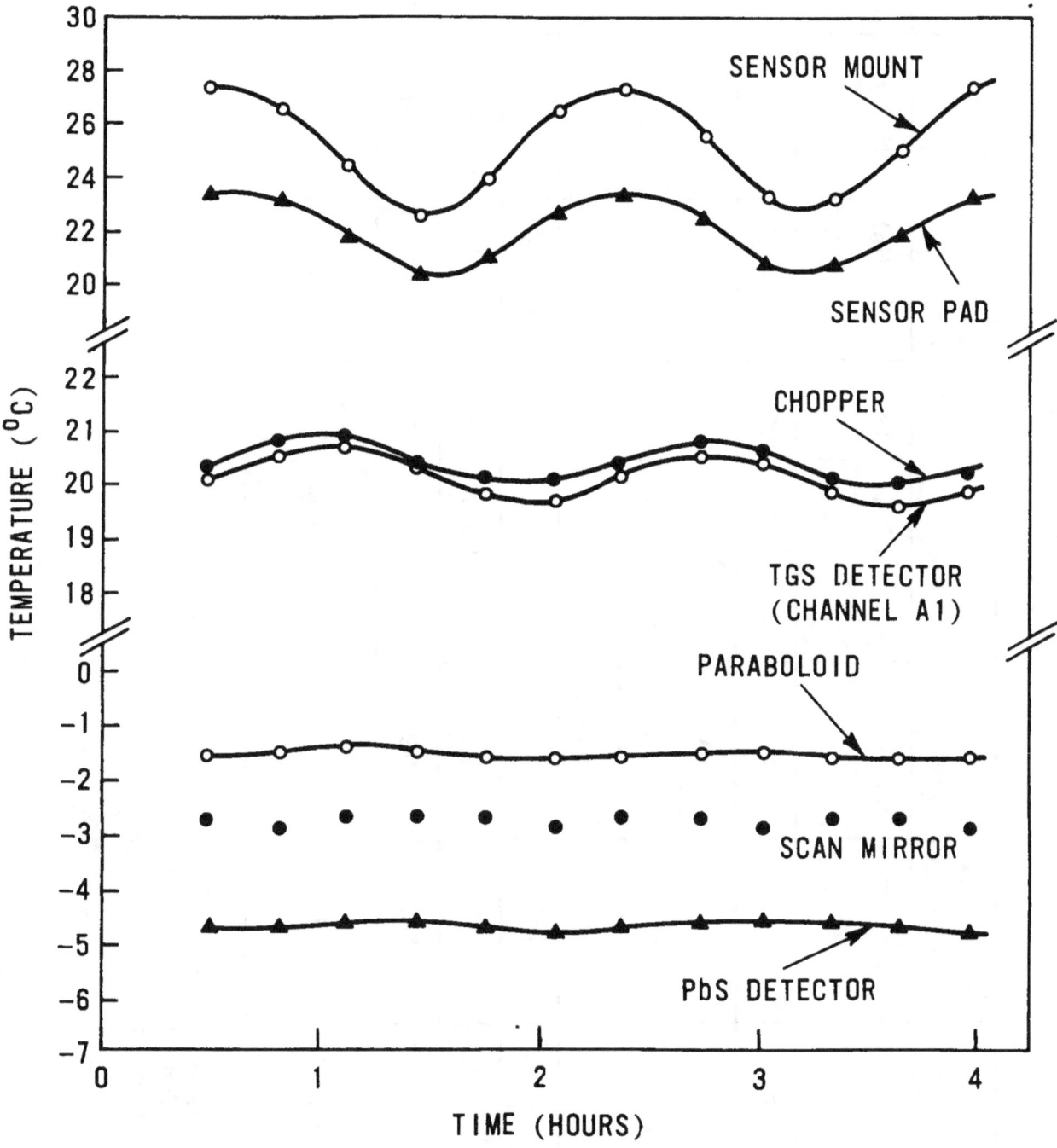

Figure 6-9. Orbital Temperature Response

Table 6-3
Signal Channel Parameters

PCM Designation	Gas	Filter Bandpass (μm)	Modulated Bandwidths		Detectors					Signal Channel Designations	
			*Black (cm^{-1})	†Line (cm^{-1})	Designation	Type	Temp	**NEP (f_B)	**NEP (f_L)	Black Modulated	Line Modulated
A1/2	CO$_2$	14.4–15.7	2.9	0.5	A1	TGS	290K	2.0 x 10^{-10}	0.9 x 10^{-10}	A1B	A1L
				6.0							A2L
A3	CO	4.1–5.4	26	1.6	A2/3/4	InSb	140K	8 x 10^{-13}	6 x 10^{-13}	A2/3/4B	A3L
A4	NO			1.0							A4L
B1/2	H$_2$O	2.5–2.6	10	0.8	B1	Pbs	260K	5 x 10^{-13}	6 x 10^{-13}	B1B	B1L
		25–100	15	0.6	B2	TGS	290K	2 x 10^{-10}	0.9 x 10^{-10}	B2B	B2L
C1	CO$_2$	14.4–15.7	2.9	0.9	C1	TGS	290K	2 x 10^{-10}	0.9 x 10^{-10}	C1B	C1L
C2	N$_2$O	7.6–7.8	1.7	3.0	C2/3	TGS	290K	2 x 10^{-10}	0.9 x 10^{-10}	C2/3B	C2L
C3	CH$_4$			0.4							C3L

*5 percent modulation depth allowed for
†mean cell pressures at highest settings
**watts for 1–8 second integrating time (i.e. single sample)

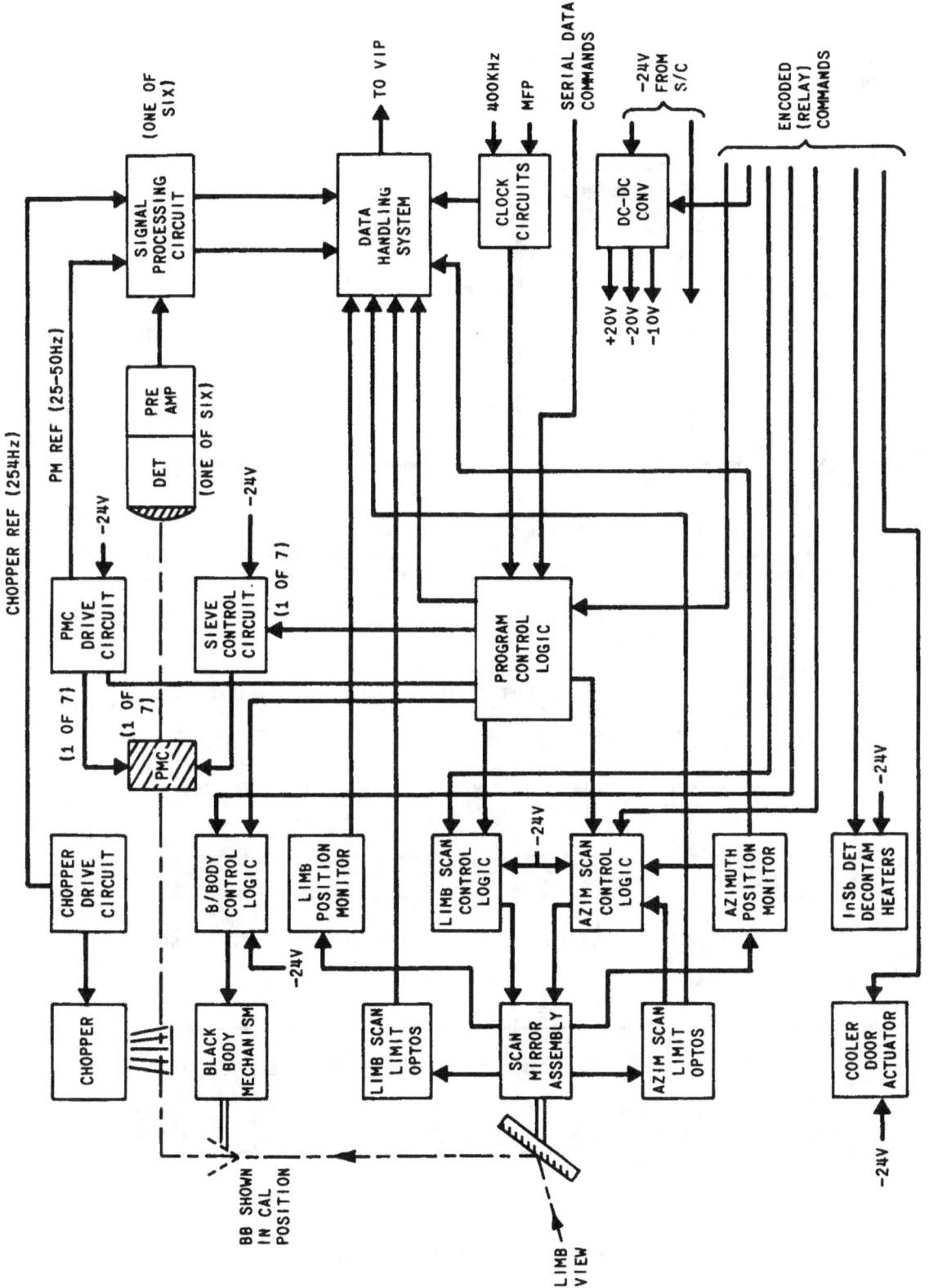

Figure 6-10. Block Schematic of SAMS Radiometer

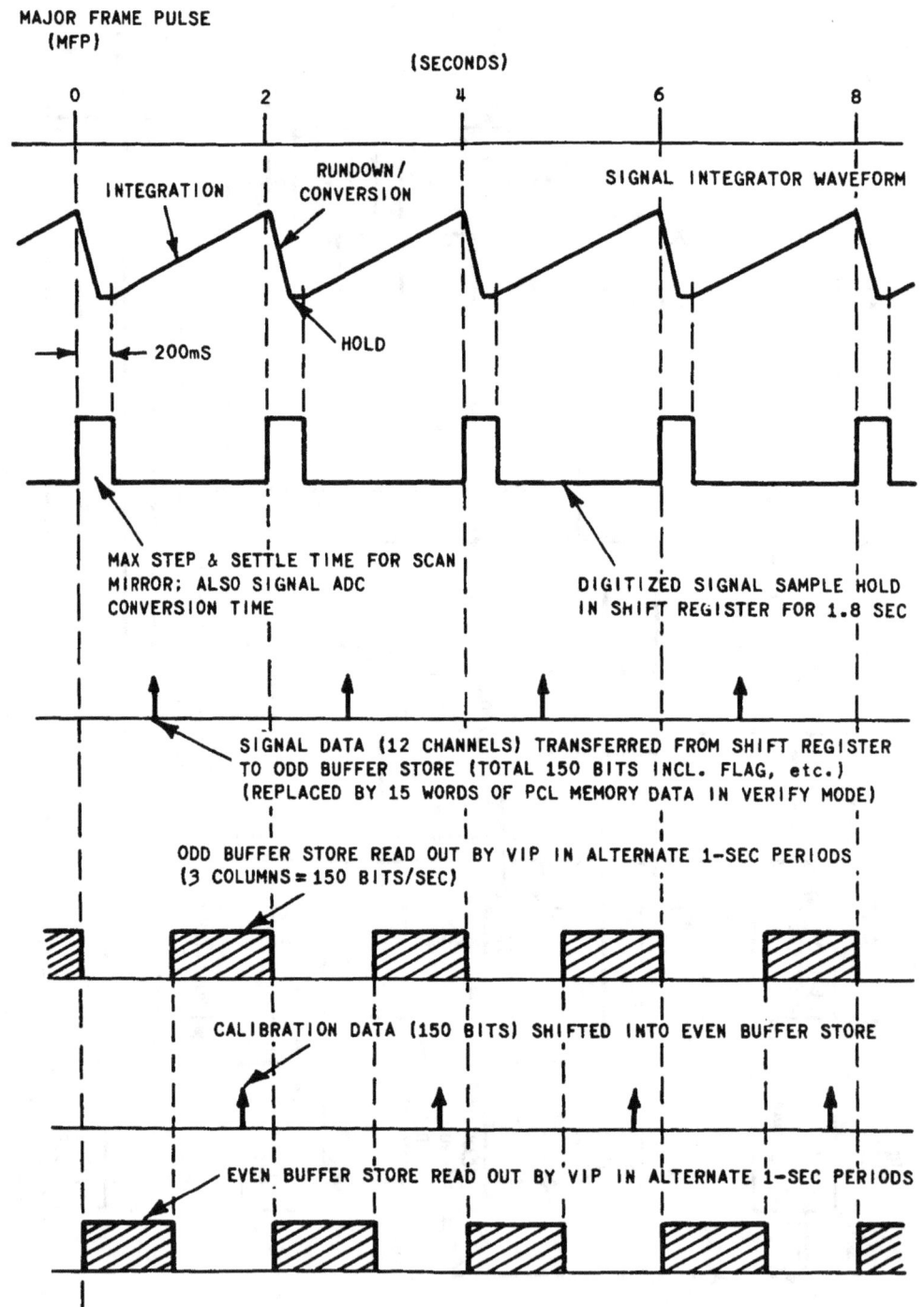

Figure 6-11. Overall Timing Diagram

The azimuth LVDT feeds a simple analog circuit giving a measurement with a resolution of 1.8 arc minutes in sight line. Unlike the limb LVDT, the output of the azimuth system is used in the position control circuit; it defines the non-scanning position in the YZ plane.

6.3.5 Blackbody

The calibration blackbody is mounted on a radius arm fixed to the shaft of a 90 degree stepping motor. In the normal energized state the arm is held out of the beam with a spring. When the motor is energized the arm moves up to an end-stop which defines the in-beam position. Other windings on the motor can be energized to assist the pull spring if this becomes necessary. Blackbody operation is normally controlled by the PCL; back-up relay commands are also provided

6.3.6 Black Chopper

The chopper consists of two etched copper grids, painted black and positioned one above the other. One is mounted on the end of a flat cantilever leaf-spring; the other is fixed to the spring mount. Oscillation is maintained at the mechanical resonant frequency of the grid/spring assembly by an amplifier having controlled limiting, and redundant pairs of piezo-electric ceramic plates, bonded to the spring leaf, which serve as drive and position-sensing elements. The drive loop supplies a phase reference to the signal processing circuits.

To minimize reaction forces on the SAMS structure the chopper baseplate is mounted on a pivot through the center of mass which allows it to move in antiphase to the spring motion.

6.3.7 Pressure Modulators

A section of a pressure modulator assembly is shown in Figure 6-12. It consists of an absorption cell coupled by drilled channels to a cylinder of 3 cms in diameter in which a piston oscillates with an amplitude of 3 mm peak to peak and a running clearance of 0.05 mm. The piston, the drive coil (part of a linear motor), and a soft iron slug are all mounted on a shaft which is carried on etched Beryllium–Copper diaphragm springs. The motor magnet is inside the cylinder, while a coil surrounding the iron slug, and which with it constitutes a differential transformer position sensor, is outside the cylinder. Oscillation is maintained at the resonant frequency and at constant amplitude by a control loop coupled to the position sensor and motor. The loop can be switched into a negative feedback mode if it is desired to inhibit piston motion. The resonant frequency of the suspended parts depends on the mean pressure of the gas in the cylinder and typically varies from 25 Hz (evacuated) to 50 Hz (at 40 mb). Piston frequency is a valuable pressure monitor and is telemetered with an accuracy of better than 1 part in 6000 for each modulator. The mean gas pressure in the the cylinder is governed by the temperature of a few grams of molecular sieve material held in a side arm, on which most of the gas is stored by absorption. The temperature is controlled by a thermostat which can be programmed from the ground via the PCL, the number of settings (two or four) depending on the enclosed gas. Before a cylinder is filled with gas it is subjected to a rigorous program of bakeout, pumping and leak testing.

6.3.8 Cooler Door Release

The cooler door is held closed by keys fixed to the shafts of two rotary solenoids which project through key plates mounted on either side of the door. On receipt of the second of two release commands, both solenoids are energized for 250 ms, the keys rotate to match slots in the plates and the door is pushed and held open by springs. The door is not recloseable in orbit.

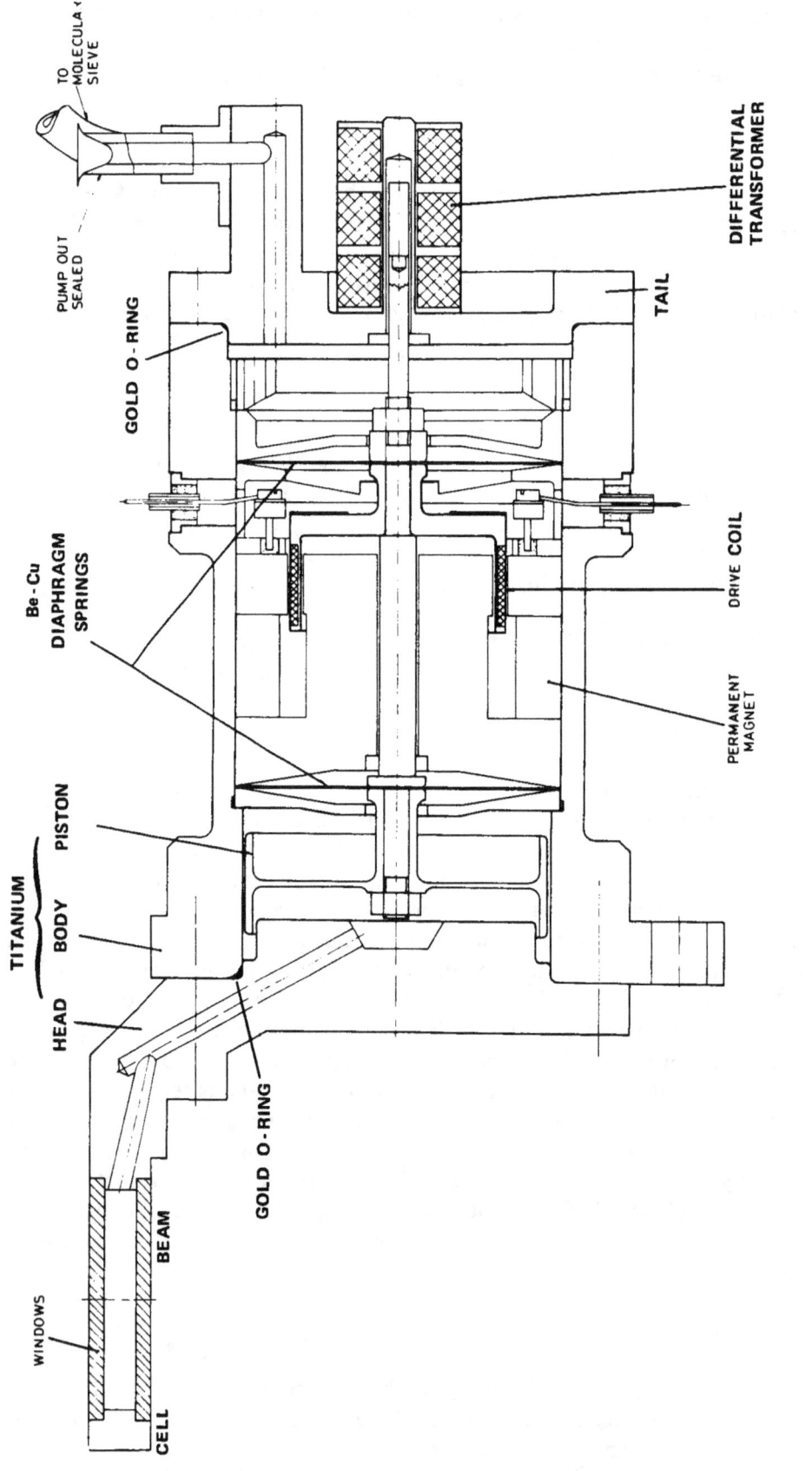

Figure 6-12. Pressure Modulator Cylinder

6.3.9 Signal Processing

The signal processing chain for one of the six pairs of line and black modulated channels is shown in Figure 6-13. Two modulators are indicated (as in the C2/3 chain) and the chopper, which is common to all six pairs, is also included. The output of the detector-preamplifier assembly, containing line and black modulated signals, is fed to active filter amplifiers having passbands of 20 Hz to 60 Hz and 180 Hz to 300 Hz respectively. Each filter amplifier is followed by a phase sensitive rectifier whose reference waveform is derived from the modulator on the chopper as appropriate, via a phase-correction network. In the former case this network is designed to operate over the whole range of frequencies (25-50 Hz) encountered for different modulator gas pressures. Selection of the correct reference waveform is controlled by the modulator enable/inhibit logic. Each phase-sensitive rectifier feeds a 12-bit dual-slope integration-type ADC, the output of which is parallel-loaded into a shift register. The ADC can recycle, giving an overrange factor of two when required (Figure 6-14 B).

The twelve registers associated with the twelve signal channels are connected in series. They are loaded simultaneously each 1.8 second measuring period and read out as a block into the data handling system. See Figure 6-15 for timing.

With the modulator pressures at the maximum settings the positions of the limb radiance and calibration signals on the overall signal transfer functions are similar to those in the chopped channels. The scale is inverted, however, on the B1 channels which are observing solar resonance fluorescence. The two cases are illustrated in Figure 6-14 (A) and (B).

6.4 Ground Calibration

6.4.1 General

An extensive program of testing and calibration has preceded launch. Measurements essential to the interpretation of signals from the instrument in orbit are of two main types:

- A calibration scheme is required for radiance which takes account of variation in operating conditions and temperatures; in particular any stray responses must be known or calculable

- The response of the instrument to emission from atmospheric paths must be calculable and a measurement therefore must be made of the response of the system to such paths of gas, so that the spectroscopic calculations by which atmospheric quantities are deduced can be verified.

The field of view of each channel must be known individually and in relation to the attitude determination channel. The limb scan angle telemetry system must also be accurately calibrated so that the attitude is known at all times during a scan sequence.

The vacuum/thermal chamber at Oxford provides thermal simulation of the spacecraft interface and of orbital temperature conditions. In addition various sources are available for calibration purposes (see Figure 6-15).

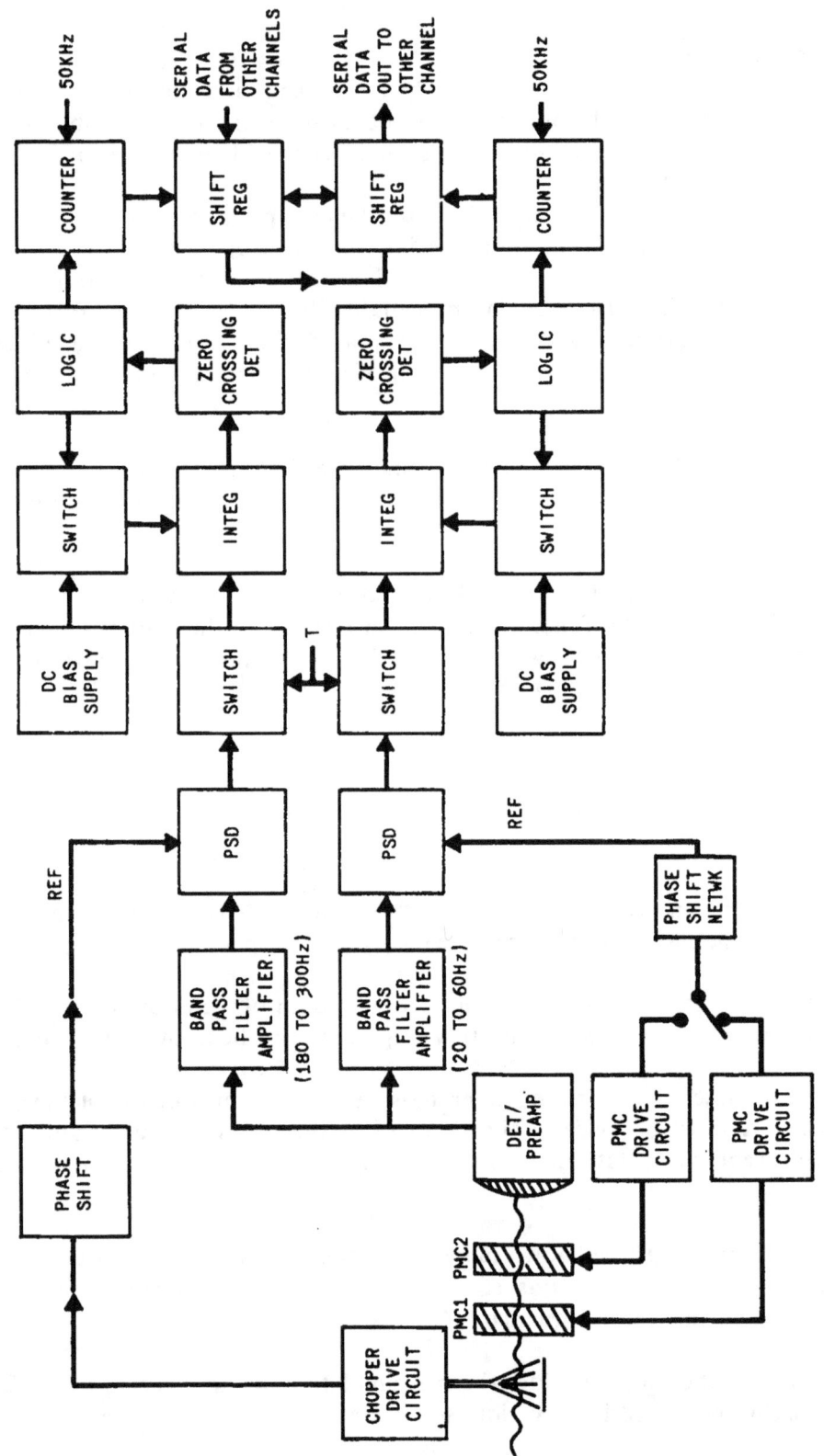

Figure 6-13. Block Schematic of Signal Electronics

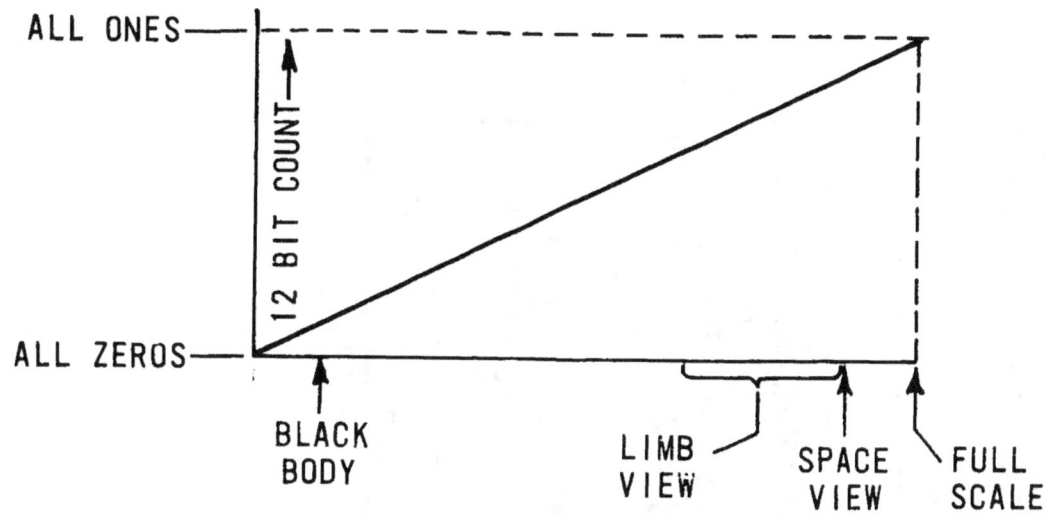

(A)

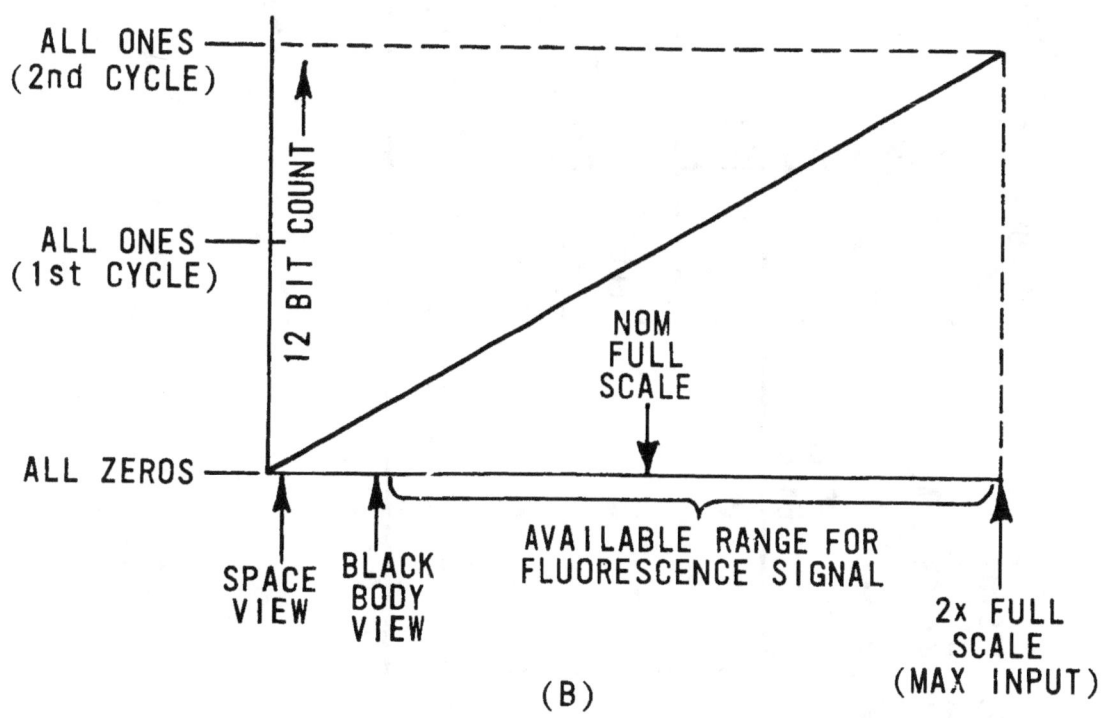

(B)

Figure 6-14. Signal Channel Transfer Characteristics for Channel B1

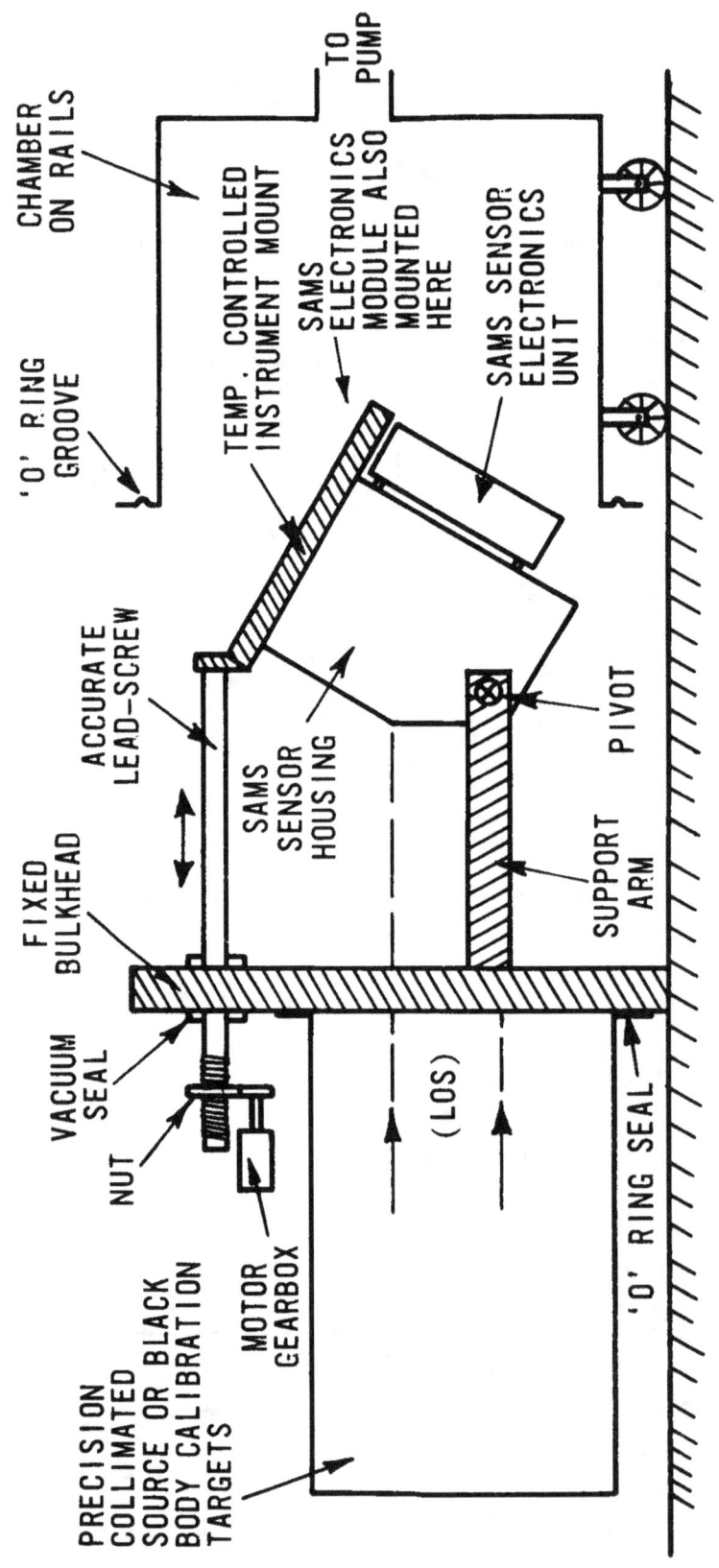

Figure 6-15. SAMS Test and Calibration Facility

6.4.2 Basic Radiance Calibration

Radiance calibration is intended to:

- Provide a basis for the model of the response of the instrument used in the in-flight calibration scheme

- Give a check of accuracy of calculations of the response of the pressure modulated (PM) channels to known radiance input

- Determine any scan angle dependence of the response

The source used for this calibration is a black cone which fills the whole instrument aperture. The temperature of this cone can be controlled at points between liquid nitrogen (LN_2 at 77K) and about +50°C. Several calibration points for known radiance inputs are obtained for each of the modulator pressure settings. During this test the temperature of the instrument is cycled at the Nimbus orbital period but with rather greater than expected excursions. In this way the temperature coefficients of response are determined dynamically; the response of the instrument is then compared with the known radiance, weighted by the calculated modulator response.

6.4.3 Scan Angle Dependence Checks

The SAMS sensor may be rotated about the axis of the scan mirror assembly by means of a high precision calibrated lead-screw (see Figure 6-15). This may be used to direct the optical axis of the instrument to the center of the cavity for all limb scan angle positions and so day variation of response with scan angle can be determined for a constant radiance input.

Targets are fitted on the axis of the LOS of the instrument, to be viewed with the mirror at the center of azimuth scan, and off axis, to be viewed at the extreme of azimuth scan. The axial target can be controlled at any temperature, but the off-axis target is maintained at LN_2 temperature. This provides a constant standard radiance throughout the extended test, and may be used in conjunction with the other target to determine any change in response at the extremes of azimuth scan.

6.4.4 Response of PM Channels to Atmospheric Paths

The response of the PM channel to gas paths is checked by measurement of the transmission of simulated atmospheric paths in the laboratory using the instrument modulator and detector subsystems before final assembly, and ones of the same type constructed specifically for this purpose. The apparatus consists of a multiple-path white cell of up to 10 meter path length. Transmissions of paths measured with this equipment are compared with calculations of response from line data, and information from both sources is combined in the radiance model used for retrieval.

6.4.5 Field of View Measurement

In orbit the LOS of the instrument is determined from the ratio of line and black modulated signals in the C1 channel (Section 6.2.9). We need, therefore, to know the LOS of the other channels with respect to this channel. In addition, the shape of the field response must be measured, particularly of the lower edge.

For this test a precision collimator with a narrow-slit hot source replaces the extended black targets. The lead-screw system is used to rotate SAMS slowly through the parallel radiation from the collimator, and the field response is thus determined.

This facility is also used for general optical checks and in particular for a calibration of the limb scan angle telemetry (LVDT) system. The lead-screw is driven slowly and continuously while the PCL is programmed to step the SAMS scan mirror so that the field of view of the instrument passes through the collimated beam once per step. By matching the respective points on the field response, the LVDT systems can be calibrated directly in terms of the lead-screw.

This test and the primary field of view measurements are carried out at several temperatures over the expected operating range.

6.5 Data Processing, Formats, and Availability

6.5.1 Data Processing

6.5.1.1 Processing Flow

SAMS data is extracted from the Nimbus 7 telemetry at GSFC and held for 24 hours. Once each day the previous day's data is transmitted to Oxford, England via NASA transmission lines. As it is received at Oxford the data undergoes quick-look processing to check the behavior of the SAMS instrument. This also allows monitoring of the stratosphere in near-real time so that an optimum operating mode can be maintained.

The image location tapes are transmitted to Oxford three to four weeks after the initial data. At this stage the data is reprocessed with the aid of more detailed and more accurate algorithms (obtained from further analyses of the original data) to produce magnetic tapes of radiances and derived SAMS parameters as described in Section 6.5.2. These tapes are mailed to the IPD at GSFC. IPD copies some of them and sends these copies to NET users. IPD uses the SAMS MATRIX tapes as input to generate the SAMS display products described in Section 6.5.3. All original tapes and microfilm are then sent to NSSDC for archiving. NSSDC makes copies of these products for users.

6.5.1.2 Summary of Processing Analysis

The quick look analysis is based on the approximation for radiance $R(v)$ at wavenumber v:

$$R(v) \cong B(v, Z_t) \, \epsilon(Z_t)$$

where $B(v, Z_t)$ is the Planck function at the tangent height Z_t and $\epsilon(Z_t)$ is the emissivity of the path through the limb. In the case of the CO_2 channels, $\epsilon(Z_t)$ is known, so that B can be derived, giving a first approximation to the temperature profile. Once this is known, B can be calculated for other gases, so that ϵ can be obtained from the measurements and hence the gas concentration at tangent height Z_t.

The final retrieval is a maximum likelihood estimator using a linearization of the full nonlinear radiance transfer equation, with the quick-look retrieval as first guess. The attitude reference provided by the spacecraft attitude control system is not of adequate accuarcy for the analysis so it is necessary to derive the tangent height from the measurements as described in Section 6.2.7. This method provides a first guess for use in the maximum likelihood estimator. Sequential estimators are used throughout to provide continuity in the solution along the tangent point track.

6.5.2 Tape Products

The following tapes are produced by Oxford, used by IPD, and then sent to the NSSDC for archiving. Brief descriptions of these tapes are as follows:

- RAT (Radiance Archive Tape)

    These tapes contain calibrated radiances versus scan angle records for all useable scans. Users may want to use these tapes to try their own methods of profile retrieval.

- MATRIX (Mapped Data Matrix Tape)

    These tapes contain northern and southern hemisphere polar map (0°-90°) and Mercator (±32°) matrices on constant pressure surfaces for all SAMS parameters to be mapped. Also on these tapes are all contoured cross-section plots. Each record contains all necessary reference and title information needed for annotation of each display.

The form and content of these is specified in a tape specification document for each tape type. The appropriate document will accompany a tape shipment to a user. See Section 1.5 of this document for details.

6.5.3 Film Products

The SAMS data are displayed on 112 different map sets and 12 different cross sections. Six parameters are mapped: temperature, water vapor, nitrous oxide, methane, carbon monoxide, and nitric oxide. (See Figures 6-16 and 6-17 for examples.) Each parameter is mapped at 16 pressure levels each month and each three months. Because of instrument operating restrictions only four parameters are mapped at six pressure levels on any given day. However, it is expected by the end of each month, data will be collected for all six parameters at all pressure levels. (Because of its day/night variability, daytime nitric oxide data is mapped separately from the nighttime nitric oxide data.) The same six parameters are averaged and displayed on 50°S to 70°N cross sections each day, month, and three months. (See Figures 6-18 and 6-19).

Table 6-4 lists the titles of all SAMS microfilm displays and the corresponding film specification number for each. All parameters are produced at more than one time scale as Table 6-4 shows. Interpreting the fourth digit from the left in each specification number gives the frequency of production of the parameter listed in the title:

   XXX1XX = produced every day,

   XXX7XX = produced every month, and

   XXX8XX = produced every three months.

All map displays contain one north and one south polar stereographic projection (pole to equator for each) and one Mercator projection (±32°). Each map contains contoured data as specified in the display title. Immediately beneath the Mercator map is contouring information giving the contour

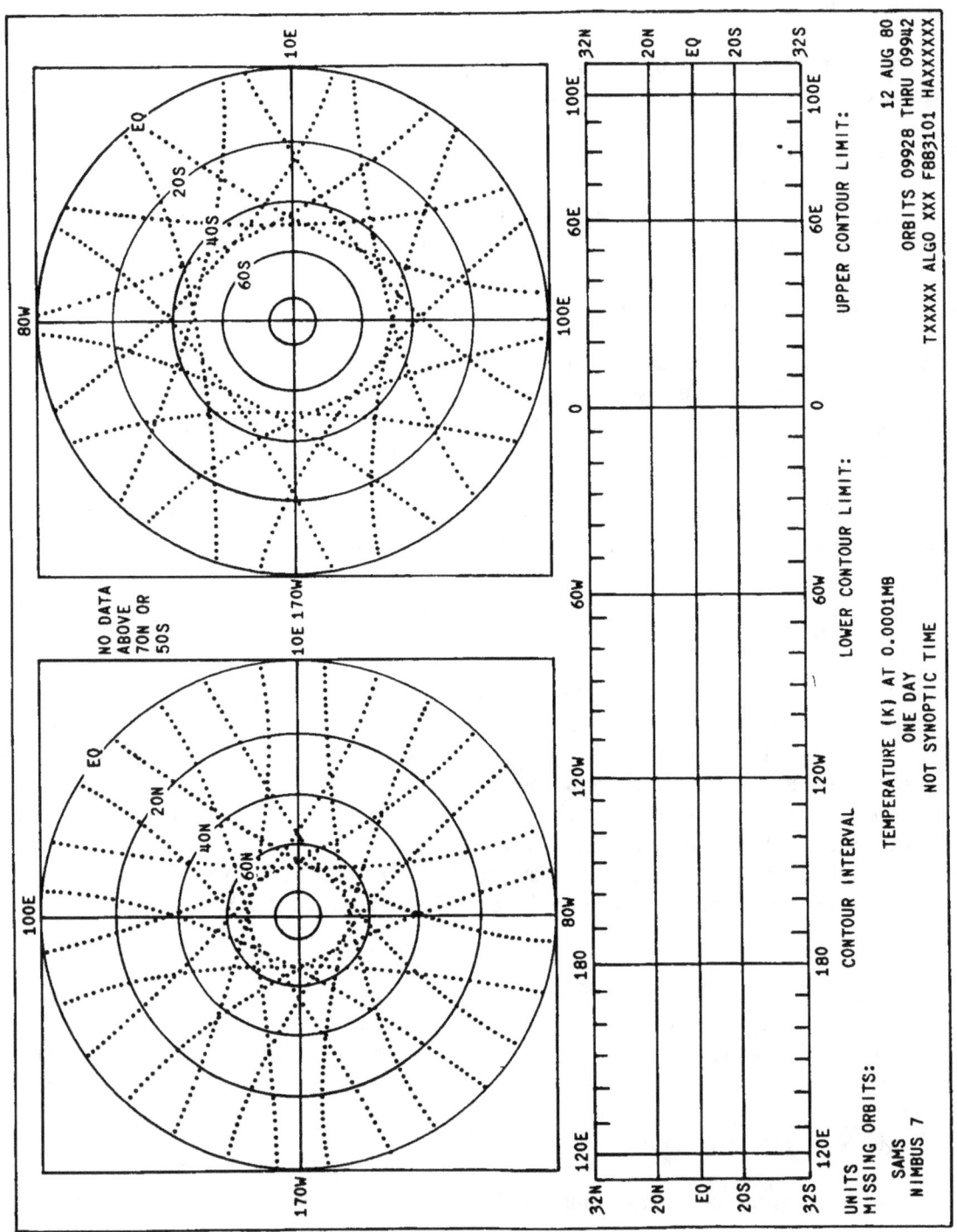

Figure 6-16. SAMS Microfilm Map Display

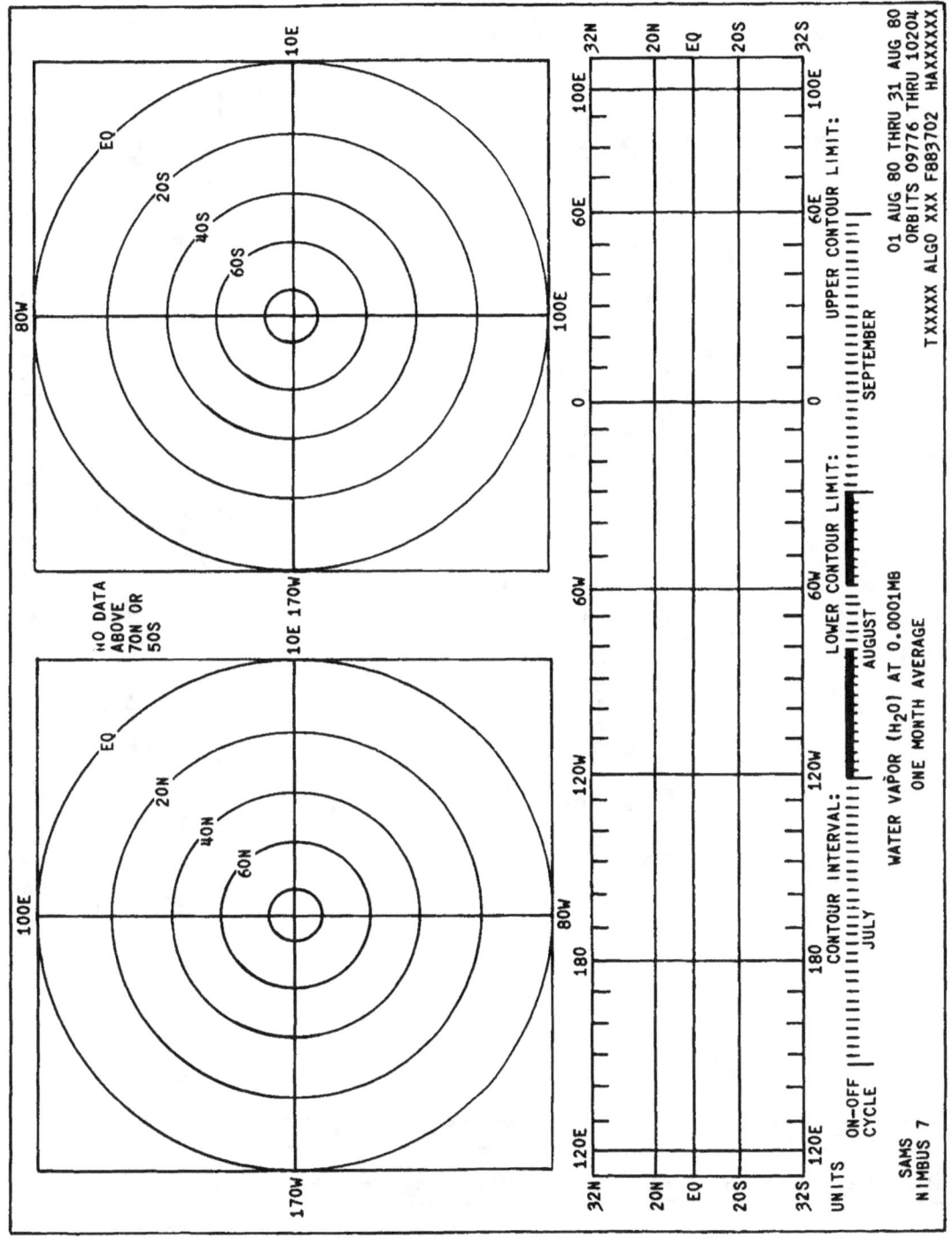

Figure 6-17. SAMS Microfilm Map Display Format for One Month and Three Month Products

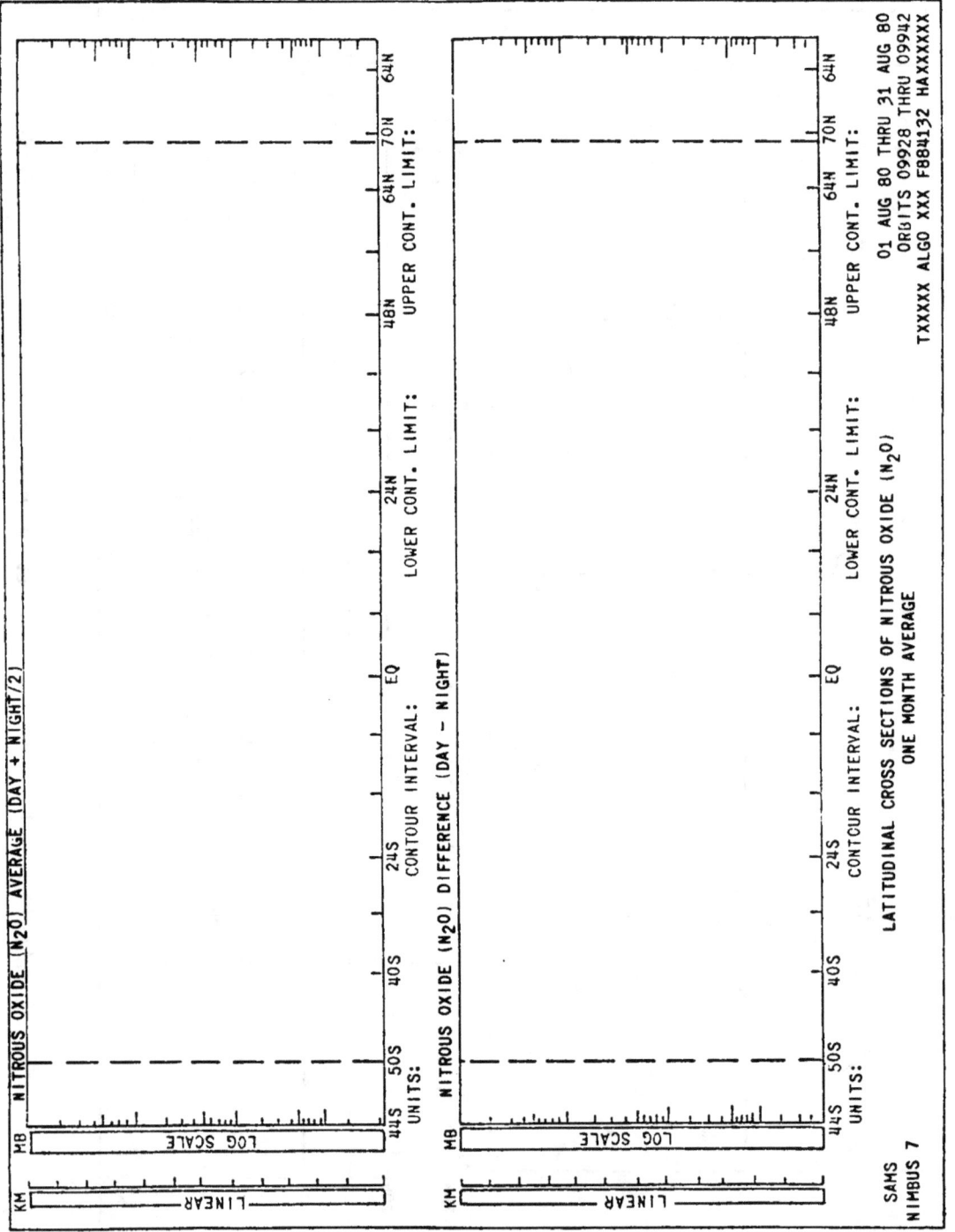

Figure 6-18. SAMS Microfilm Cross Section Display Format for all Parameters except Nitric Oxide

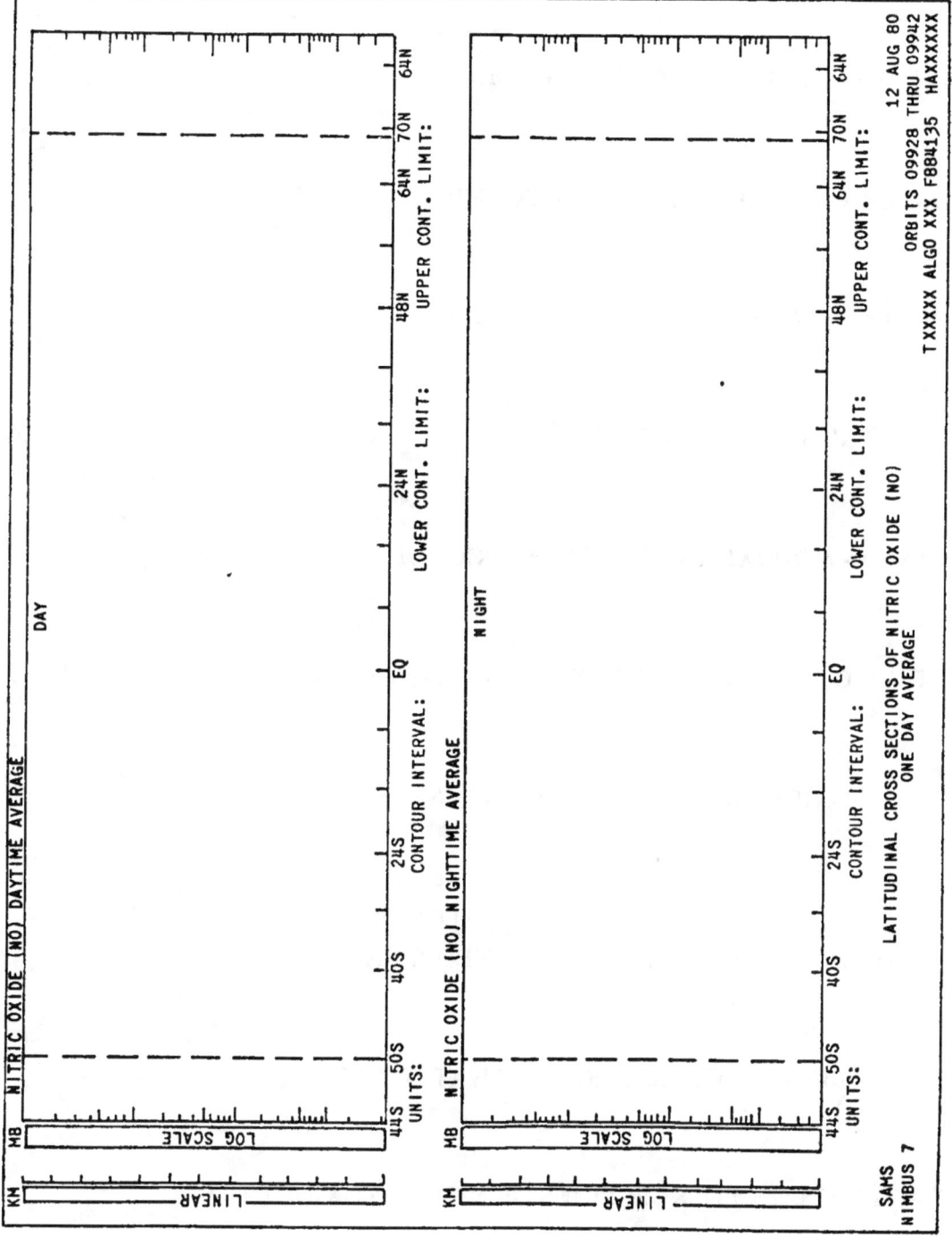

Figure 6-19. SAMS Microfilm Cross Section Display Format for Nitric Oxide

Table 6-4
Titles and Film Specification Numbers for SAMS Microfilm Products

Film Spec Number	Film Product Title
	MAPS
883101	TEMPERATURE (K) AT XXX.XXXX*MB
883701	
883801	
883102	WATER VAPOR (H_2O) AT XXX.XXXX MB
883702	
883802	
883103	NITROUS OXIDE (N_2O) AT XXX.XXXX MB
883703	
883803	
883104	METHANE (CH_4) AT XXX.XXXX MB
883704	
883804	
883105	CARBON MONOXIDE (CO) AT XXX.XXXX MB
883705	
883805	
883106	NITRIC OXIDE (NO) AT XXX.XXXX MB (daytime data)
883706	
883806	
883107	NITRIC OXIDE (NO) AT XXX.XXXX MB (Nightime data)
883707	
883807	
	CROSS SECTIONS
884130	LATITUDINAL CROSS SECTIONS OF TEMPERATURE (K)
884730	
884830	
884131	LATITUDINAL CROSS SECTIONS OF WATER VAPOR (H_2O)
884731	
884831	
884132	LATITUDINAL CROSS SECTIONS OF NITROUS OXIDE (N_2O)
884732	
884832	
884133	LATITUDINAL CROSS SECTIONS OF METHANE (CH_4)

Table 6-4 (Continued)

Film Spec Number	Film Product Title
884733	
884833	
884134	LATITUDINAL CROSS SECTIONS OF CARBON MONOXIDE (CO)
884734	
884834	
884135	LATITUDINAL CROSS SECTIONS OF NITRIC OXIDE (NO)
884735	
884835	

*The one month and three month maps are contoured and displayed at 16 pressure levels. These are 300, 100, 70, 30, 10, 5, 2, 1, 0.5, 0.1, 0.03, 0.01, 0.003, 0.001, 0.0003, and 0.0001 millibars. Because of instrument operating constraints, for any particular set of one-day maps all pressure levels are not possible. However, a large collection of one-day map sets should have all pressure levels as shown above.

units, interval between contour lines, and the maximum and minimum value contoured. All map displays contain an indicator of the quantity of data within a display. On the one day displays there is a "missing orbits per day" code specifying how many orbits of data are missing from that day's input to the map (See Figure 6-16). On the one month and three month maps (Figure 6-17) there is an "on-off cycle" scale specifying the days during the display period when the instrument was on and off. If a day on the scale is "filled-in" the instrument was on, the data was collected, interpreted, and used in the contouring. If a day is not filled in, the instrument was either off or the data was unuseable for some reason.

The polar stereographic maps on the one day displays (Figure 6-16 is an example) contain "dot tracks" with each dot representing a tangent point data value location used to construct the contours on the maps.

Figures 6-18 and 6-19 are examples of the SAMS cross section displays. Figure 6-18 is typical for all parameters except nitric oxide. Figure 6-19 is typical for nitric oxide. All parameters (except nitric oxide) have cross section displays of the average of the day and night data and the difference between the day and night data. Because of the large day-night difference the nitric oxide displays have one cross section with only daytime data and the other with only nighttime data. Data are referenced vertically by pressure and standard atmosphere altitudes, and horizontally by latitudes.

Title and reference information at the bottom of all displays is mostly self-explanatory. The right half of the last line, however, requires explanation. This information is mainly used for cataloging and information control. These items are: the physical tape number the data is stored on (TXXXXX), the algorithm reference number used in processing the data (ALGO XXX), the film specification number (F88XXXX), the project data format code (HA), and the film frame number (XXXXXX).

6.5.4　Data Availability

The SAMS experimental data consisting of the magnetic tapes described in Section 6.5.2 and the 16 mm microfilm described and illustrated in Section 6.5.3 are archieved at the NSSDC. It is anticipated that the first data sets will not arrive at NSSDC until at least three to six months after launch. Users requesting SAMS data should read Section 1.5 of this document for general tape and film ordering information.

6.6　Reference

1. Curtis et al. The Pressure Modulator Radiometer. Proc. Roy. Soc. London; A $\underline{337}$, pp. 135-150, 1974.

SECTION 7

THE SOLAR BACKSCATTER ULTRAVIOLET (SBUV) AND TOTAL OZONE MAPPING SPECTROMETER (TOMS) EXPERIMENT

by

Donald Heath and Arlin J. Krueger
National Aeronautics and Space Administration
Goddard Space Flight Center

and

Hongwoo Park
Systems and Applied Sciences Corporation
Riverdale, Maryland

7.1 Introduction

The Solar Backscatter Ultraviolet and Total Ozone Mapping Spectrometer (SBUV/TOMS) experiment is an expanded and improved version of the Backscatter Ultraviolet (BUV) experiment on Nimbus 4, and is composed of two essentially independent instruments. The Solar Backscatter Ultraviolet (SBUV) subsystem consists of a double Ebert-Fastie spectrometer and a filter photometer similar to the BUV. Both channels simultaneously view identical fields of solar radiation scattered by the terrestrial atmosphere in the nadir of the solar flux scattered from the instrument diffuser plate which is deployed on command. The spectrometer serially monitors 12 selected narrow wavelength bands in the spectral region from 250 nm to 340 nm, or continuously scans the wavelength range from 160 nm to 400 nm, while the photometer measures the light in a fixed band centered at 343 nm. This instrument is intended for use in determining the total ozone and its vertical distribution above the ozone maximum for measuring the ultraviolet solar spectral irradiance.

The Total Ozone Mapping Spectrometer (TOMS) subsystem employs a single monochromator whose IFOV is scanned through the subsatellite point and perpendicular to the orbital plane. The backscattered radiation is sampled at six wavelengths from 312.5 nm to 380 nm sequentially in three degree steps in the ±51 degrees cross scan from the nadir. This scanning creates a contiguous mapping of the total ozone since the scans of consecutive orbits overlap.

The polar orbit of this satellite makes possible the ozone measurements on a global basis and the expected instrumental lifetime should permit the long-term monitoring of ozone and the ultraviolet solar flux. Recent advances in the standard spectral irradiance sources used in the calibration will provide better accuracy in the absolute solar flux measurement.

7.2 Scientific Objectives

The SBUV/TOMS is designed to measure the extraterrestrial ultraviolet solar irradiance and the solar ultraviolet radiation backscattered from the earth and its atmosphere. Methods to recover the ozone information from backscattered ultraviolet measurements are described in References 1, 2, 3, and 4.

The objectives of the SBUV/TOMS experiment may be summarized as follows:

SBUV

- Determine the total amount of atmospheric ozone in a vertical column above the subsatellite point
- Determine the vertical profile of ozone above the ozone maximum
- Measure the ultraviolet solar spectral irradiance and monitor its temporal variability over the wavelength range from 160 to 400 nm, with a spectral resolution of 1 nm.

TOMS

- Obtain contiguous mapping of total ozone.

7.3 Instrumentation

The SBUV/TOMS consists of three units: two sensor modules and one electronic module (ELM). The sensor modules house the optical components of each subsystem, the high voltage power supplies, and the first stages of the signal processing electronics. The electronics module houses the bulk of the signal processing electronics plus the circuitry required to support the whole system; the command receiving and decoding relays, low voltage power supplies, interface circuits for spacecraft clocks, telemetry readout and control circuits. Figure 7-1 shows a functional block diagram for the SBUV/TOMS. A preliminary design of the SBUV/TOMS has been reported in Reference 5.

7.3.1 SBUV Subsystems

Figure 7-2 shows an optical diagram of the SBUV. The SBUV is composed of a double monochromator, a photometer and supplementary optical mechanical and electronic equipment. Two Ebert-Fastie type monochromators are used in tandem and connected with reflective transfer optics and an intermediate slit. Each monochromator has a single spherical collimating mirror of a 250 mm focal length coated with aluminum and a halographic diffraction grating having a 52 mm x 52 mm rules area and 2400 grooves per mm.

The optical design of the monochromator was aided by computer-plotted ray tracing. The spatial shape of the exit slit was also determined by ray tracing. The spectrometer has fixed entrance and exit slits which are 30 mm long. Their widths are equivalent to a 1 nm spectral bandwidth near 300 nm.

The SBUV monochromator scans 12 discrete wavelengths ranging from 250 nm to 340 nm and 5 wavelength calibration steps centered near 253.7 nm, or scans continuously the wavelength range from 160 to 400 nm in 0.2 nm steps. The wavelength in the continuous scan mode is determined from the cam position telemetered by a high resolution encoder mounted on the cam shaft. The two gratings are rigidly mounted to a single casting. The casting provides the function of a bearing axis, a wavelength drive arm, and support for the cam follower bearing.

The SBUV used three detectors: one photomultiplier tube (PMT) and one photodiode for the monochromator, and one photodiode for the photometer. The PMT is an ITT F4090 type with a bialkali photocathode and a fused silica window. The window material was selected for transmittance at 160 nm and minimum florescence resulting from the high energy particle in the trapped radiation belts. The photodiodes are vacuum sealed and have the same photocathode characteristic as the P

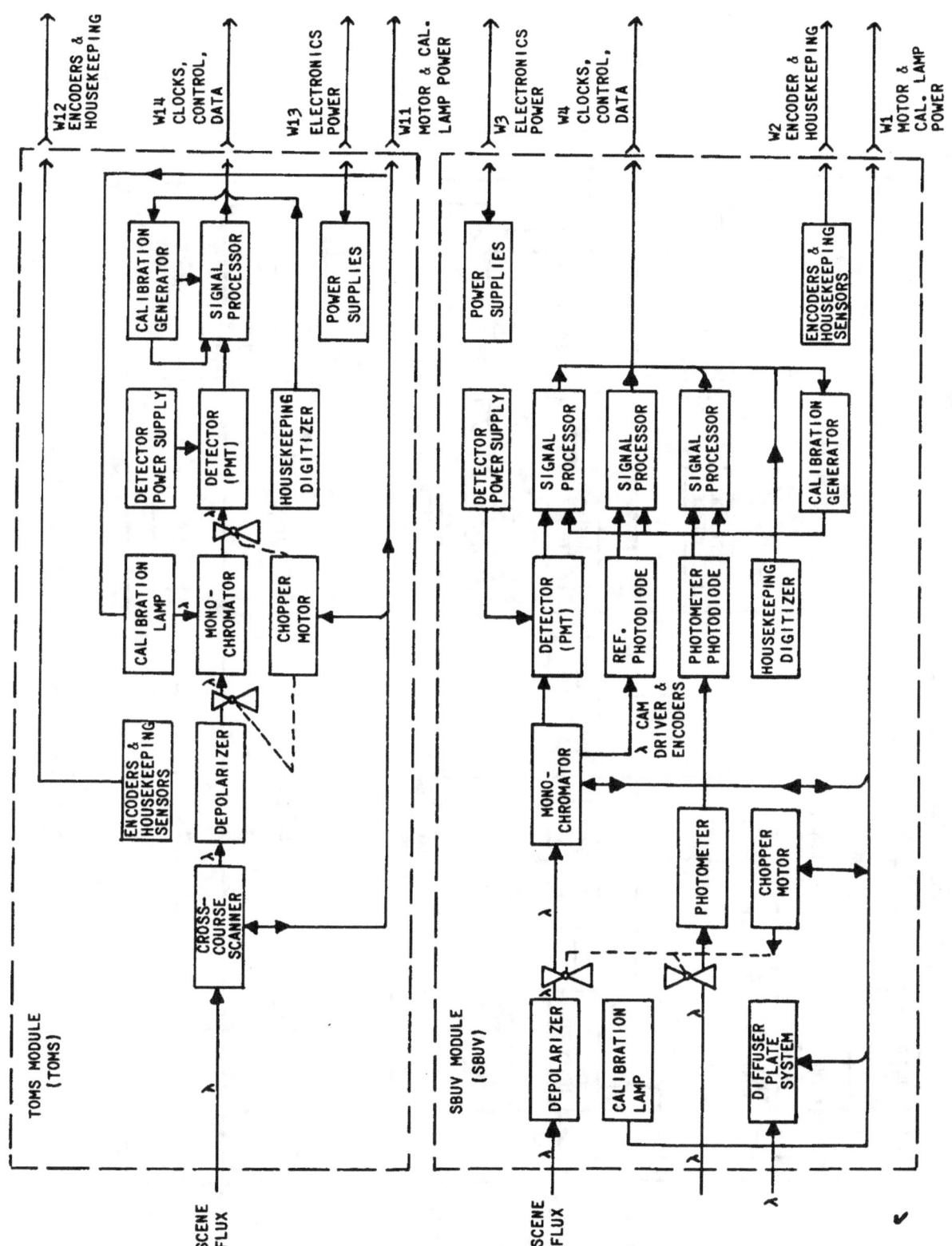

Figure 7-1 Functional Block Diagram of SBUV/TOMS Modules

177

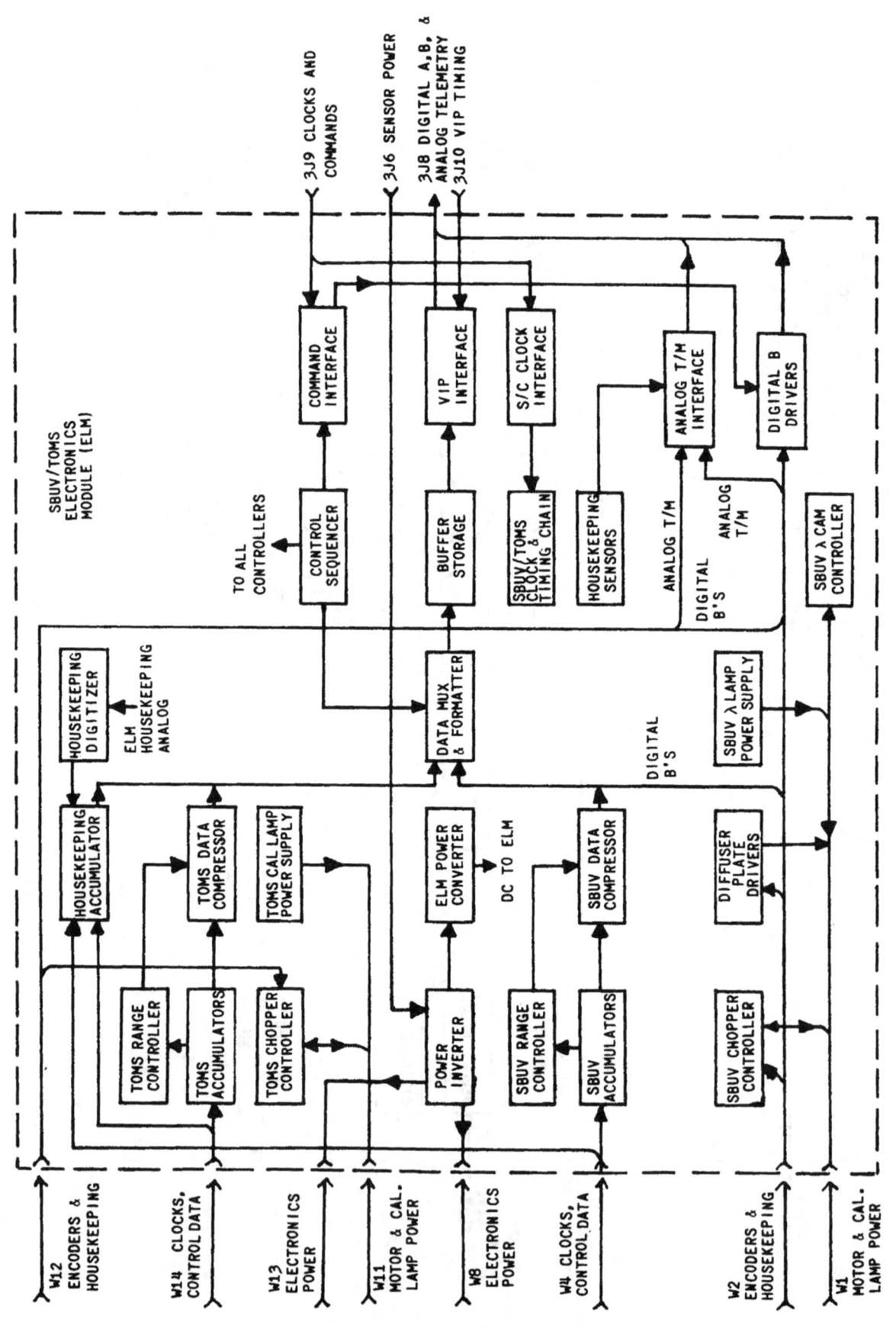

Figure 7-1a Functional Block Diagram of SBUV/TOMS Modules

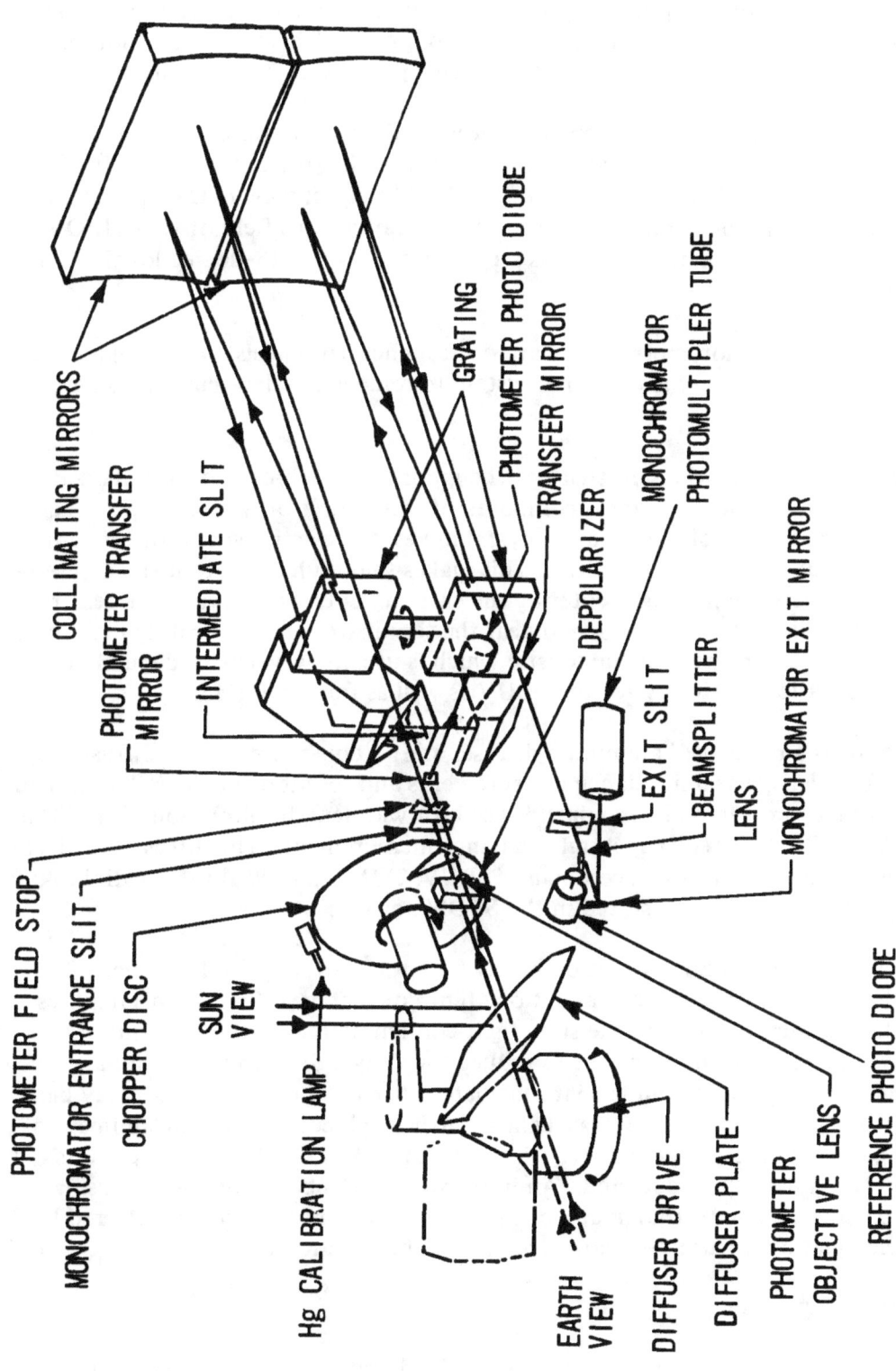

Figure 7-2 SBUV Optics Diagram

A focusing mirror system immediately following the exit slit images the gratings on a rectangular optical stop immediately in front of the PMT. This optical stop matches the grating image in both size and location and thus prevents light rays that scatter from walls and structural parts of the monochromator from reaching the photomultiplier. A portion (about ten percent) of the exit beam is directed to the reference photodiode by a small reflective beam divider. This reference photodiode is used at high light levels to monitor the gain of the PMT during prolonged operation in space.

The photometer of the SBUV, which is designed to measure the ground reflectivity, employs a photodiode, an interference filter, a field lens and transfer mirrors. It has a 3 nm bandwidth (full width at half maximum) centered at 343 nm and an 11.3° x 11.3° field of view (FOV) which matches and is co-linear with that of the double monochromator. The instantaneous field of view (IFOV) traces 200 km wide swaths on the ground. The swaths are separated by the 26-degree longitude interval between successive orbits.

The monochromator and the photometer each have optical chopper wheels. The chopper operates at a frequency of 25 Hz, enabling the detecting system to remove the dark current as described in Section 7.3.4.

A depolarizer is used to eliminate the sensitivity of the grating monochromator to polarization of the backscattered radiation. The depolarizer is composed of four pieces of wedge-shaped quartz so the optical activity characteristic of the quartz crystal mixes the electric vector of the radiation within the 1 nm bandwidth along the slit height. This makes the resultant output signal insensitive to the state of initial polarization of light entering the entrance slit. This device is necessary because the backscattered sunlight is highly polarized near the terminator and the grating monochromator may have a very strong polarization characteristic which is usually wavelength dependent. Figure 7-3 shows the polarization sensitivity for the SBUV as well as for the TOMS.

The SBUV also contains a roughened aluminum diffuser and a stepper motor. The diffuser plate is used to view the sun for solar spectral irradiance measurements and to intercept the radiation from the onboard low-pressure mercury-argon lamp which is used for wavelength calibration. The TOMS shares the diffuser with the SBUV for solar spectral irradiance measurement. The diffuser is driven by a stepper motor to one of three positions on command: SBUV, TOMS, STOW. In the nadir looking earth radiance measurement, the diffuser is stored in the STOW position.

The absorption cross section of ozone varies by nearly five orders of magnitude between 250 nm and 340 nm, and the solar flux at 250 nm is about one tenth of that at 340 nm. This requires a high stray light rejection capability to avoid the stray light contamination by the more intense solar flux at longer wavelengths. The prime means to reduce the stray light is to use a double monochromator. The second monochromator and the intermediate slit reduces the unwanted stray light emerging from the first monochromator. The stray light is further reduced by forming an image of the grating on a rectangular optical stop immediately in front of the detector face. This permits only the light reflected from the grating to be detected. An improvement has been obtained by using gratings ruled by photorestive technique (halographic gratings). These produce lower scattered light than a mechanically ruled grating. Figure 7-4 shows the stray light rejection capability of the SBUV.

7.3.2 TOMS Subsystem

The TOMS is a single Ebert-Fastie spectrometer with a fixed grating and an array of exit slits. Figure 7-5 is the optics diagram of the TOMS. The TOMS has a 3° by 3° instantaneous field of view and measures six discrete wavelengths ranging from 312.5 to 380 nm with a 1 nm bandwidth.

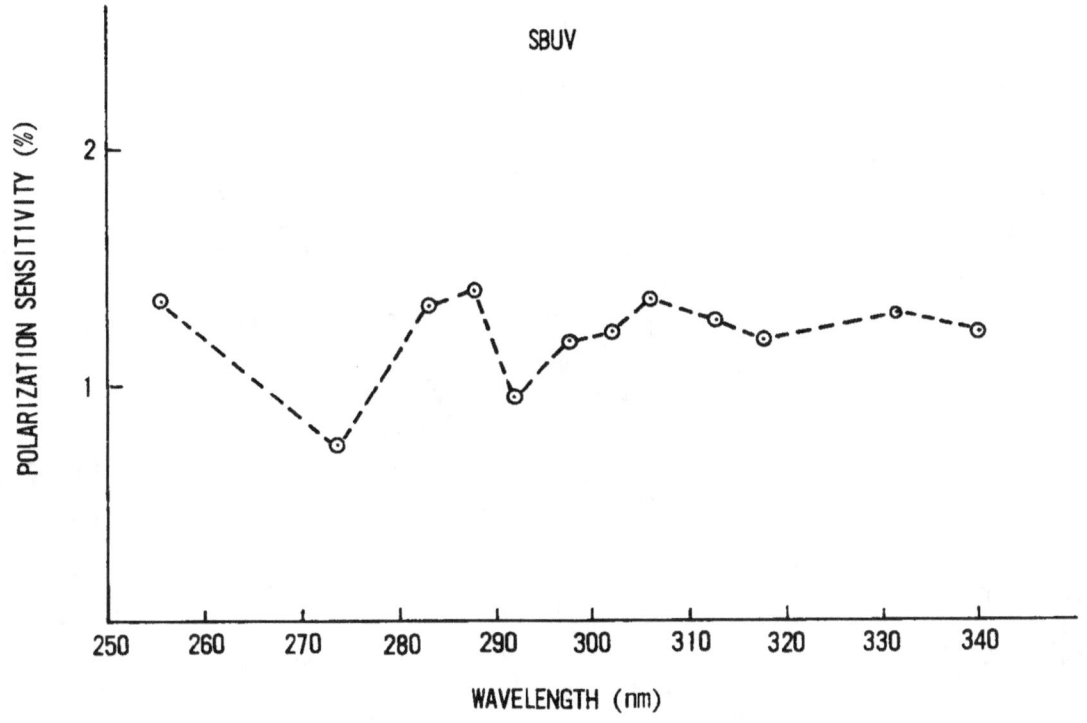

a. Polarization Sensitivities for SBUV

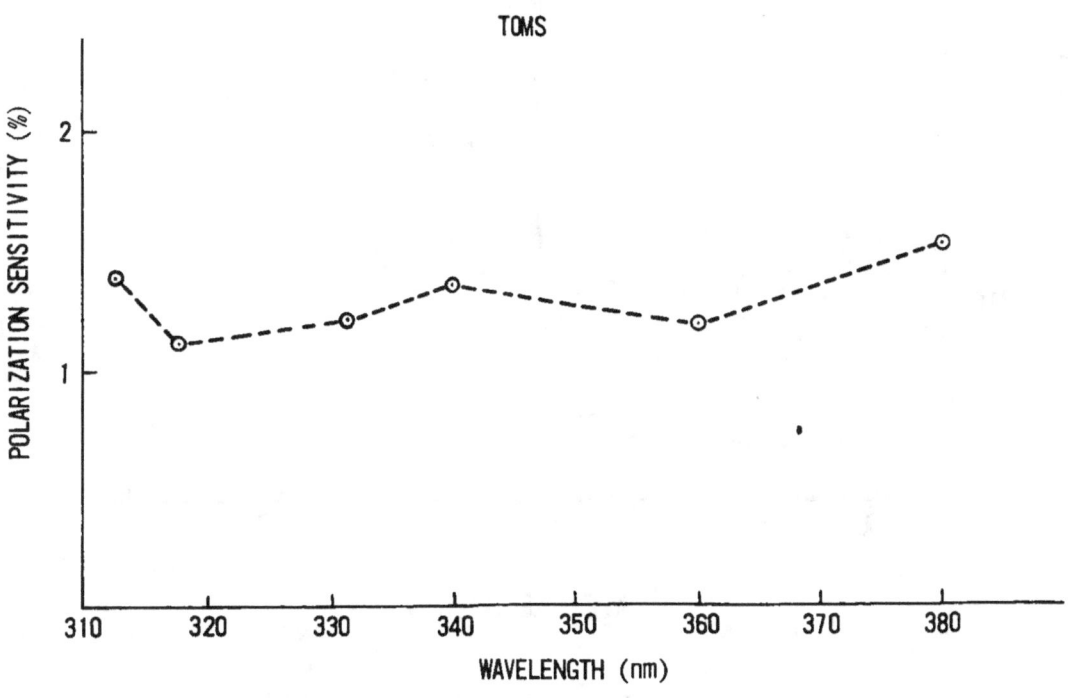

b. Polarization Sensitivities for TOMS

Figure 7.3 Polarization Sensitivities for SBUV/TOMS

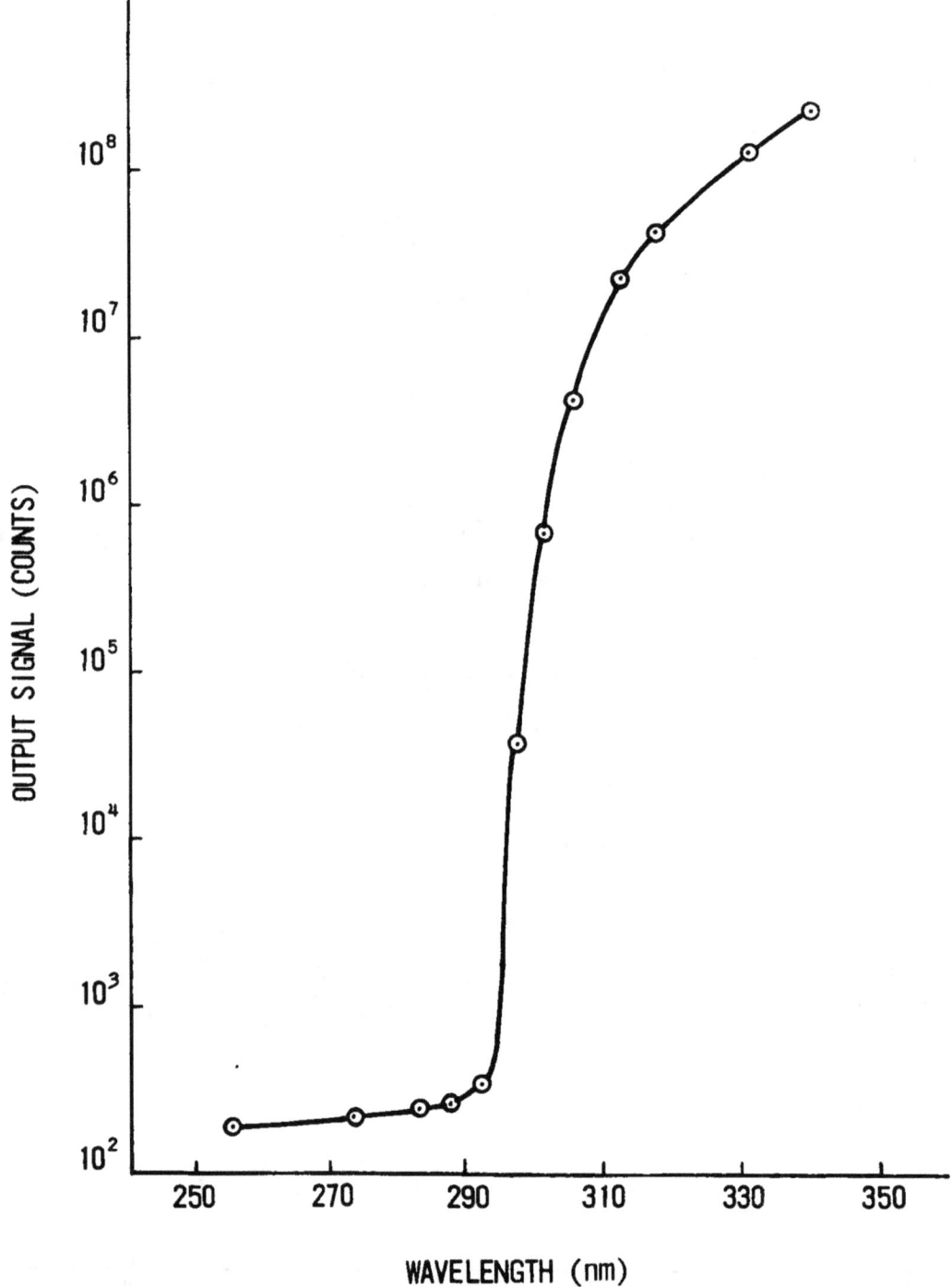

Figure 7-4 Stray Light Rejection Capability of SBUV

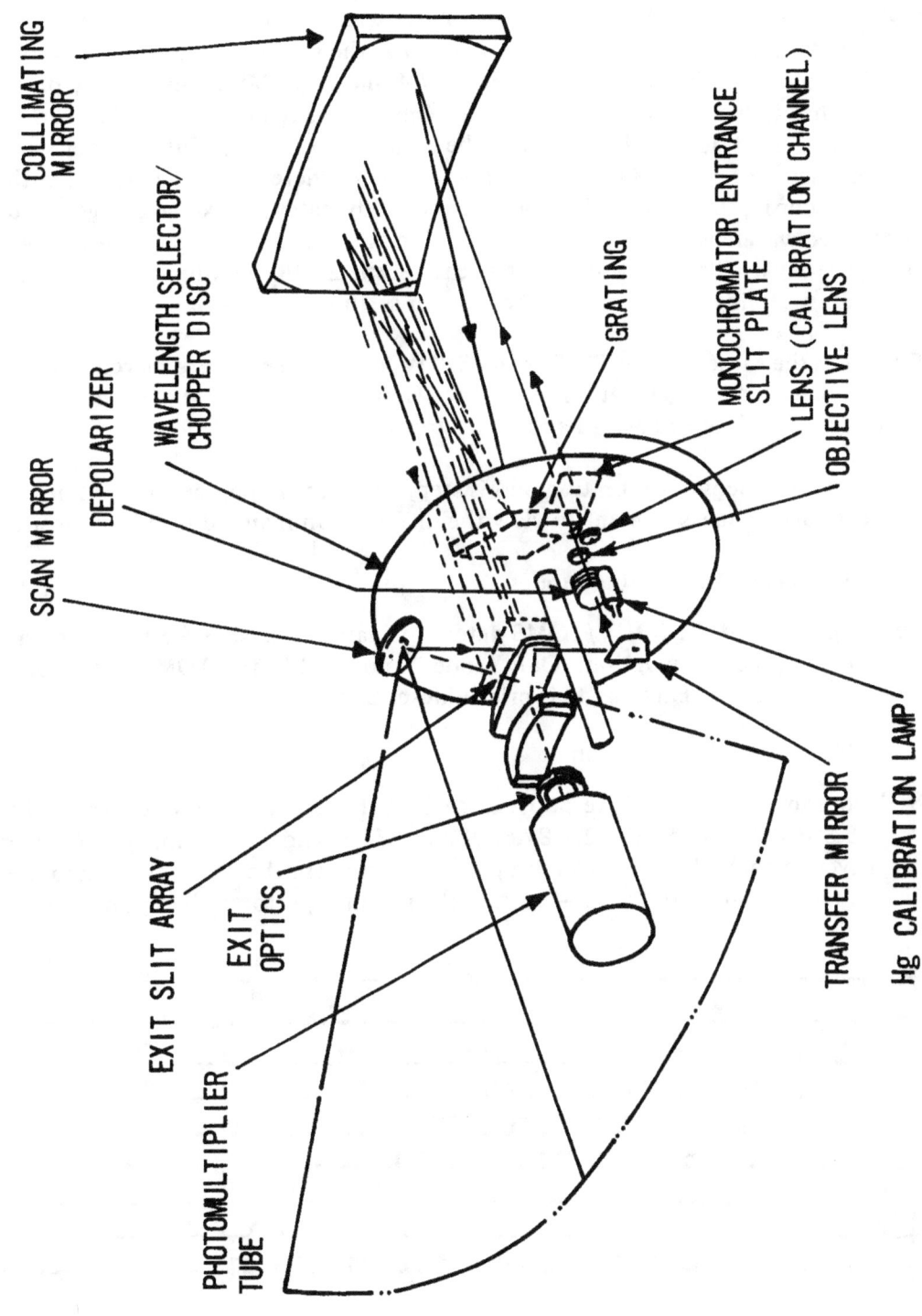

Figure 7-5 TOMS Optics Diagram

183

The PLL is synchronous to the Nimbus clock frequency. The disc also serves as chopper to allow synchronous modulation of the optical signal. Figure 7-6 shows a timing diagram of the TOMS chopper sequence at each scene. The TOMS light chopping frequency is 143 Hz. The wavelength selector disc also has a set of entrance slits for wavelength calibrations. A set of holes (exit slits for the 317.5 nm band) in the wavelength selector serve as a fixed exit during wavelength calibration.

The TOMS is a cross-course scanning instrument. A scanning mirror scans across the track ±51 degrees from the nadir in 3 degree steps. Figure 7-7 shows the instantaneous field of view (IFOV) of the SBUV/TOMS in the nadir direction and Figure 7-8 shows TOMS fields of view projected on a plane tangent to the earth at the subsatellite point. One complete cross scan takes eight seconds, including one second for retrace, while recording the data for 35 scenes. The encoder senses the position of the scan mirror on a six-bit code word which is telemetered to the ground with the scene data. At each scene during stepping, the chopper sequentially gates all six wavelengths four times such that the selected wavelengths are gated in succession and the order of gating is reversed in next sampling (see Figure 7-6). The total time spent at each scene is 200 ms which includes 168 ms for six wavelengths data sampling and 32 ms for the scanner settling.

The TOMS uses the same type of PMT as the SBUV, and has a separate mercury-argon lamp for wavelength calibration and a separate depolarizer. However, as noted above, the TOMS shares the diffuser with the SBUV for solar irradiance measurement.

The TOMS scanner is length calibration and is stepped up to six degrees over a horizontal position during solar irradiance measurement to view the diffuser commanded to TOMS position.

7.3.3 SBUV/TOMS Operating Modes

Five operating modes of the SBUV/TOMS determine data processing sequences, data formats, and the SBUV wavelength cam operation. Only Mode 2 is related to the TOMS operation. The TOMS has five scanner modes. Each mode is briefly described.

7.3.3.1 SBUV Mode 1 (Step Scan)

The SBUV measures photometric response at each of the following wavelengths 339.8, 331.2, 317.5, 312.5, 305.8, 301.9, 297.5, 292.2, 287.6, 283.0, 273.5, and 255.5 nm. The wavelength cam is sequentially driven to each of the wavelength positions listed (in the order given) and upon reaching each position, the cam stops and a one-second integration is performed. Upon completion of

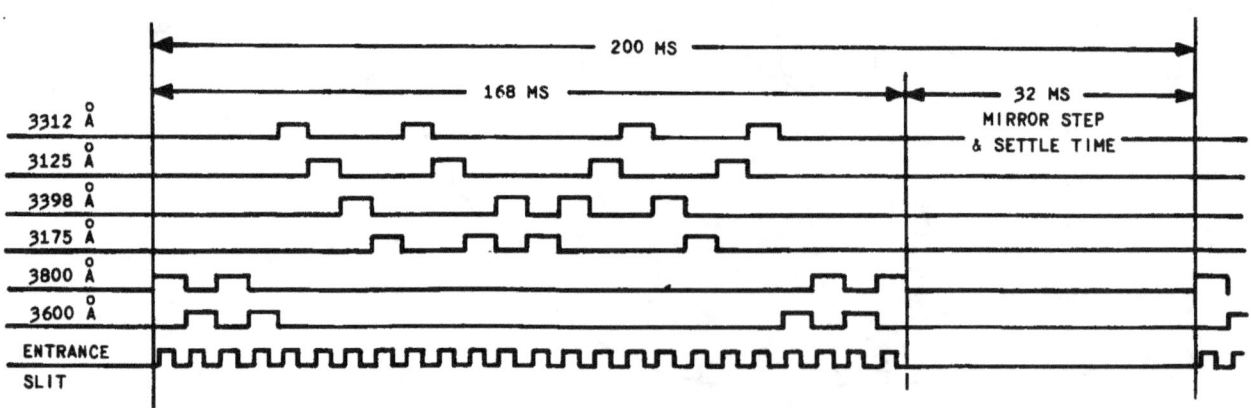

Figure 7-6 TOMS Chopper Timing Sequence

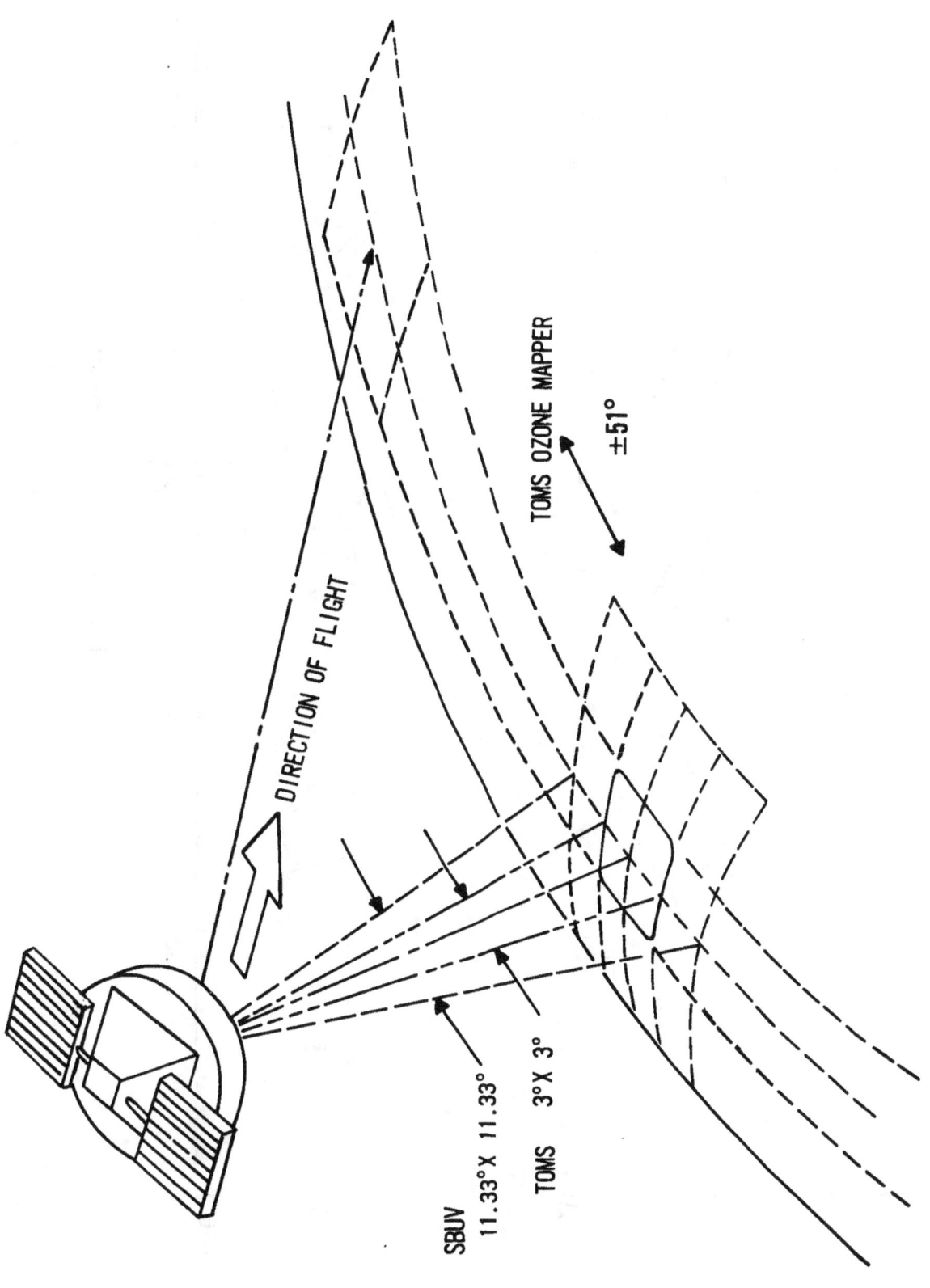

Figure 7-7 IFOV of SBUV/TOMS in the Nadir

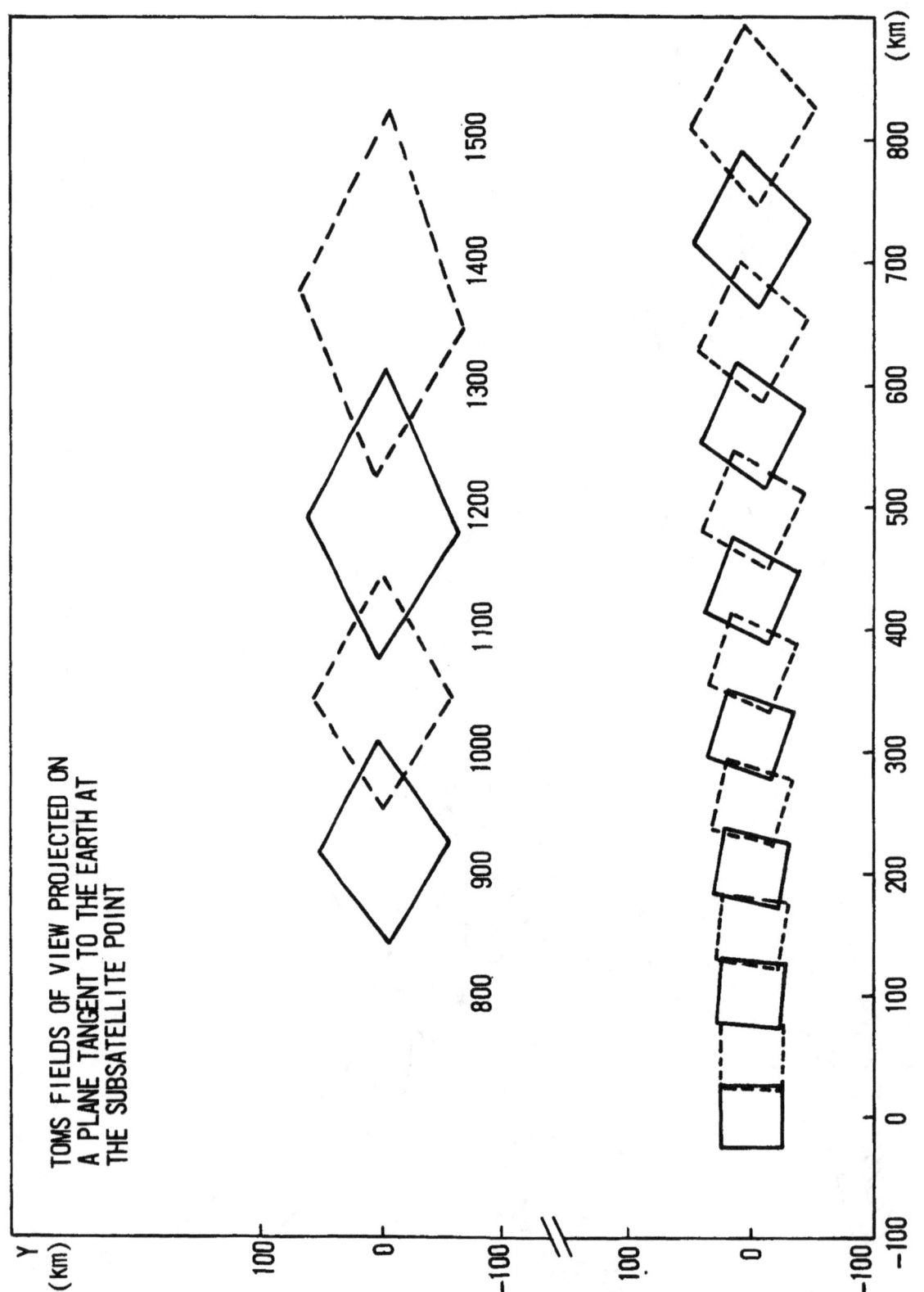

Figure 7-8 TOMS Projected Fields of View in a Cross-Scan

the sampling period, the cam advances to the next position. This cycle repeats until all 12 positions have been sampled. At the completion of the last sample (255.5 nm), the cam continues until the cage position is reached. This sequence requires 32 seconds (2 VIP major frames) and is repeated until a different mode is commanded.

7.3.3.2 SBUV Mode 2 (Wavelength Calibration)

Mode 2 is similiar to mode 1 except different wavelengths are sampled. The wavelength positions are 254.7, 254.2, 253.7, 253.2, and 252.7 nm in order. The SBUV remains in this mode until commanded otherwise. Each measurement sequence cycle requires 32 seconds (two VIP major frames). During mode 2, the TOMS samples response at four calibration wavelengths: 297.5, 297.0, 296.5, and 296.0 nm.

7.3.3.3 SBUV Mode 3 (Cage Cam)

The cage cam command causes the SBUV wavelength cam to move and stay in the cage position. During this mode, the SBUV monochromator, photometer, and reference channels sample data at one second intervals at 408.8 nm.

7.3.3.4 SBUV Mode 4 (Continuous Scan)

During mode 4, the SBUV scans from 160 nm to 400 nm sampling data at 80 millisecond intervals in 0.2 nm increments. One complete wavelength scan in mode 4 requires seven VIP major frames (112 seconds). During the first major frame the SBUV wavelength cam advances to the beginning of the linear scan region and the data sampling (160 nm to 400 nm) occurs during the next six major frames.

7.3.3.5 SBUV Mode 5 (Scan Off)

The command of mode 5 causes the SBUV wavelength cam motion to cease. During this mode, the SBUV monochromator, photometer, and reference channels sample data at one-second intervals at the wavelength determined by the current cam position.

7.3.3.6 TOMS Scanner Modes

The five scanner modes for TOMS are: scan off mode, single step mode, normal scan mode, stowed mode, and view diffuser mode.

In single step mode, the scanner stops and responds to actuation of the momentary relay connected to the TOMS single step command line. In normal scan mode, the scanner scans 35 positions, retraces to the first scene and starts another scan of 35 scenes.

In stowed mode, the scanner slews to the stowed position and stops. This mode is required for the TOMS wavelength calibration. In view diffuser mode, required in solar spectral irradiance measurement, the scanner slews to the view diffuser position and stops.

7.3.4 Electronic System

Each sensor module houses a detector power supply, the first stages of the signal processing, a calibration generator, and its own optical and mechanical systems. The calibration generator is a constant current source used for the in-flight electronic calibration. The electronics module accommodates all the supporting electronic circuits including the data processing and command logic.

The SBUV/TOMS has four data output channels: one for the monochromator, one each for the reference and the photometer for the SBUV, and one for the TOMS. The signal processing is very similar in each channel, though not identical. A distinct feature in the SBUV/TOMS detecting electronics compared to the original Nimbus 4 BUV experiment is the use of the optical chopper to eliminate the dark current contribution in the output by the up/down counting technique.

The signal detected in the SBUV/TOMS monochromator channel is processed as following. The chopper optical signal from the exit slit falls on the PMT generating a current in the detector. This chopper detector current is amplified by the multi-range electrometer. Three electrometer outputs in different gain ranges are fed to three voltage-to-frequency converters (VFC), respectively. The VFC outputs are routed through the interface cable to digital accumulators in the ELM. The accumulator is controlled and gated in synchronism with the light chopper so it counts up when the light is admitted to the PMT and counts down when the chopper blocks the light. Thus the accumulator works as an integrator and synchronous demodulator, and rejects any unchopped signals. The accumulated data are compressed (for mode 4), formated and sent to the VIP telemetry through a buffer storage. The integrated period is one second in step scan and 80 milliseconds in continuous scan. The SBUV monochromator channel has three parallel outputs, each of which has full scale 175,000 counts in step scan mode (14,000 counts in continuous scan) and covers different parts of the 10^6 dynamic range the SBUV measures with better than one percent resolution accuracy.

The signal processing for the reference and photometer channels is identical to that for the monochromator except that only one VFC and accordingly one output is employed in each channel. The data are sampled in a one-second period and the output has a full scale of 175,000 counts.

The TOMS also uses the optical chopping and up/down counting system in the signal processing. Even though the TOMS has four gain ranges, four VFC's and four accumulators, only one output is chosen by the TOMS range controller and telemetered through the VIP. The highest gain range output among the unsaturated ranges is selected for transmission.

The SBUV/TOMS telemetry system used three VIP channels: digital A, digital B, and VIP analog. The digital A channel transmits the sensory data, status bits and housekeeping data that require resolution and accuracy greater than can be provided by analog telemetry. Status bits indicate various conditions within the sensors. The digital B telemetry verifies proper operation of command relays or proper operation of circuits and mechanisms, which are initiated via the command relays. The VIP analog telemetry consists of eleven temperatures from thermistors and three ac power supply monitor voltages.

7.4 Calibration

7.4.1 Prelaunch Calibration

The radiometric calibration of the SBUV/TOMS is composed of three parts: radiance calibration, irradiance calibration, and system linearity. The SBUV/TOMS measures the ultraviolet solar irradiance on the top of the earth's atmosphere and the earth radiance due to the solar illumination. The irradiance measurement mode uses only one additional optical element (diffuser) in instrumentation compared to the radiance measurement mode, but quite different calibration techniques are required. A principal shared in both calibrations is the determination of the instrument response to a known source.

The spectral range of the SBUV/TOMS cannot be covered by a single standard in calibration. Standard spectral irradiance used in various wavelength ranges were the 1000 watt tungsten quartz halogen, the argon mini arc and the deuterium (D_2) arc which were obtained from the National Bureau of Standards (NBS). A standard spectral radiance source which can fill the field of view of the instrument is obtained by combining as irradiance source with a reflective diffuser whose diffuse reflectance is known. Several fresh coated and aged $BaSO_4$ diffuser plates and one aluminum diffuser plate have been used in the radiance calibration. The aluminum diffuser is used to monitor the stability of the $BaSO_4$ diffuser reflectance. The reflectance of all the diffuser plates have been measured periodically at NBS.

The irradiance calibration is achieved by three operations: ambient calibration, vacuum calibration and goniometric calibration. The ambient calibration includes the calibration runs in the ambient for the wavelength range longer than 200 nm. The tungsten quartz halogen lamp isused from 250 nm to 400 nm while the argon mini arc and deuterium lamp are used from 200 nm to 250 nm. A vacuum calibration was performed in an ion pumped vacuum chamber for the wavelength range from 160 nm to 200 nm with an argon mini arc. In the ambient and vacuum calibration, the experimental arrangement is such that the instrument axis coincides with the vertical direction and the calibration source illuminates the center of the deployed diffuser in the horizontal direction as the SBUV/TOMS views the sun in the terminator. The third operation, the goniometric calibration, is required since the incident angle of the sun on the diffuser continuously changes as the satellite orbits the earth. Therefore, the angular response of the SBUV/TOMS is determined in the goniometric calibration.

Radiance calibration requires a radiance source which fills the field of view of the instrument. This is achieved by illuminating a $BaSO_4$ diffuser in normal incidence with a standard irradiance source. For the normal incidence, the reflected beam off the $BaSO_4$ diffuser is nearly Lambertian. The radiance calibration is carried out by measuring the response of the instrument viewing the diffuser plate. To ensure the accuracy in radiance calibration, several $BaSO_4$ and one aluminum diffuser have been used and their reflectance has been measured periodically. The measurement shows a reproducibility within one percent.

The system linearity test is a necessary complementary part of the prelaunch calibration. The SBUV/TOMS has a wide dynamic range of operation and the radiometric calibration described above provides the instrument response only near one signal level. The system linearity test has been completed by using two different techniques. For the lower part of the dynamic range, the instrument is held at a fixed distance from a $BaSO_4$ diffuser plate and views the diffuser. The radiance of the $BaSO_4$ plate is controlled by varing the distance from a tungsten quartz halogen lamp to the diffuser while the response of the instrument is monitored. For the upper part of dynamic range, the output of the SBUV monochromator channel is compared to that of the reference diode channel which is assumed to have a linear response. The sun, guided by a heliostat, is used as a source to drive the SBUV at higher signal level. The solar flux is attenuated by placing uniform neutral density screens in the light path. The linearity of the lower part of the dynamic range was checked independently by using a set of standard neutral density filters with a tungsten quartz halogen lamp.

7.4.2 In-flight Calibration

7.4.2.1 Electronics Calibration

The gain stability of the signal processing is checked if electronics calibration (ECAL) is commanded in the scan off mode. A precise simulated chopped signal is injected into the input of each electrometer amplifier and data are accumulated as in the radiance measurements. Several signal levels and modulation combinations are automatically sequenced to fully test each electrometer range.

7.4.2.2 Wavelength Calibration

The purpose of in-flight wavelength calibration is to detect wavelengths shifts in the monochromators which could be caused by excessive temperature differentials or mechanical displacement of the wavelength determining components due to shock or vibration.

The SBUV uses the same wavelength calibration scheme used by the BUV on Nimbus 4. It consists of five precise cam steps centered about at 253.7 nm, which is the wavelength of the strongest mercury emission line. To perform wavelength calibration, the SBUV wavelength calibration lamp is warmed up for at least five minutes and the diffuser is moved to the SBUV position. Upon the wavelength calibration command, the SBUV cam scans through the five steps which are separated by 0.5 nm, measuring the radiation from the mercury lamp reflected off the diffuse plate. A triangle is drawn from the output since the spectrometer has a triangular slit fucntion of a 1 nm bandwidth. Any shift in wavelength is deduced from the position of the peak of the triangle. Table 7-1 lists the actual wavelengths (λ_o) for the 12 SBUV steps determined from the ground calibration. The true wavelengths (λ_t) in space will be determined from the in-flight wavelengths calibration using an equation:

$$\lambda_t = \lambda_o + (253.725 - \lambda p - 0.113)$$

where λp is the peak wavelength of the triangle of the output in vacuum. The calibration lamp spectrum obtained in the continuous scan mode can also provide a wavelength calibration check.

The TOMS in-flight wavelength calibration works on the same pricniple as that of the SBUV, even though the TOMS has a different wavelength scanning mechanism and uses a different mercury emission line (296.7 nm). The TOMS has four special entrance slits in the wavelength selector which function as the SBUV cam wavelength steps. Table 7-2 shows the actual wavelengths for the TOMS from the ground calibration. Any change in wavelength scale can be corrected by using a formula:

$$\lambda_t = \lambda_o + (296.814 - \lambda p + 0.120)$$

where λ_t, λ_o and λ_p have the same meaning as in the SBUV wavelength calibration.

7.5 Data Processing, Formats, and Availability

7.5.1 Data Processing

Figures 7-9a and 7-9b present an overview of the SBUV/TOMS data reduction sequence from initial processing in the MetOCC through user image generation by the IPD at GSFC. The MetOCC produces an SBUV/TOMS User Formatted Output (UFO) tape and an Image Location Tape (ILT). These tapes, together with terrain and THIR cloud information are used by SACC at GSFC to produce the tapes described in Section 7.5.2. SACC sends these tapes to the IPD. IPD copies some and sends these copies to NET users. IPD uses other tapes (as shown in Figure 7-9b) as input to generate the SBUV/TOMS microfilm and montage displays described and listed in Section 7.5.3. All original tapes and microfilm are then sent to NSSDC for archiving. NSSDC makes copies of these products for users.

Table 7-1

Actual Wavelengths of SBUV Wavelength Steps

Wavelength Step (nm)	Prelaunch Actual Wavelength (nm)
339.8	339.871
331.2	331.240
317.5	317.540
312.5	312.544
305.8	305.851
301.9	301.951
297.5	297.565
292.2	292.268
287.6	287.681
283.0	283.078
273.5	273.587
255.5	255.631

NOTE: Actual wavelength in space can be determined from inflight wavelength calibration using the following equation:

$$\lambda_t \text{ (nm)} = \lambda_o + (253.725 - \lambda_p - 0.113)$$

where λ_t is true wavelength, λ_o prelaunch wavelength, and λ_p the peak wavelength from inflight wavelength calibration. The term of 0.113 nm appears because the wavelength steps are not centered exactly at 253.725 nm.

Table 7-2

Actual Wavelengths of TOMS

Wavelengths Step (nm)	Prelaunch Actual Wavelength (nm)
380.0	379.956
360.0	359.904
339.8	339.803
331.2	331.195
317.5	317.454
312.5	312.456

NOTE: Actual wavelength in space can be determined from inflight wavelength calibration using the following equation:

$$\lambda_t \text{ (nm)} = \lambda_o + (296.814 - \lambda_p + 0.120)$$

where λ_t is true wavelength, λ_o prelaunch wavelength, and λ_p the peak wavelength from inflight wavelength calibration. The term of 0.120 nm appears because the wavelength steps are not centered exactly at 296.814 nm.

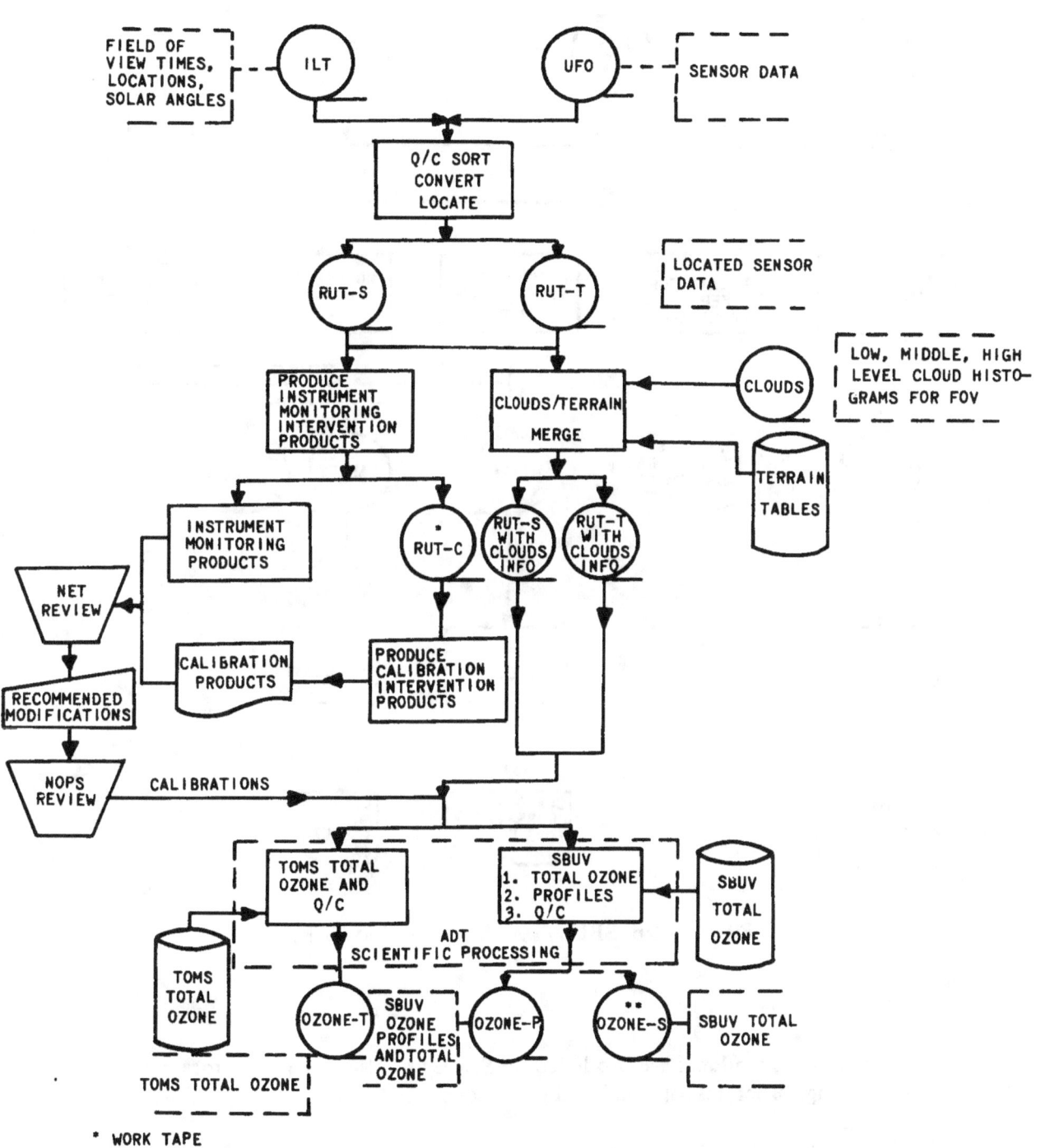

Figure 7-9a SBUV/TOMS Process Flow Chart

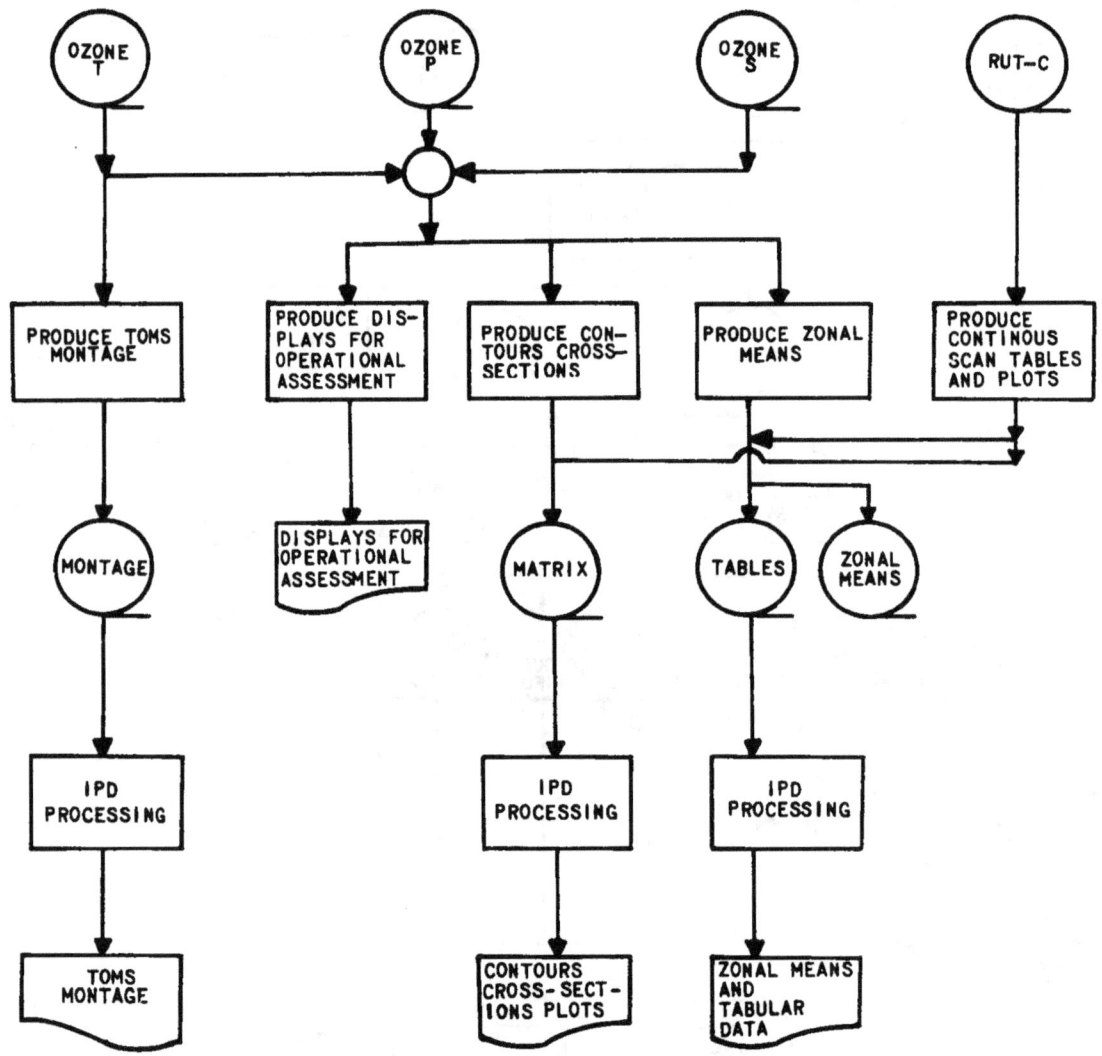

Figure 7-9b SBUV/TOMS Process Flow Chart

7.5.2 Tape Products

The following tape products are produced by SACC and used by IPD before being sent on to NSSDC for archiving. Brief descriptions of these tapes are as follows:

- RUT-S (Raw Units Tape for SBUV)
  Contains the most elementary and complete form of the SBUV data. Each tape record contains one of four formats depending on the instrument operating mode. Each record contains essentially the same solar satellite and earth reference data, plus housekeeping data. The number and type of monochrometer and photometer values included in each record type is determined by the instrument mode.

- RUT-T (Raw Units Tape for TOMS)
  Contains the most elementary and complete form of the TOMS data. It contains solar, satellite, and earth reference data, plus housekeeping data. It also contains the TOMS radiance values at specified wavelengths for each IFOV along each orbit.

- OZONE-S (SBUV Ozone Tape)
  Contains SBUV high-level ozone profiles and total atmospheric profiles.

- OZONE-T (TOMS Ozone Tape)
  Contains TOMS total ozone profiles.

- ZMT (Zonal Means Tape)
  Contains in computer-compatible format all SBUV and TOMS of wavelengths, solar irradiances, high-level ozone mass mixing ratios, and total ozone values.

- TABLES (all SBUV/TOMS tables for microfilm display)
  Contains SBUV and TOMS data formatted from production of all tables on microfilm (See Section 7.5.3 for film examples of these tables.)

- MATRIX (Map Data Matrix Tape)
  Contains daily, monthly and three month world grids of both SBUV and TOMS data plus northern and southern hemisphere polar stereographic map matrices containing the contour values of the parameters to be displayed on microfilm. These tapes also contain SBUV cross-section records of contour values, and SBUV plots to be on microfilm displays. Each record contains all necessary reference and title information needed for annotation of each of these displays. (See Section 7.5.3 for film examples of these.)

- MONTAGE (Montage Tape for TOMS)
  Contains all TOMS data and all title and reference information needed for annotation and display of each TOMS one-day montage. (See Section 7.5.3 for a film example of this.)

The form and content of each of these tapes is specified in a tape specification document for each tape type. The appropriate document will accompany a tape shipment to a user. See Section 1.5 of this document for details.

7.5.3 Display Products

The SBUV/TOMS data are displayed on 18 different map sets, one cross section, 18 tables, three plots, and one montage. TOMS and SBUV are contoured and displayed as total ozone maps. (See Figure 7-10 for an example.) SBUV data is also contoured and mapped at 14 high-altitude pressure levels. (See Figure 7-11). The orbital high-level SBUV values are also presented in a cross-section format (Figure 7-12). There are SBUV tables of solar irradiances (Figures 7-13 and 7-14), total ozone (Figure 7-15), and high-level ozone mass mixing ratio (Figure 7-16). There is one TOMS table showing total ozone (Figure 7-17). The three plot displays (Figures 7-18, 7-19, and 7-20) show SBUV values of solar irradiance, terrestrial radiance, and terrestrial albedo, respectively. The TOMS data are also presented in one-day montages showing each orbital swath of total ozone as shades of gray. (Figure 7-21).

Table 7-3 lists the titles of all SBUV/TOMS microfilm and montage displays and the corresponding film specification number for each. Most parameters are produced at more than one time scale

Table 7-3
SBUV/TOMS Film Products

Film Spec Number	Film Product Title
	MAPS
633101	TOTAL OZONE (TOMS)
633701	
633801	
633102	TOTAL OZONE (SBUV)
633702	
633802	
633103	OZONE MASS MIXING RATIO AT XX.X* MB (SBUV)
633703	
633803	
	CROSS SECTIONS
634030	HIGH LEVEL X-SECTIONS OF OZONE MASS MIXING RATIO
	MONTAGE
632140	TOTAL OZONE MAPPING SPECTROMETER DAILY MONTAGE OF DAY-TIME DATA FOR (date)
	TABLES
636060	SOLAR IRRADIANCE IN XX ANGSTROM STEPS BETWEEN 1600A AND 4000A
636161	TOTAL OZONE (TOMS Zonal means)
636561	
636761	
636861	
636162	TOTAL OZONE (SBUV Zonal means)
636562	
636762	
636862	
636163	HIGH LEVEL OZONE MASS MIXING RATIO (SBUV Zonal means)

Table 7-3 (Continued)

Film Spec Number	Film Product Title
636563	
636763	
636863	
636764	SOLAR IRRADIANCE IN 10 ANGSTROM STEPS BETWEEN 1600A AND 4000A
	PLOTS
637780	PLOTS OF THE RATIO OF ORBITAL MEASUREMENTS OF SOLAR IRRADIANCE TO THE 27 DAY MEAN
637781	PLOTS OF TERRESTRIAL RADIANCE VS WAVELENGTH
637782	PLOTS OF TERRESTRIAL ALBEDO VS WAVELENGTH

*The maps for each time period showing SBUV high-level ozone mass mixing ratios are contoured and displayed at 14 pressure levels. These are: 40, 30, 20, 15, 10, 7, 5, 4, 3, 2, 1.5, 0.7, and 0.4 millibars.

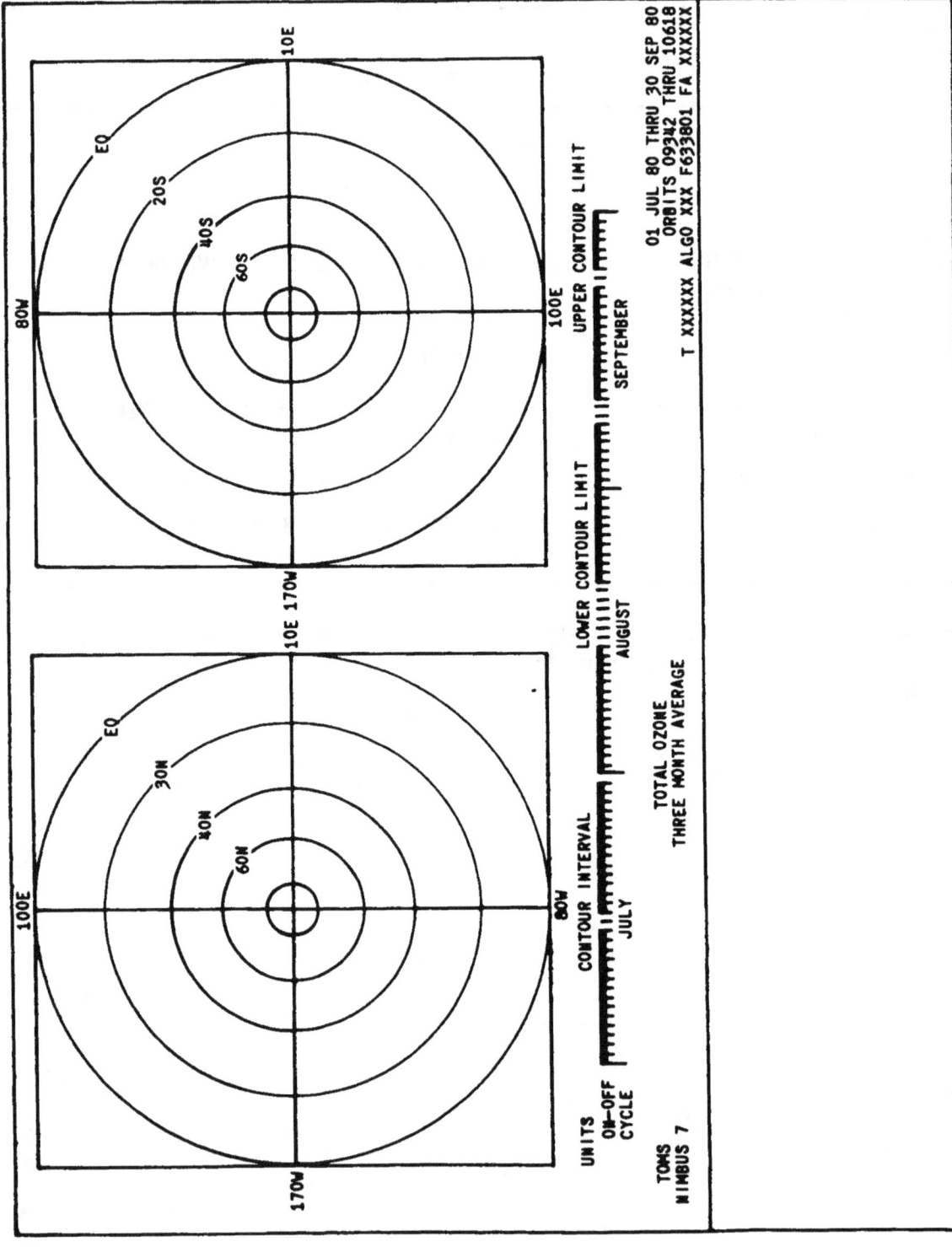

Figure 7-10 SBUV/TOMS Microfilm Map Format for One Month and Three Months Displays

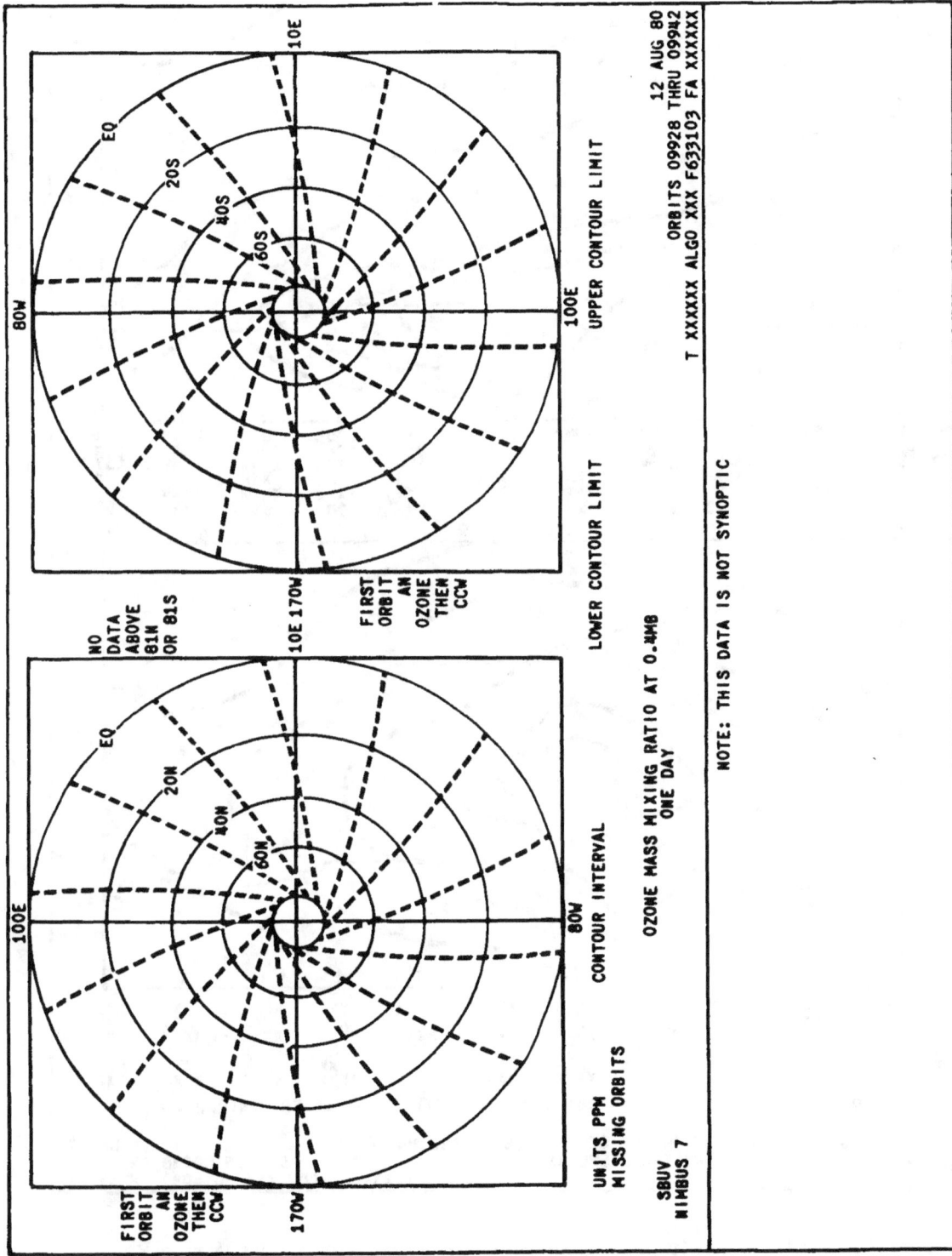

Figure 7-11 SBUV Microfilm Map Format for One Day Displays

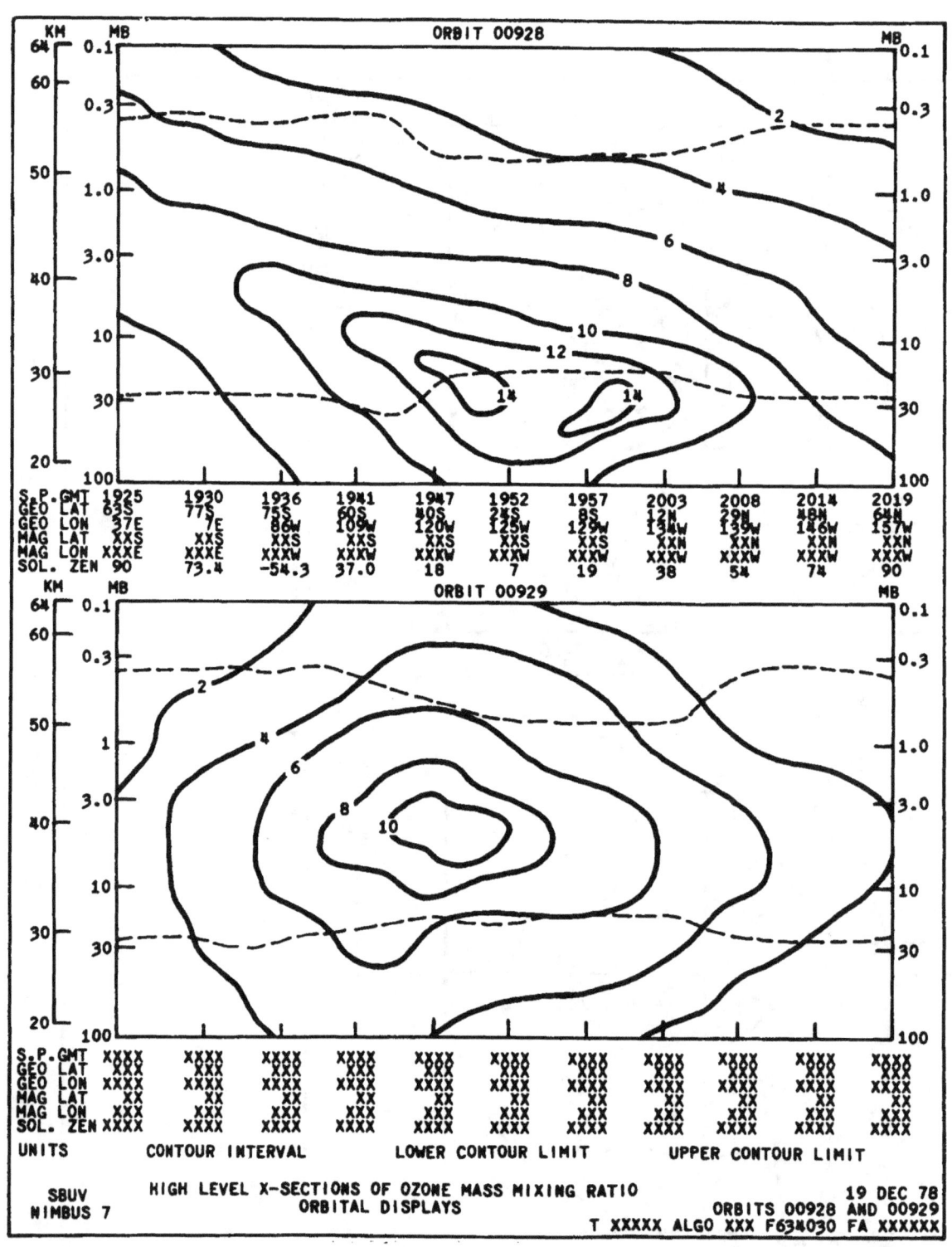

Figure 7-12 SBUV Microfilm Cross Section Format of Ozone Mass Mixing Ratio

```
          SOLAR IRRADIANCE IN 02 ANGSTROM STEPS BETWEEN 1600A AND 4000A
                               ONE ORBIT MEASUREMENTS

          SBUV                                                        12AUG80
          NIMBUS 7                                                ORBIT 09931

                     GMT    LATITUDE LONGITUDE  SOLAR AZ  SOLAR ZEN
      SCAN START 142629     79.855    131.6W      124        78        WAVELENGTH (λ) IS IN ANGSTROMS
      SCAN STOP  142805     76.355    171.5W      119        76        IRRADIANCE (F) IS IN W/M*2 X10-12

                        WAVELENGTH SPAN ON THIS PAGE FROM 1600A TO 2400A

      F         F         F         F         F          F         F          F          F          F
1601 XXX 1681 XXX 1781 XXX 1841 XXX 1921 XXX  2001 XXX 2081 XXX 2161  XXX  2141  XXX  2321  XXX
1603 XXX 1683 XXX 17 3 XXX 1843 XXX 1923 XXX  2003 XXX 2083 XXX 2163  XXX  2143  XXX  2323  XXX
  |   |    |   |    |   |    |   |    |   |     |   |    |   |    |    |     |    |     |    |
  v   v    v   v    v   v    v   v    v   v     v   v    v   v    v    v     v    v     v    v
1677 XXX 1777 XXX 1837 XXX 1917 XXX 1977 XXX  2077 XXX 2157 XXX 2237  XXX  2317  XXX  2397  XXX
1679 XXX 1779 XXX 1839 XXX 1919 XXX 1999 XXX  2079 XXX 2159 XXX 2239  XXX  2319  XXX  2399  XXX

                                       TXXXXX ALGO XXX F636060 FBXXXXXX
```

Figure 7-13 SBUV Microfilm Table Format of Solar Irradiance

SOLAR IRRADIANCE IN 10 ANGSTROM STEPS BETWEEN 1600A AND 4000A
(27 DAY MEANS AT ONE A.U.)

SBUV   BARTELS NUMBER   01AUG80 THRU 27AUG80
NIMBUS 7   2004   ORBITS 09776 THRU 10148

```
        SOLAR AZIMUTH ANGLES AT SATELLITE                WAVELENGTH (λ) IS IN ANGSTROMS
DAY     01AUG  07AUG  14AUG  21AUG  27AUG                IRRADIANCE (F) IS IN W/M*2 X10-12
ANGLE   XXX    XXX    XXX    XXX    XXX
                              WAVELENGTH SPAN ON THIS PAGE FROM 1600A TO 2800A

CNTR  MEAN  STD  MAX  MIN       CNTR  MEAN  STD  MAX  MIN       CNTR  MEAN  STD  MAX  MIN
       F    DEV  VAL  VAL              F    DEV  VAL  VAL              F    DEV  VAL  VAL
1605  XXX   XX   XXX  XXX       2005  XXX   XX   XXX  XXX       2405  XXX   XX   XXX  XXX
1615  XXX   XX   XXX  XXX       2015  XXX   XX   XXX  XXX       2415  XXX   XX   XXX  XXX
 |     |    |    |    |          |     |    |    |    |          |     |    |    |    |
 |     |    |    |    |          |     |    |    |    |          |     |    |    |    |
 v     v    v    v    v          v     v    v    v    v          v     v    v    v    v
1985  XXX   XX   XXX  XXX       2385  XXX   XX   XXX  XXX       2785  XXX   XX   XXX  XXX
1995  XXX   XX   XXX  XXX       2395  XXX   XX   XXX  XXX       2795  XXX   XX   XXX  XXX
                        TXXXXX ALGO XXX F636764 FBXXXXXX
```

Figure 7-14   SBUV Microfilm Table Format of Solar Irradiance

```
                              TOTAL OZONE
                          ONE MONTH ZONAL MEANS
SBUV
                                                    01AUG80 THRU 31AUG80
NIMBUS 7                                            ORBITS 09776 THRU 10204

        GEODETIC                          GEOMAGNETIC
        LATITUDE   TOTAL OZONE            LATITUDE   TOTAL OZONE
          ZONE    MEAN      STD    POP      ZONE    MEAN      STD    POP
        (CENTER) (M-ATM-CM) DEV   FLAG    (CENTER) (M-ATM-CM) DEV   FLAG
          80N      XXX      XX.X    *       80N      XXX      XX.X    *
          70N      XXX      XX.X    *       70N      XXX      XX.X    *
          60N       |        |      |       60N       |        |      |
          50N       |        |      |       50N       |        |      |
          40N       |        |      |       40N       |        |      |
          30N       |        |      |       30N       |        |      |
          20N       |        |      |       20N       |        |      |
          10N       |        |      |       10N       |        |      |
           0        |        |      |        0        |        |      |
          10S       |        |      |       10S       |        |      |
          20S       |        |      |       20S       |        |      |
          30S       |        |      |       30S       |        |      |
          40S       |        |      |       40S       |        |      |
          50S       |        |      |       50S       |        |      |
          60S       |        |      |       60S       |        |      |
          70S       V        V      V       70S       V        V      V
          80S      XXX      XX.X    *       80S      XXX      XX.X    *

                                     TXXXXX ALGOXXX F636767 FBXXXXXX
```

Figure 7-15 SBUV Microfilm Table Format of Total Ozone

as Table 7-1 shows. Interpreting the fourth digit from the left in each specification number gives the frequency of production of the parameter listed in the title:

    XXX0XX = produced orbitally, or more often,
    XXX1XX = produced every day,
    XXX5XX = produced every seven days,
    XXX7XX = produced every month (or 27 days equalling one Bartel period), and
    XXX8XX = produced every three months.

All map displays contain one north and one south polar stereographic projection (pole to equator for each). Each map contains contoured data as specified in the display title. Immediately beneath the maps is contouring information giving the contour units, interval between contour lines, and the maximum and minimum value contoured. All map displays contain an indicator of the quantity of data within a display. On the one day display there is a "missing orbits per day" code specifying how many orbits of data are missing from that day's input to the map (See Figure 7-10). On the one month and three month maps (Figure 7-11) there is an "on-off cycle" scale specifying the days during the display period when the instrument was on and off. If a day on the scale is "filled in" the instrument was on, the data was collected, interpreted, and used in the contouring. If a day is not filled in, the instrument was either off or the data was unuseable for some reason.

The polar stereographic maps on the one day displays (Figure 7-10 is an example) contain "dot tracks" with each dot representing a target point data value location used to construct the

HIGH LEVEL OZONE MASS MIXING RATIO (PPM)
ONE DAY ZONAL MEANS

SBUV 12AUG80
NIMBUS 7 ORBITS 09928 THRU 09942

GEOMAGNETIC LATITUDE OZONE (CENTER)	40 MB MEAN STD DEV POP	39 MB MEAN STD DEV POP	20 MB MEAN STD DEV POP	15 MB MEAN STD DEV POP	10 MB MEAN STD DEV POP	7 MB MEAN STD DEV POP	5 MB MEAN STD DEV POP
80N	XXX XX.X XXX	XXX XX.X XXX	XXX XX.X XXX	XXX XX.X XXX	XXX XX.X XXX	XXX XX.X XXX	XXX XX.X XXX
70N							
60N							
50N							
40N							
30N							
20N							
10N							
0							
10S							
20S							
30S							
40S							
50S							
60S							
70S							
80S	XXX XX.X XXX	XXX XX.X XXX	XXX XX.X XXX	XXX XX.X XXX	XXX XX.X XXX	XXX XX.X XXX	XXX XX.X XXX

GEOMAGNETIC LATITUDE OZONE (CENTER)	4 MB MEAN STD DEV POP	3 MB MEAN STD DEV POP	2 MB MEAN STD DEV POP	1.5 MB MEAN STD DEV POP	1 MB MEAN STD DEV POP	0.4 MB MEAN STD DEV POP
80N	XXX XX.X XXX	XXX XX.X XXX	XXX XX.X XXX	XXX XX.X XXX	XXX XX.X XXX	XXX XX.X XXX
70N						
60N						
50N						
40N						
30N						
20N						
10N						
0						
10S						
20S						
30S						
40S						
50S						
60S						
70S						
80S	XXX XX.X XXX	XXX XX.X XXX	XXX XX.X XXX	XXX XX.X XXX	XXX XX.X XXX	XXX XX.X XXX

TXXXXX ALGO XXX F636163 FBXXXXXX

Figure 7-16 SBUV Microfilm Table Format of High Level Ozone Mixing Ratio

TOTAL OZONE
ONE MONTH ZONAL MEANS

TOMS
NIMBUS 7

01AUG80 THRU 31AUG80
ORBITS 09776 THRU 10204

GEODETIC LATITUDE ZONE (CENTER)	TOTAL OZONE MEAN (M-ATM-CM)	STD DEV	POP FLAG	GEOMAGNETIC LATITUDE ZONE (CENTER)	TOTAL OZONE MEAN (M-ATM-CM)	STD DEV	POP FLAG
POLE	XXX	XX.X	*	POLE	XXX	XX.X	*
80N	XXX	XX.X	*	80N	XXX	XX.X	*
75N				75N			
70N				70N			
65N				65N			
60N				60N			
55N				55N			
50N				50N			
45N				45N			
40N				40N			
35N				35N			
30N				30N			
25N				25N			
20N				20N			
15N				15N			
10N				10N			
5N				5N			
0				0			
5S				5S			
10S				10S			
15S				15S			
20S				20S			
25S				25S			
30S				30S			
35S				35S			
40S				40S			
45S				45S			
50S				50S			
55S				55S			
60S				60S			
65S				-65S			
70S				70S			
75S				75S			
80S				80S			
POLE	XXX	XX.X	*	POLE	XXX	XX.X	*

TXXXXX ALGO XXX F636761 FBXXXXXX

Figure 7-17 TOMS Microfilm Table Format of Total Ozone

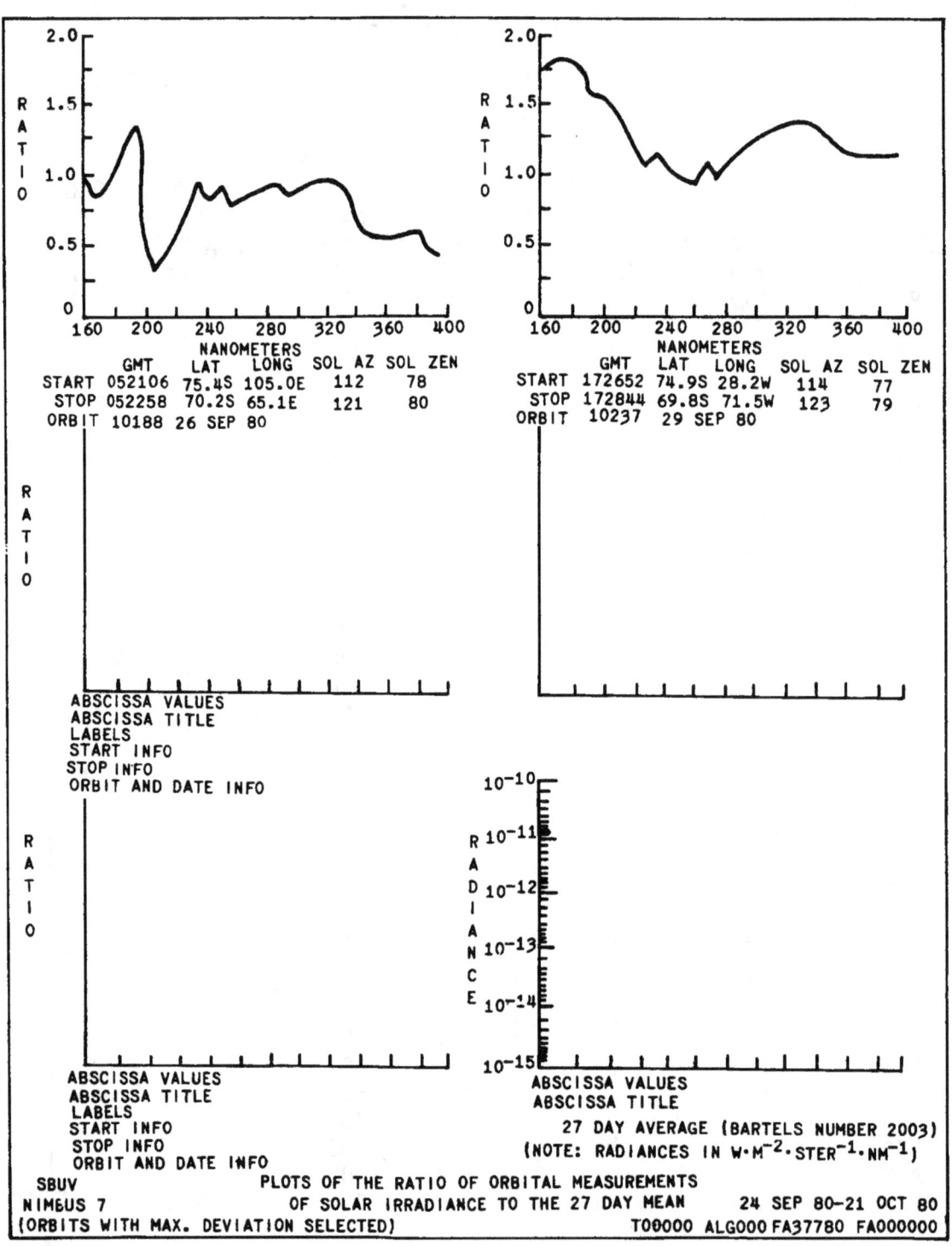

Figure 7-18 SBUV Microfilm Format of Solar Irradiance Plots

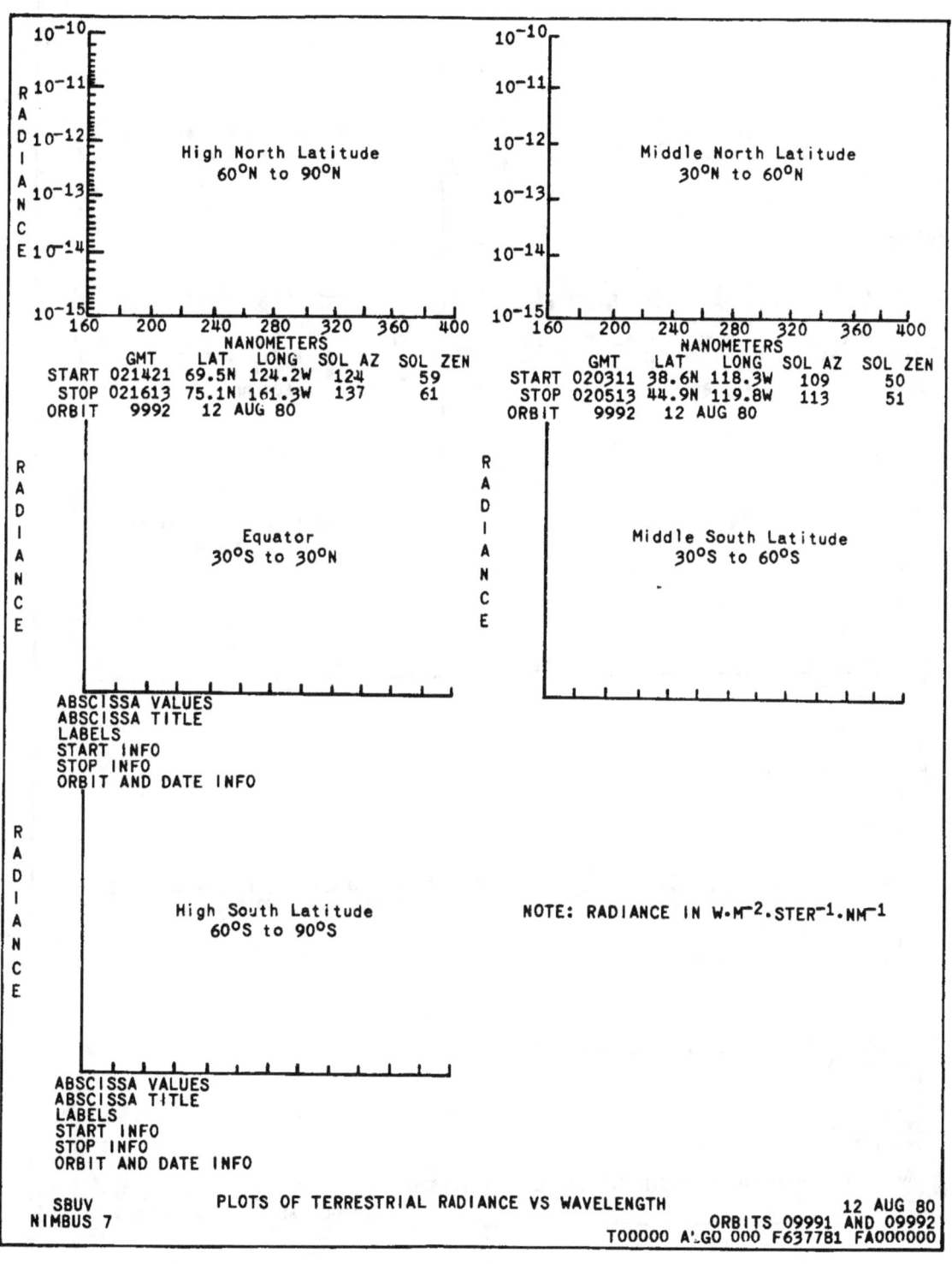

Figure 7-19 SBUV Microfilm Format of Terrestrial Radiance Plots

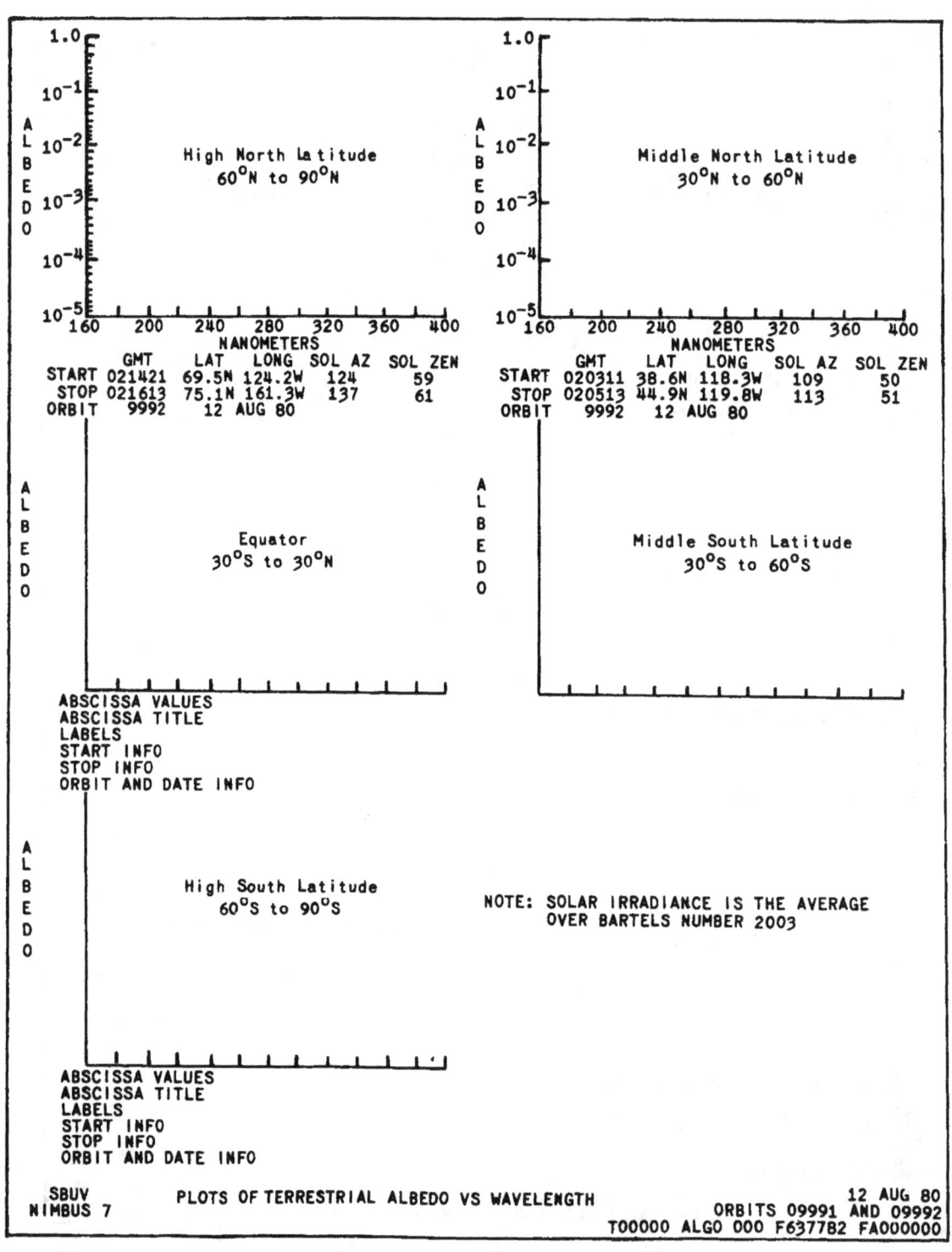

Figure 7-20 SBUV Microfilm Format of Terrestrial Albedo Versus Wavelength Plots

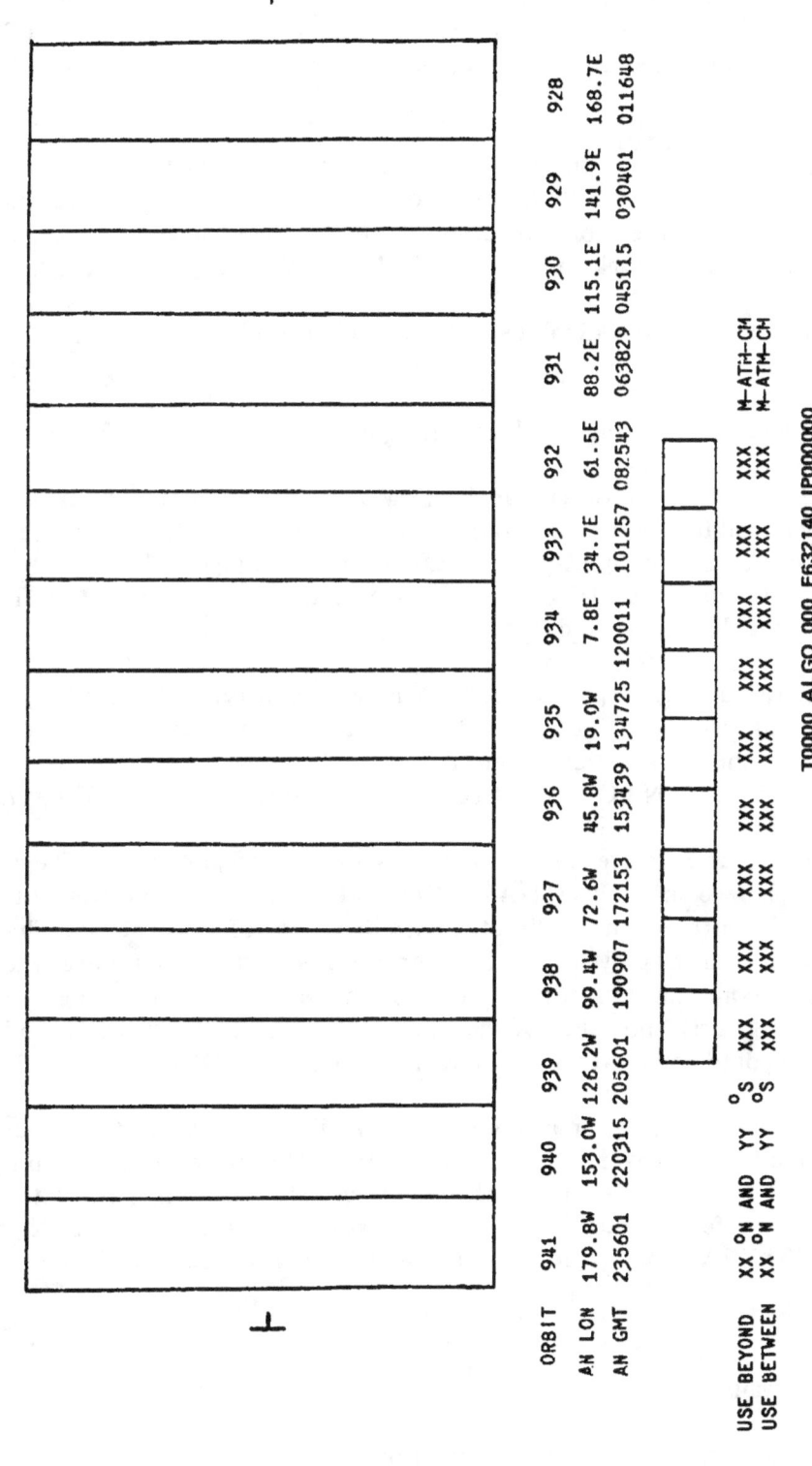

Figure 7-21 TOMS Montage Format

contours on the maps. The dots also represent the subpoint tracks and, together with the orbit numbers given as reference in the lower right corner, they can be used to assign orbit numbers to each track. The labeling adjacent to each map "first orbit A.N. Zone . . ." plus the horizontal lines over to each maps equator circle define the longitudinal zone of the ascending node (A.N.) of the first orbit given as reference in the lower right. Orbit numbers then increase clockwise (CW) on the left map (northern hemisphere) and the same orbit numbering increases counterclockwise (CCW) on the right map.

Figure 7-12 is an example of the SBUV orbital cross section displays. Two orbits are on each display. Data are referenced vertically by pressure and standard atmosphere altitudes, and horizontally at 11 locations by satellite subpoint values of GMT, geocentric and geomagnetic latitudes and longitudes (GEO LAT, GEO LON, and MAG LAT, MAG LON), and solar zenith angle (SOL ZEN).

Examples of the SBUV/TOMS tables on microfilm are shown in Figures 7-13 through Figure 7-17.

The SBUV plots on microfilm are shown in Figures 7-18, 7-19, and 7-20.

Figure 7-21 illustrates the format for the TOMS montage. The individual swaths of TOMS data are electronically stored until a day of data is assembled. Then the data is exposed on 241 mm by 241 mm (9.5 inch) black and white film as a TOMS world montage. The format and reference fiducial are identical to the daytime THIR montages (see Section 9.4.1) so that the THIR cloud data can be compared with the TOMS ozone data.

Each display contains appropriate title information identifying the satellite, the experiment, and the date the data was recorded. The 13 or 14 orbits of data are in the center of each display with a reference fiducial on the left and right. Beneath each orbital swath is its data orbit number plus an ascending node longitude (AN LON) and ascending node Greenwich Mean Time (AN GMT).

Beneath the data orbit reference information is a nine-step gray scale. The gray scale is calibrated in milli-atmosphere-centimeters (M-ATM-CM) and allows a user to estimate this quantity within each data swath. If all of a swath is calibrated with one range of M-ATM-CM values, only one set of numbers appears beneath the gray scale. However, if it is shown that data near the equator should be enchanced, then a second range of values is shown. The second (bottom) range is for all data within certain latitude limits north and south of the equator (USE BETWEEN XX °N and XX °S). The first (top) range is for all data beyond these latitude limits (USE BEYOND XX °N and XX °S).

Title and reference information at the bottom on all displays is mostly self-explanatory. The right half of the last line, however, requires explanation. This information is mainly used for cataloging and information control. These items are: the physical tape number the data is stored on (TXX XXX), the algorithm reference number used in processing the data (ALGO XXX), the film specification number (F63XXXX), the Project Data Format Code (FA, FB, or FC), and the film frame number (XXXXXX).

7.5.4 Data Availability

The SBUV/TOMS experimental data consisting of the magnetic tapes described in Section 7.5.2 and the 16 mm microfilm and 241 mm montage displays listed and illustrated in Section 7.5.3 are archived at the NSSDC. It is anticipated the first data sets will not arrive at NSSDC until at least six months after launch. Users requesting SBUV/TOMS data should read Section 1.5 of this document for general tape and film ordering information.

7.6 References

1. Singer, S. F., and R. C. Wentworth: "A Method for the Determination of the Vertical Ozone Distribution from a Satellite", J. Geophys. Res., Vol. 62, No. 2 (June 1957), pp. 299-308.

2. Twomey, Sean: "On the Deduction of the Vertical Distribution of Ozone by Ultraviolet Spectra Measurements from a Satellite.", J. Geophys. Res., Vol. 66, No. 7 (July 1961), pp 2153-62.

3. Twomey, S. and H. B. Howell: "A Discussion of Indirect Sounding Methods with Special Reference to the Deduction of Vertical Ozone Distribution from Light Scattering Measurements", Mon. Weather Rev., Vol. 91, No. 10-12 (October-December), pp. 659-64.

4. Kaplan, Lewis D.: "On the Determination of Upper-Atmosphere Composition from Satellite Measurements" in Chemical Reactions in the Lower and Upper Atmosphere. New York, Interscience, 1981, pp. 269-74.

5. Heath, D. F., A. J. Krueger, H. R. Roeder and B. D. Henderson: "The Solar Backscatter Ultraviolet and Total Ozone Mapping Spectrometer (SBUV/TOMS) for Nimbus G" Optical Engineering, Vol. 14, No. 4, pp. 323-331.

6. "Final Report for SBUV/TOMS Calibration" in preparation for National Aeronautics and Space Administration by Systems and Applied Sciences Corporation, Riverdale, Maryland.

7. a) "Requirement Specifications for Nimbus-G SBUV/TOTAL Ozone (R-SAD 1/78-2)",
   b) "Requirement Specifications for Nimbus-G SBUV/Vertical Ozone profile (R-S AB 6178-66)",
   c) "Requirement Specifications for Nimbus-G TOMS Ozone Algorithm (R-SAD 12/77-35)", all prepared for National Aeronautics and Space Administration by Systems and Applied Sciences Corporation, Riverdale, Maryland.

SECTION 8

THE SCANNING MULTICHANNEL MICROWAVE RADIOMETER (SMMR) EXPERIMENT

by

Dr. Per Gloersen
National Aeronautics and Space Administration
Goddard Space Flight Center
Greenbelt, Maryland 20771

and

Len Hardis
OAO Corporation
50/50 Powder Mill Road
Beltsville, Maryland 20705

8.1 Introduction

The Scanning Multichannel Microwave Radiometer (SMMR) was conceived in order to obtain sea surface temperature and near-surface winds, two very important parameters required by oceanographers for developing and testing global ocean circulation models and other aspects of ocean dynamics (Reference 26). The design was based on experience gained from a wide variety of experiments carried out in the laboratory, in the field, on board aircraft, and on board spacecraft using microwave radiometers over a wide wavelength range.

Several microwave radiometers have flown on previous Nimbus satellites. The Electrically Scanned Microwave Radiometer (ESMR) operated at 1.55 cm wavelength on Nimbus 5 and at 0.81 cm on Nimbus 6, and provided very useful surface observation data. The Nimbus E Microwave Spectrometer (NEMS) on Nimbus 5 and the Scanning Microwave Spectrometer (SCAMS) on Nimbus 6 were significant atmospheric-observing experiments. Much of the radiometer hardware technology developed for the NEMS and SCAMS was directly applied to the SMMR. The earliest attempt at the SMMR concept was the Passive Multichannel Microwave Radiometer (PMMR) proposed for the Earth Observatory Satellite (Reference 27). That instrument consisted of ten channels, similar to SMMR, and utilized phased arrays as antennae.

Earlier work (References 1 and 4) has demonstrated that variations of sea surface wind give rise to variations in the observed microwave brightness temperatures — even at wind speeds of less than seven meters per second where foam is not present. More recently, the spectral nature of this variation has been studied (Reference 5) and found to be separable from microwave brightness temperature changes caused by atmospheric and sea surface temperature variations (Reference 6). These studies provide the basis for extracting sea surface winds and temperatures from the SMMR data.

Other geophysical parameters are extracted from the SMMR data. These include: sea ice parameters, a mesoscale soil wetness index, snow accumulation rates over continental ice sheets, subsurface physical temperatures in snow cover, and atmospheric parameters over open ocean water of total water vapor, total non-precipitating liquid water, and rainfall rate.

The SMMR channel wavelengths are centered at 0.8 cm, 1.4 cm, 1.7 cm, 2.8 cm, and 4.6 cm. Polarization components of the microwave radiation are extracted for each channel. The smallest cell resolution is about 20 km for the 0.8 cm channel.

8.2 Scientific and Technical Objectives

The scientific objectives of the experiment are to:

- Extract geophysical parameters from the multispectral microwave radiances

- Verify the extraction algorithms

- Utilize the extracted parameters in climate modeling and assessment

- Support ongoing and new operational maritime uses (Fleet Weather Facility-USN/FWF, Fleet Numerical Weather Control-FNWC)

- Identify new observables

8.3 Instrument Description

The SMMR is a ten-channel instrument delivering orthogonally polarized antenna temperature data at the five microwave wavelengths indicated in Table 8-1. A simplified block diagram is shown in Figure 8-1. A summary table of sensor design characteristics is given in Table 8-2.

Six conventional Dicke-type radiometers are utilized. Those operating at the four longest wavelengths measure alternate polarizations during successive scans of the antenna; the others, at the shortest wavelength, operate continuously for each polarization. A two-point reference signal system is used, consisting of an ambient RF termination and a horn antenna viewing deep space. A switching network of latching ferrite circulators selects the appropriate polarization or calibration input for each radiometer.

The most novel feature of the instrument is the antenna subsystem: A 42-degree offset-parabolic reflector focuses the received power into a single feedhorn covering the entire range of operating wavelengths provides coaxial antenna beams for all channels.*

The design of the feed utilizes a ridge-loaded corrugated conical horn with peripheral slot couplers, mode transducers, and filters. Scanning is achieved by oscillating the reflector about an axis coincident with the axis of the feedhorn. The instrument is installed on the spacecraft in such a manner that this axis is parallel to the local vertical, resulting in a conical scan pattern with the angle of incidence constant on the surface of the earth near 50 degrees. The reflector is supported on a hexapod attached to a ring surrounding the feedhorn. This ring, in turn, is supported on three peripheral roller bearings and is driven through two cogged belts by a preprogrammed servo system with position and velocity feedback. The entire mechanism is caged during launch; release is by redundant pyrotechnic devices.

---

*Mr. C. R. Loughlin (GSFC) and Dr. Kurt Richter (as a NAS Fellow at GSFC) collaborated on a design study for a Passive Multichannel Microwave Radiometer (PMMR). They developed a single antenna dish receiver concept for the PMMR design. This design was adapted to SMMR and was a key factor in making SMMR accepted as part of the Nimbus payload.

Table 8-1
SMMR Performance Characteristics

Parameter	Channel				
	1	2	3	4	5
Wavelength (cm)	4.54	2.8	1.66	1.36	0.81
Frequency (GHz)	6.6	10.69	18.00	21.00	37.00
R-F Bandwidth (MHz)	250	250	250	250	250
Integration Time (ms) (approximate)	126	62	62	62	30
I-F Frequency Range (MHz)	10-110	10-110	10-110	10-110	10-110
Dynamic Range (°K)	10-330	10-330	10-330	10-330	10-330
Absolute Accuracy (°K rms)	<2.0	<2.0	<2.0	<2.0	<2.0
Temperature Resolution, $\Delta T rms$ (°K) (per IFOV)*	0.9	0.9	1.2	1.5	1.5
Antenna Beam Width (±0.2°)	4.2	2.6	1.6	1.4	0.8
Antenna Beam Efficiency (percent)	87.0	87.0	87.0	87.0	87.0
Scan Cycle ±0.4 rad (±25°)/second**	4.096	4.096	4.096	4.096	4.096
Double Sideband Noise (dB) (maximum)	5.0	5.0	5.0	5.0	5.0

*IFOV are remapped to form equal sized cells (150, 90, 50km) across the swath prior to retrieval of geophysical parameters; the ΔT rms's are correspondingly lower.

**Add 2 ms (used for integer dump) for complete IFOV cycle time.

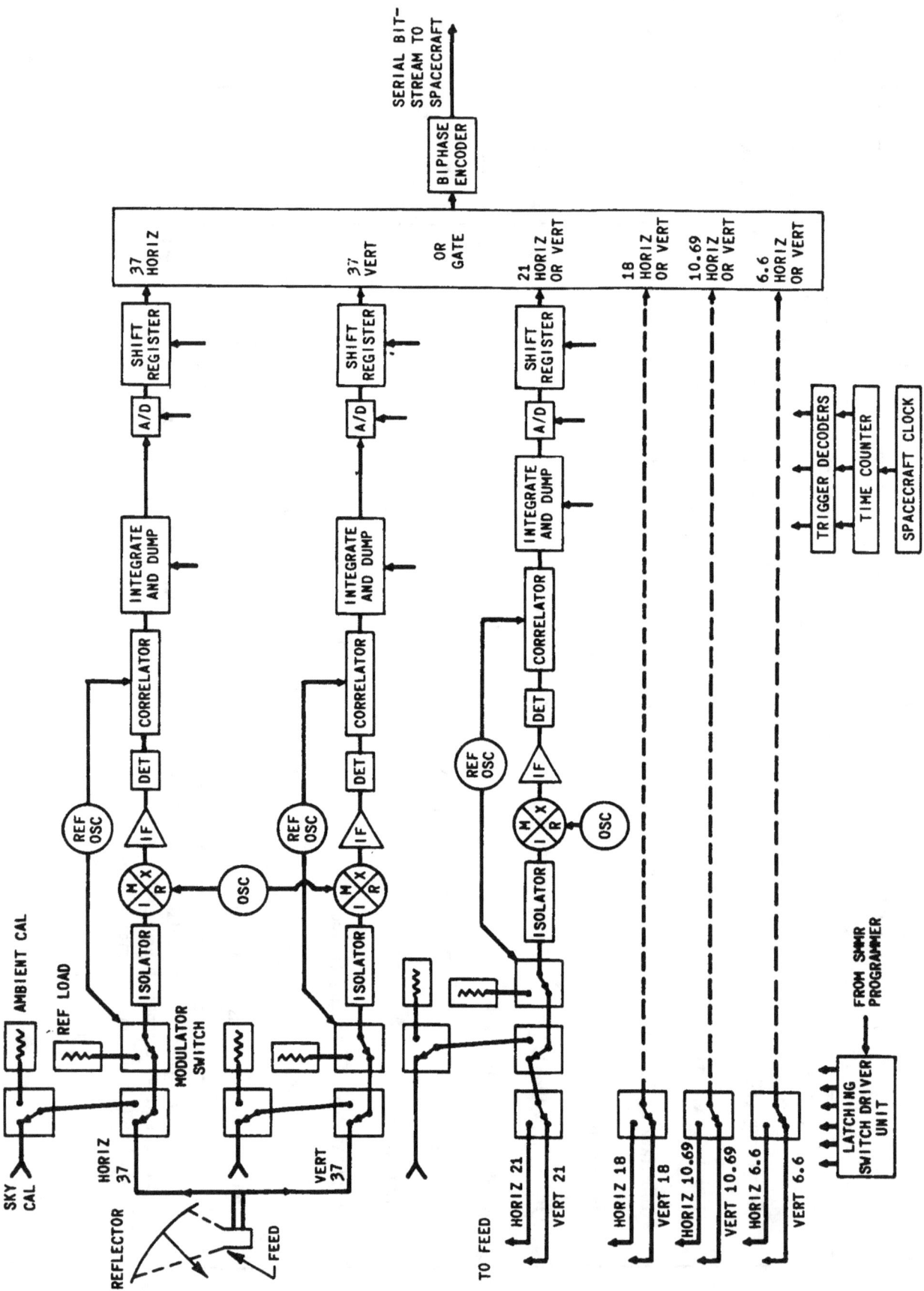

Figure 8-1. Block Diagram of the SMMR Electronics.

Table 8-2
SMMR Sensor Design Characteristics

Item	Characteristics
Detectors:	RF diode – Dicke – Superheterodyne
Size:	Two 15.3- by 33.0- by 20.4-cm modules (two Nimbus bays)
	One 15.3- by 16.5- by 20.4-cm module (one-half Nimbus bay)
	Parabolic section antenna, 80 cm in diameter
	Multifrequency antenna feed
Weight:	52.3 Kg
Power:	60 Watts
Commands:	12
Data:	DAPS – 2 kbs[1]
Telemetry:	Digital B – 9[2]
	Analog – 19
Clock:	Time code
	Strobe
	1 Hz
	10 kHz
	1.6 MHz

(1) Nimbus 7 data processing
(2) Lower data rate in DAPS for collecting digital words

The remainder of the instrument, including the radiometers, control electronics, power supply, data and programmer subsystems, is a derivative of the Nimbus 5 NEMS and Nimbus 6 SCAMS instruments with minor modifications to take advantage of improved state-of-the-art-components.

Physically, the SMMR instrument consists of five hardware elements:

- The antenna assembly consisting of the reflector, fabricated of graphite epoxy, and the feedhorn

- The scan mechanism, including momentum compensation devices

- An RF module containing the input and reference switching networks, the mixer-IF preamplifiers, and the Gunn local oscillators

- An electronics module containing the main IF amplifiers, all the post-detection electronics, and the power supplies for the scan and data subsystems

- A power supply module which contains the dc-to-dc converters and regulators for the rest of the instrument

The antenna, scan mechanism, RF module, and sky horn cluster are mounted on a bridgelike platform which is then installed as a preassembled, aligned and calibrated unit on the spacecraft. The electronics and power supply modules are mounted separately and are cabled to the instrument and spacecraft through connectors. Figures 8-2 through 8-5 contain various views of the instrument. Its overall size can be visualized by noting that the elliptical antenna reflector is approximately 110 cm x 80 cm. Total instrument weight is about 50 kg, its power consumption 60 watts, and its digital data output rate 2 kbs.

On Nimbus 7, the SMMR scan pattern is forward viewing and scans equally (see Table 8-1) to either side of the orbital track so the swath is centered on that track. With a subsatellite velocity of about 6.5 kilometers per second and a scan period of 4.096 seconds, overlap coverage is provided at all wavelengths.

To conserve power, the scan is sinusoidal. Part of the time spent at the scan extremeties is utilized for reading the radiometer internal and space horn references. The dwell time of all the SMMR channels are integral multiples of and synchronous with the 0.81 cm channel dwell time of 32 ms. Concurrent dwell time facilities multispectral data analyses on various geometric scales.

Conversion of the raw radiometric readings to microwave brightness temperatures involves correcting for actual antenna patterns, including sidelobe effects, as well as separating out the horizontal and vertical polarization components of each of ten channels of radiometric data. These equations will be available after launch.

8.4 Calibration

8.4.1 Prelaunch Calibration

The instrument is calibrated by the following equation for each wavelength and polarization:

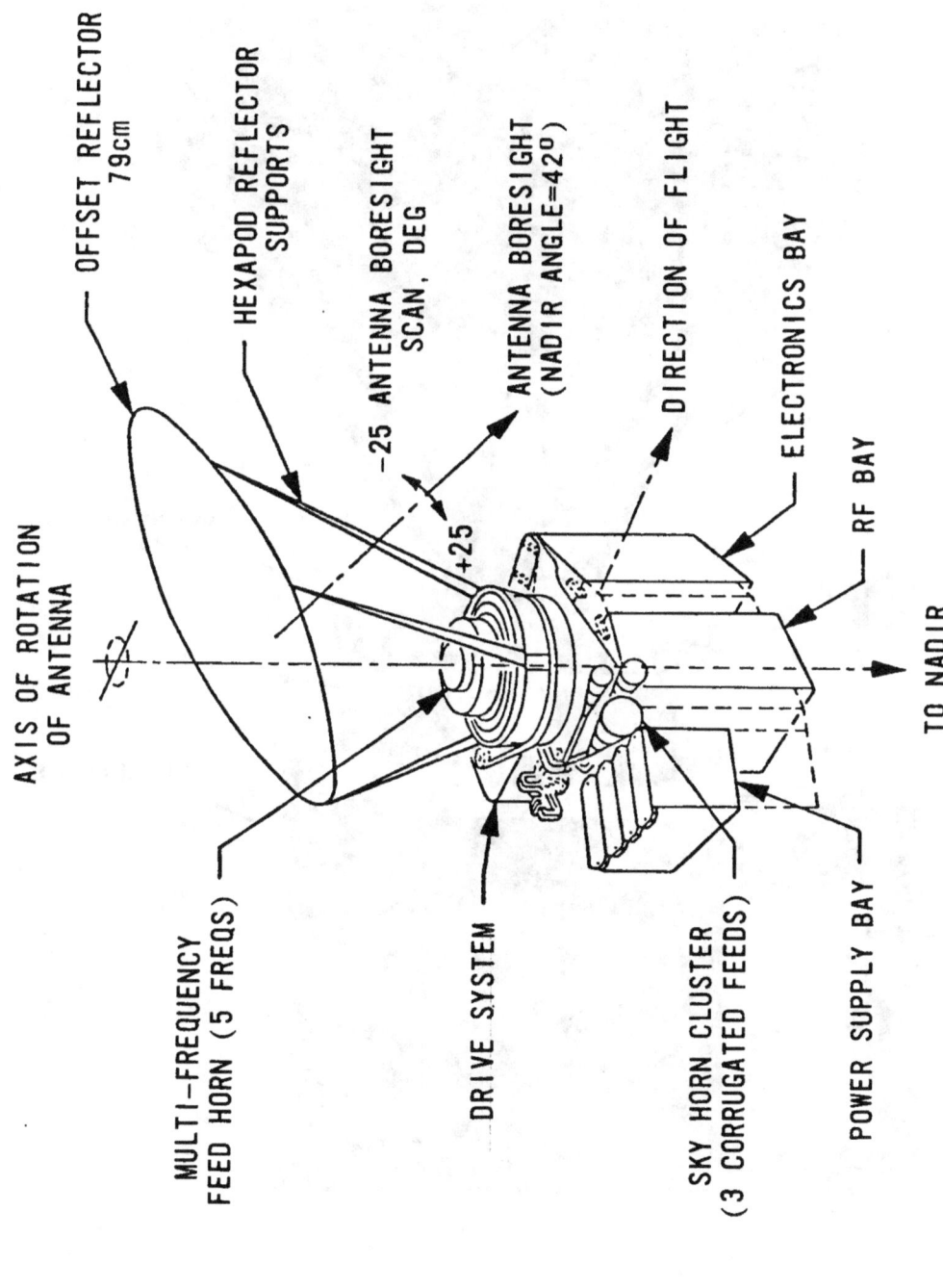

Figure 8-2. SMMR Instrument Configuration Showing Antenna, Feed Horn Drive Assembly, and Electronic Boxes (configured to fit the Nimbus 7 sensor ring bays)

Figure 8-3. SMMR Instrument Showing Front View

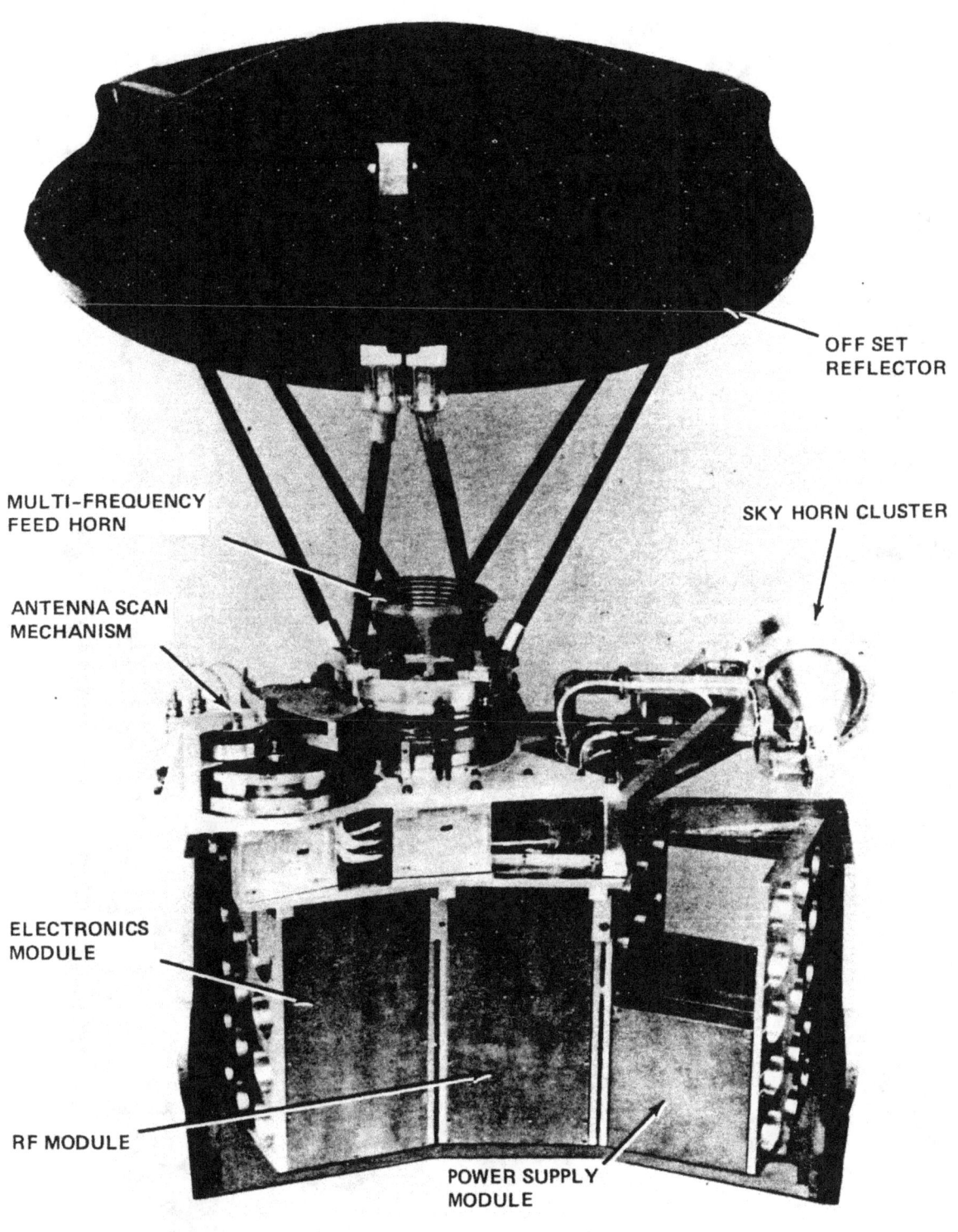

Figure 8-4. SMMR Instrument Showing Rear View

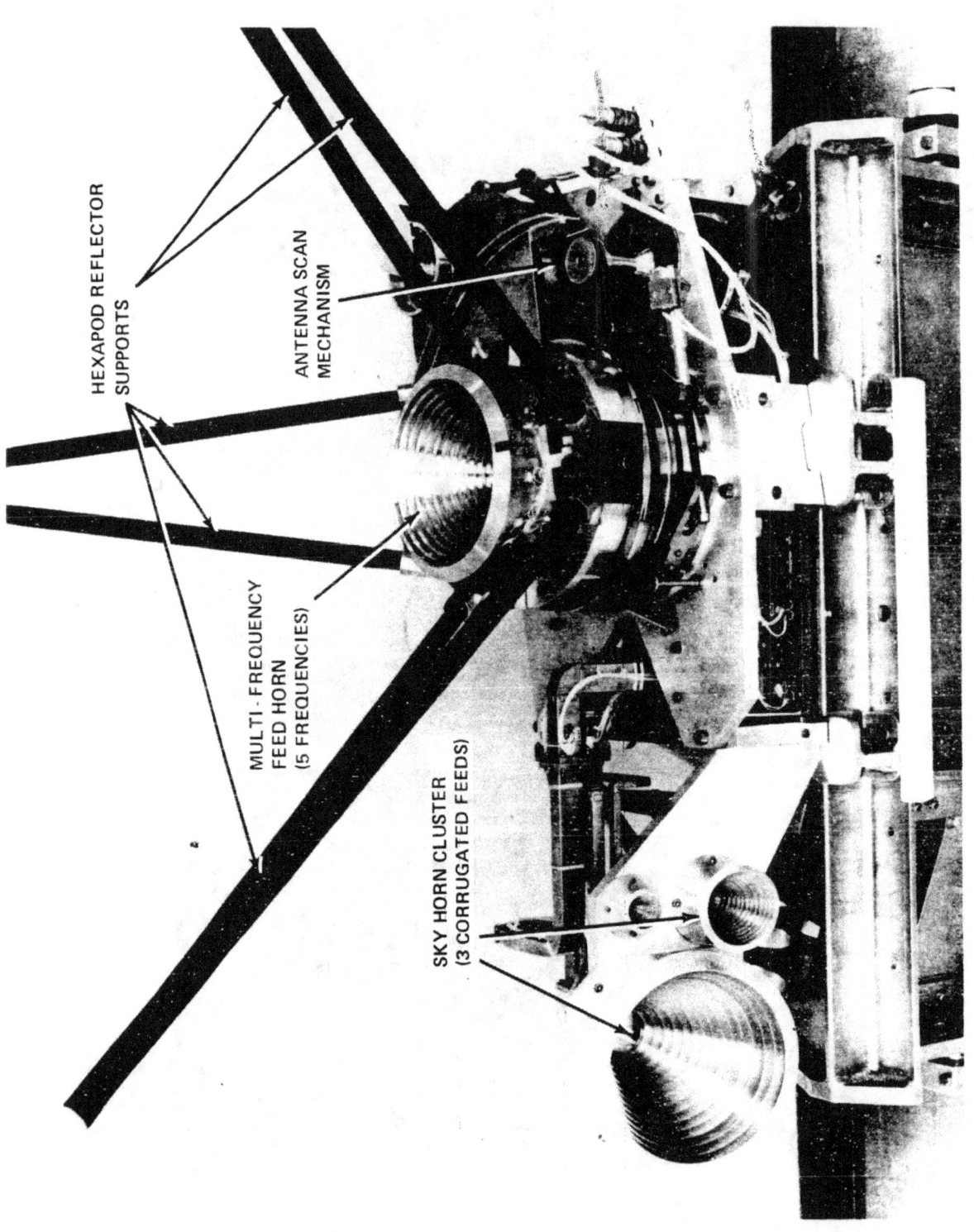

Figure 8-5. SMMR Instrument Configuration Close-up Drive and Feed Details

$$T_{Ai} = a_{i00} + a_{i01}(T_I - T_{10}) + a_{i02}(T_I - T_{10})^2 + a_{i10}N + a_{i11}(T_I - T_{10})N + \ldots + a_{i12}(T_I - T_{10})^2 N \quad (1)$$

where a_{ijk} will be supplied by launch based on thermal-vacuum chamber and laboratory tests and updated after launch during the validation period,

i = 1 through 10 (radiometer channel number)

T_A is the total brightness temperature, uncorrected for polarization mixing and sidelobes

T_I is the instrument temperature

T_{10} is a mean, or convenient intermediate value of N

$N = \dfrac{C_A - C_H}{C_C - C_H}$ is the normalized counts

C_A is the digital counts from the radiometer, when viewing the earth

C_C is the digital counts from the radiometer, when reading the space horn

C_H is the digital counts from the radiometer, when reading internal warm surface

8.4.2 Post-launch Calibration

After launch, the prelaunch constants (Equation 1) will be updated by checking against earth targets of known properties — open, calm sea water with clear skies or light clouds, and consolidated first-year sea ice. The T_B's will be verified by comparsion with T_B's obtained with an airborne radiometer with all SMMR channels during Nimbus 7 underflights. The underflights are particularly important, since extrapolation from the laboratory cold reference of 100°K to the postlaunch value of 30°K cannot be done with complete confidence.

8.5 Operational Modes

The SMMR has been constrained to a 50 percent duty cycle due to spacecraft power limitations. This duty cycle is achieved by operating on alternate days. Even so, the SMMR maps the entire earth every six days. The overlap resulting from 100 percent duty cycle is absent, and the interval between observations of a given point is increased as a result of the 50 percent duty cycle operations.

The "alternate-day" operating pattern must be maintained throughout the lifetime of the SMMR to ensure the integrity of the time series analyses of the data. Should extra observing time be made available (i.e., short periods of 100 percent duty cycle), care must be taken to return to the original alternate day schedule at the end of the 100 percent duty cycle period.

8.6 Data Processing, Formats, and Availability

8.6.1 Data Processing

Throughout the SMMR data processing system, a modular software design is planned in order to facilitate program changes and algorithm refinement during the instrument validation phase. The many geophysical parameters are derived from linear combinations of the ten measured radiances, or functions of the radiances. The initial algorithms supplied during the prelaunch period are based on data acquired in a series of flight measurements with a SMMR simulator flown on the CV-990 aircraft with correlative surface truth, on data from other microwave radiometers (i.e., Nimbus 5 and 6) and on theory. Most of the prelaunch analyses and field expeditions outlined in Section 8.7 deal with the development of key algorithms. The schedule for their development reflects the amount of analysis effort needed following completion of the required CV-990 flights.

The SMMR data stream processing has been separated into three distinct categories. The initial flight data are received by the Meteorological Operations Control Center (MetOCC). Figure 8-6 contains the flow chart for the data processing performed in the MetOCC.

The user formatted output tape from MetOCC is then tranferrred to and processed by the Science and Applications Computer Center (SACC). SACC derives the required geophysical parameters from the radiometric data.

The algorithms necessary for these conversions are presently in various stages of development. A list of these algorithms is given in Table 8-3. It should be noted that all constants used in the algorithms are subject to revision in the post-launch period. The SMMR flow charts for data processing at SACC are given in Figures 8-7 through 8-13.

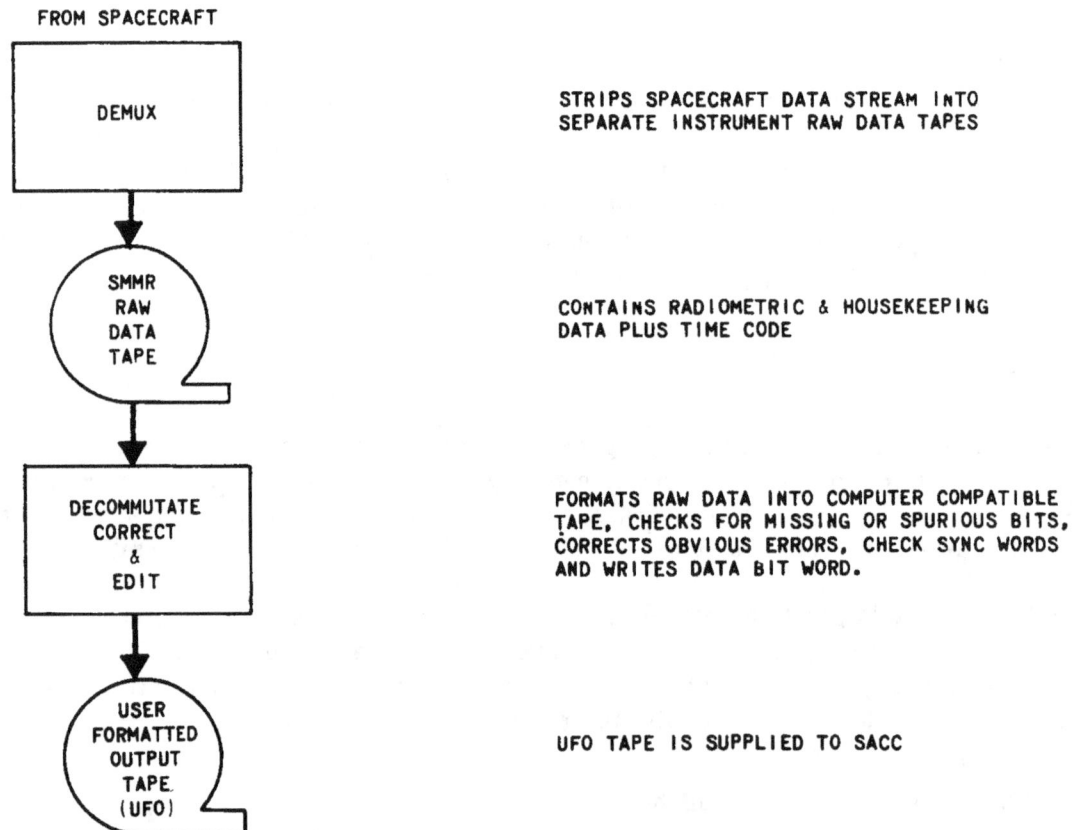

Figure 8-6. SMMR Data Processing Flow Chart in the Meteorological Operations Control Center

Table 8-3
SMMR Science Algorithms

Algorithm	Status			Remarks
	Delv'd	Due	Open	
A. Ocean				
1. Sea Ice Temperature	X			
2. Sea Surface Wind Speed	X			
3. Atmospheric Water Vapor	X			
4. Atmospheric Liquid Water	X			
5. Rain Rate	X			
B. Sea Ice				
1. Ice Concentration	X			1 Algorithm approach delivered. However, mathematical model studies have shown results to be "noisy" based on constants that are on hand. The algorithm is being "fine-tuned" and the constants re-evaluated.
2. Ice Surface Temperature	X			
3. Multiyear Ice Fraction	X			
4. Thin First Year Ice Fraction	1			
5. Temperature Difference Between Thick and Thin Ice	1			
6. Atmospheric Contribution	X			
C. Land				
1. Rain (Yes/No)	X			2 A post-launch algorithm development effort contingent on sufficient A/C flight time to obtain satisfactory ground truth data.
2. Soil Wetness Index		2,	3	
3. % Open Water Over Land		3,		
4. Land Surface Temperature		3,		
				3 A tentative functional form may be delivered at launch with highly preliminary constants for initial research purposes.
D. Snow				
1. Dry Snow (Yes/No)		3,	4	4 Considered a possible post launch algorithm effort. Contingent upon funds to perform snow underflights.
2. Snow Layer Water Equivalent		3,	4	
3. Snow Surface Temperature		3,	4	
4. Snow Sub-surface Temperature.		3,	4	
E. Ice Sheet				
1. 1.7 cm 0.6 $T_V - T_H$	X			5 Results from comparison of recent A/C data with mathematical model shows that the model has anomalies. Because of higher priority of algorithms B-4 and B-5, effort has been held in abeyance.
2. 2.8 cm 0.6 $T_V - T_H$	X			
3. Surface Temperature	5			
4. Sub-surface Temperature	5			

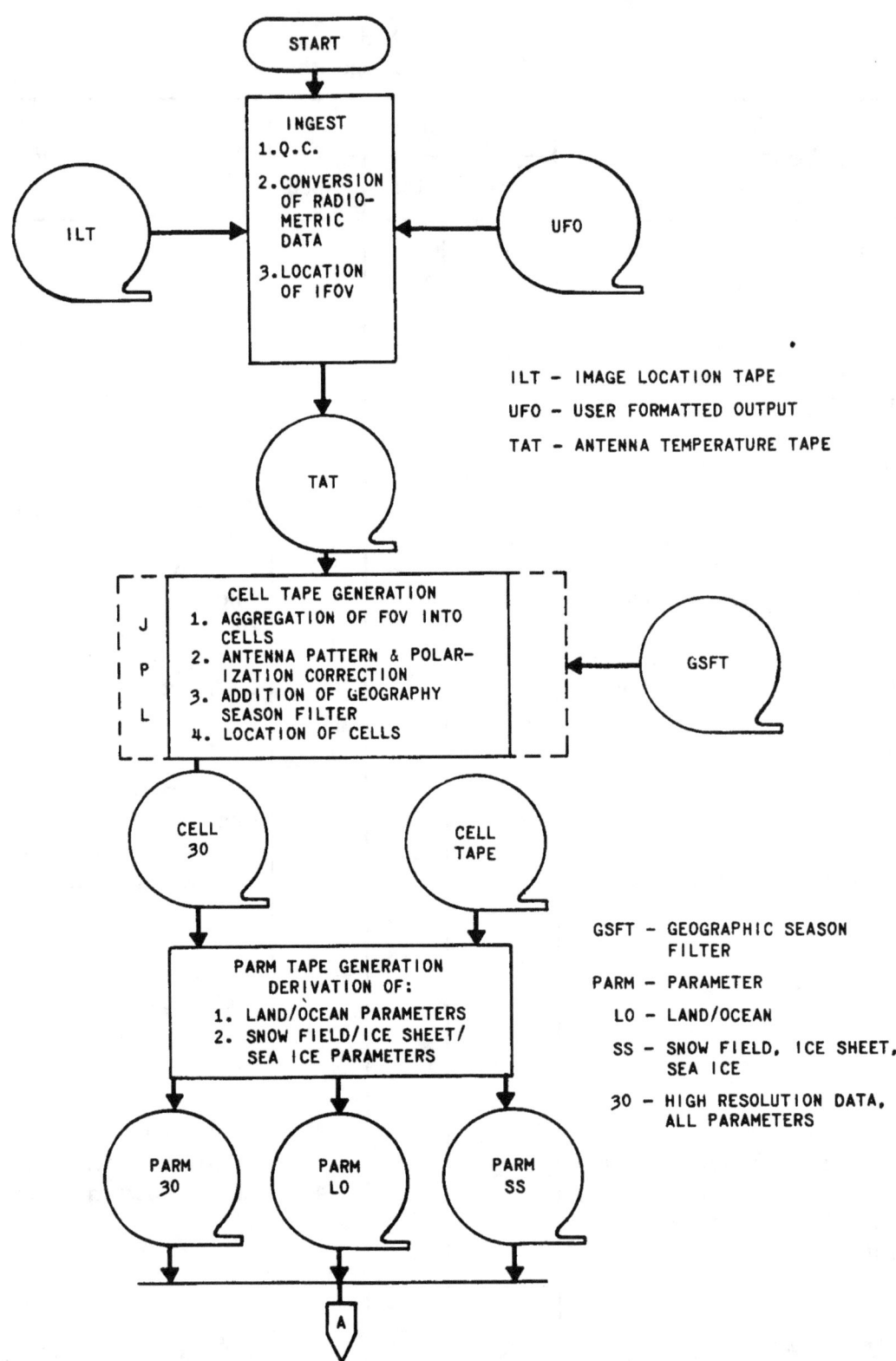

Figure 8-7. SMMR Data Processing Flow Chart

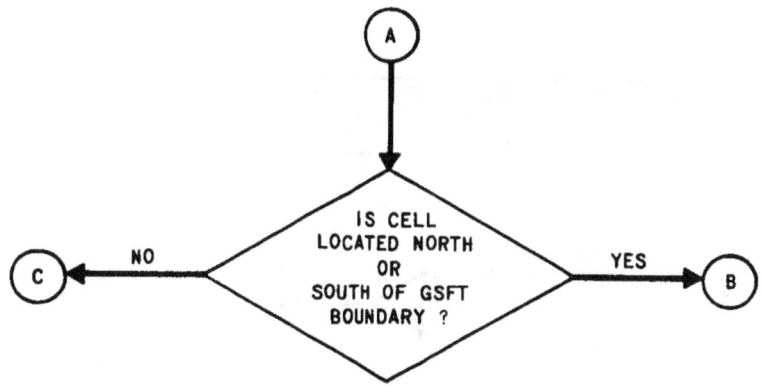

Figure 8-8. SMMR Data Processing Flow Chart

For the purpose of obtaining brightness temperatures (T_B's) corrected for polarization mixing and sidelobes from the antenna temperatures (T_A's), and for the processing algorithms for the various geophysical parameters, the T_A's from the IFOV's are remapped into four different equal-sized square cells. All but the 4.6 cm channel are in the 97.5 km cells, all but the 4.6 and the 2.8 cm channels are in the 60 km cells, and only the 0.8 cm channel is in the 30 km cells. The cells serve to take advantage of the overlap in IFOV's to give lower T_B rms values than those in Table 8-1. Thus the spatial resolutions peculiar to a given geophysical parameter retrieval depends on which SMMR channels are utilized in that retrieval. The data processing flow charts for the generation of the parameter tapes are shown in Figure 8-12.

To complete the process in SACC, the parameter (PARM) tapes are processed by using map matrix information on discs to generate matrix tapes. These matrix tapes contain the color coding for any parameter at each located (latitude and longitude) grid intersection. The data stream processing for this phase is shown in Figure 8-13.

The Information Processing Division (IPD) at GSFC then receives the matrix tapes, merges them with MAP tapes, and produces the final color image products as shown in the right half of Figure 8-13.

8.6.2   Tape Products

The following tapes are produced by SACC and used by IPD before being sent to the NSSDC for archiving. Brief descriptions of these tapes are as follows:

- TAT (Antenna Temperature Tape)

    Contains calibrated antenna temperatures and earth locations for each IFOV for each polarization. Also contains ephemeris, attitude, and SMMR housekeeping information. This is the most basic form of the SMMR data available to users.

- CELL-ALL

    Contains horizontal and vertical polarization brightness temperatures and seasonal geographic filters for each of the five channels at 150 km resolution (as discussed in Section 8.6.1), for all but the 4.6 cm channel at 97.5 km resolution, for all but the 4.6 cm and 2.8 cm channels at 60 km resolution, and for only the 0.8 cm channel at 30 km resolution. Data are

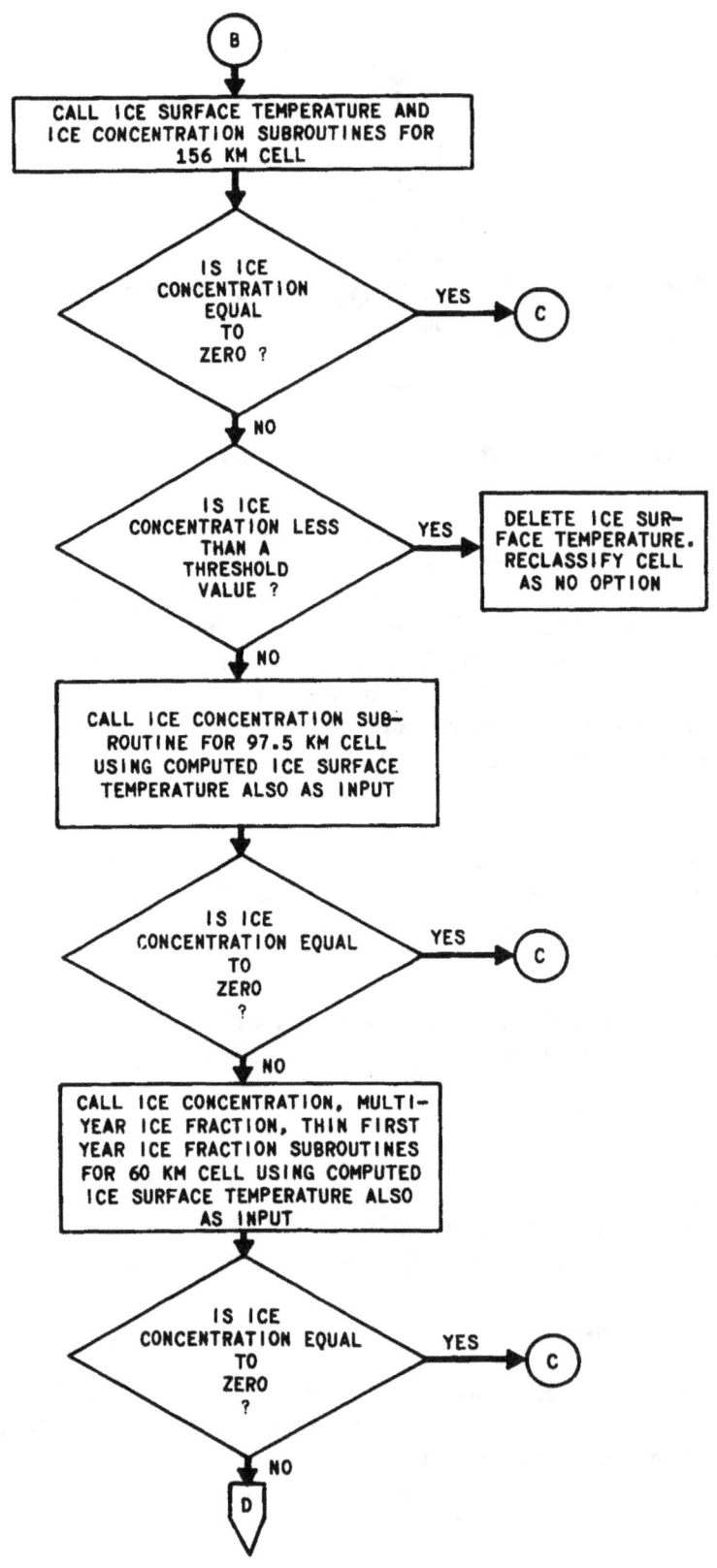

Figure 8-9. SMMR Data Processing Flow Chart

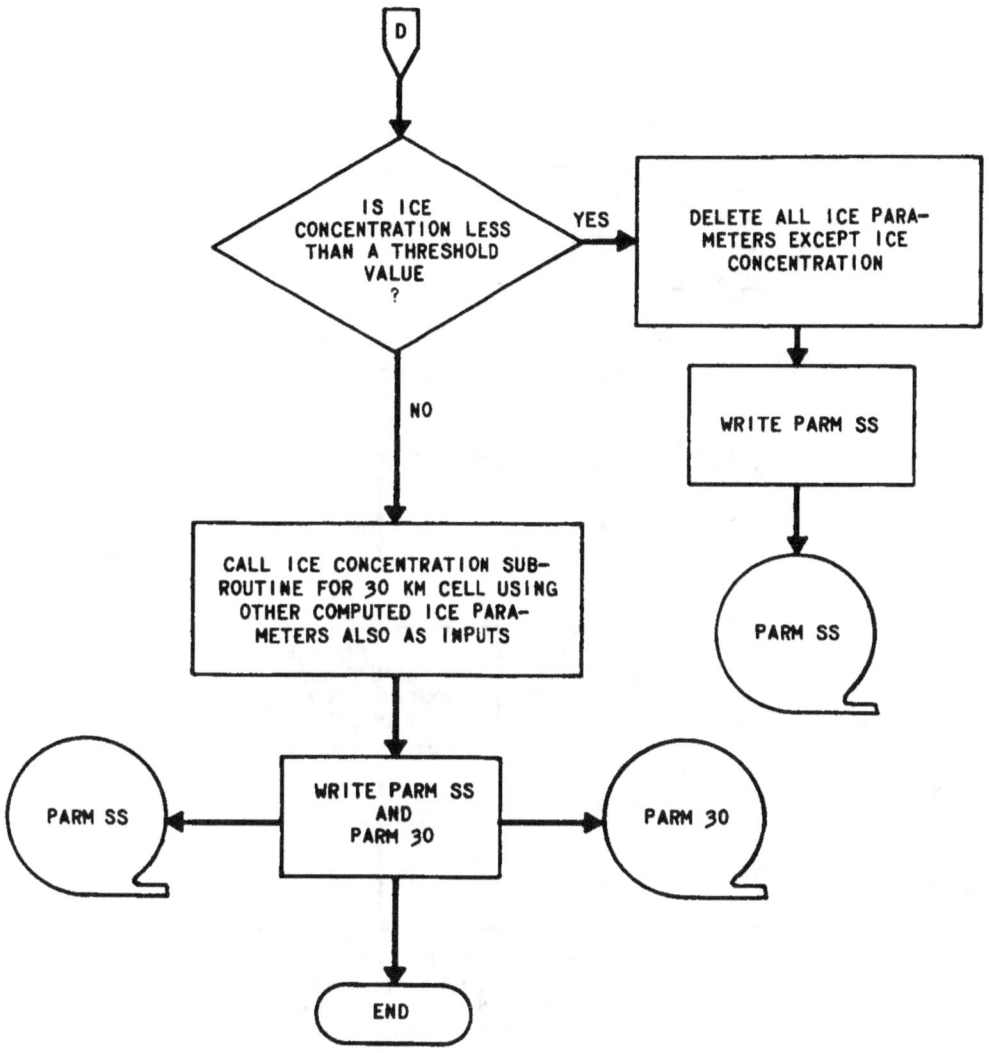

Figure 8-10. SMMR Data Processing Flow Chart

grouped by cells and bands of various sized, but each combination of cells and bands equals 780 km². Location coordinates are given for each cell and band.

- PARM LO (Parameters of Land–Ocean Tape)

- PARM SS (Parameters of Sea Ice and Snow and Ice on Land Tape)

- PARM 30 (Parameters of 37 GHz channel Tape)

Each tape type contains derived parameters for each IFOV as specified by the tape titles.

- MAP LO (Mapped Parameters of Land–Ocean Data Tape)

- MAP SS (Mapped Parameters of Sea Ice and Snow and Ice on Land Tape)

- MAP 30 (Mapped Parameters of 37 GHz channel Data Tape)

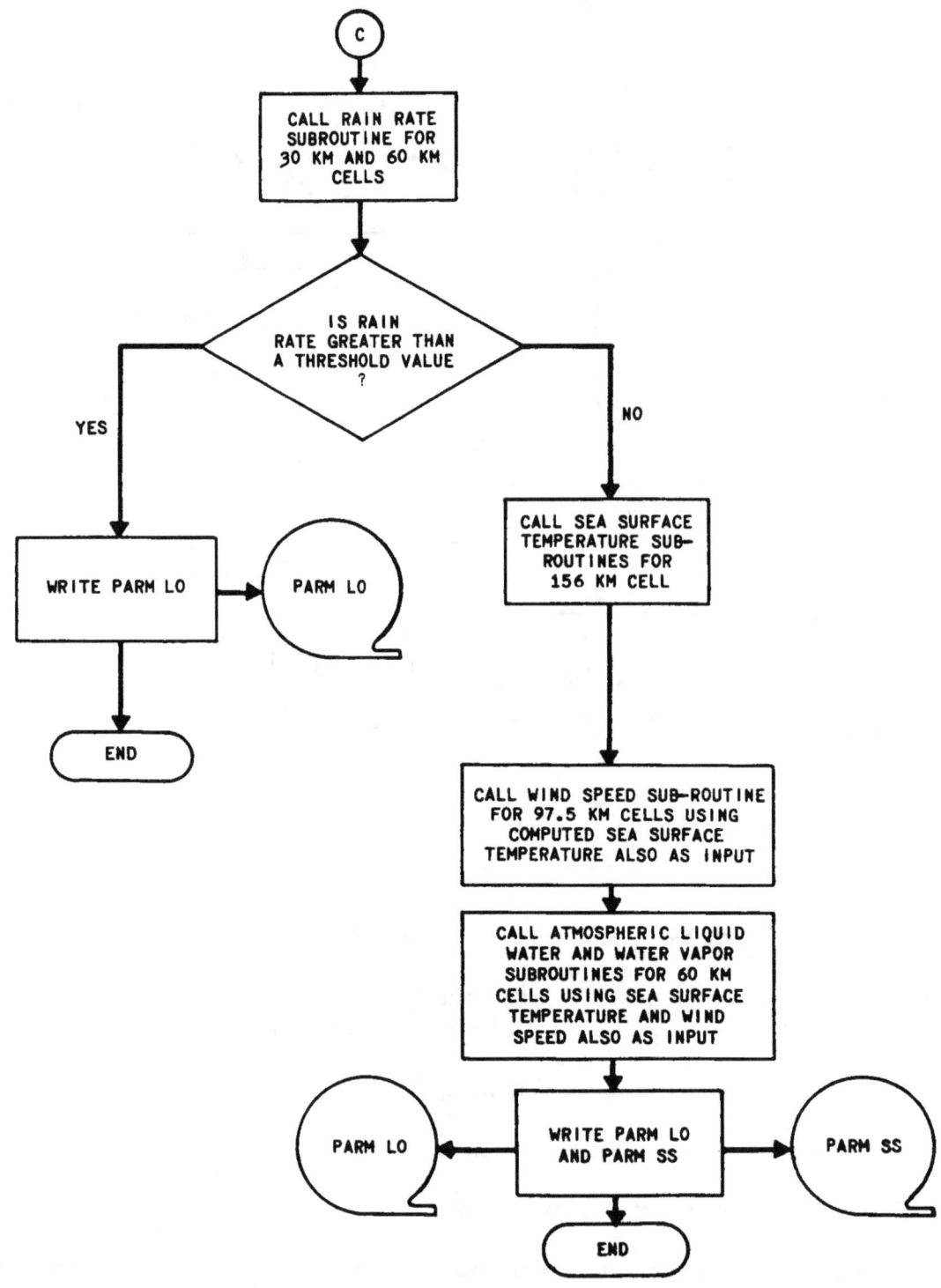

Figure 8-11. SMMR Data Processing Flow Chart

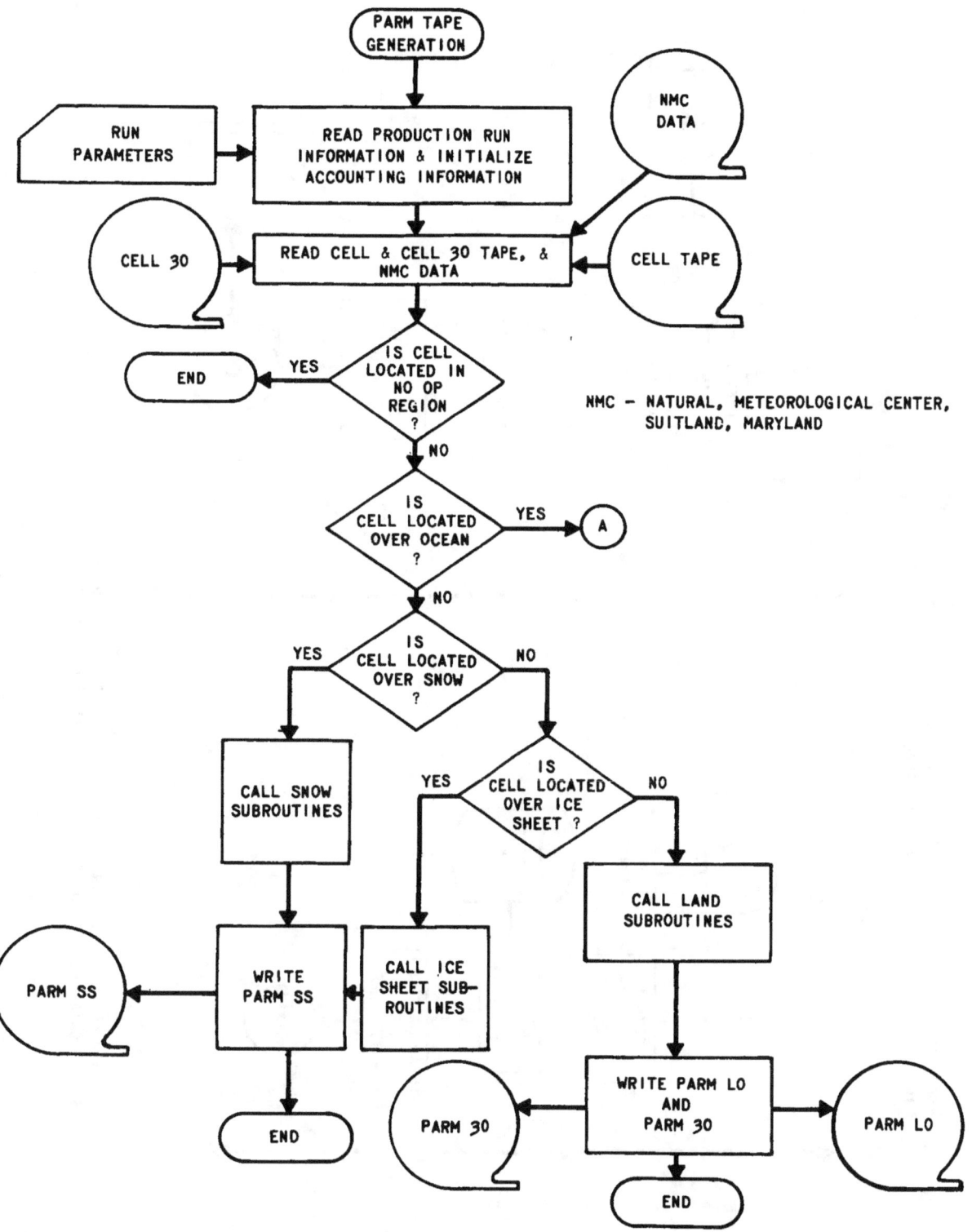

Figure 8-12. SMMR Data Processing Flow Chart

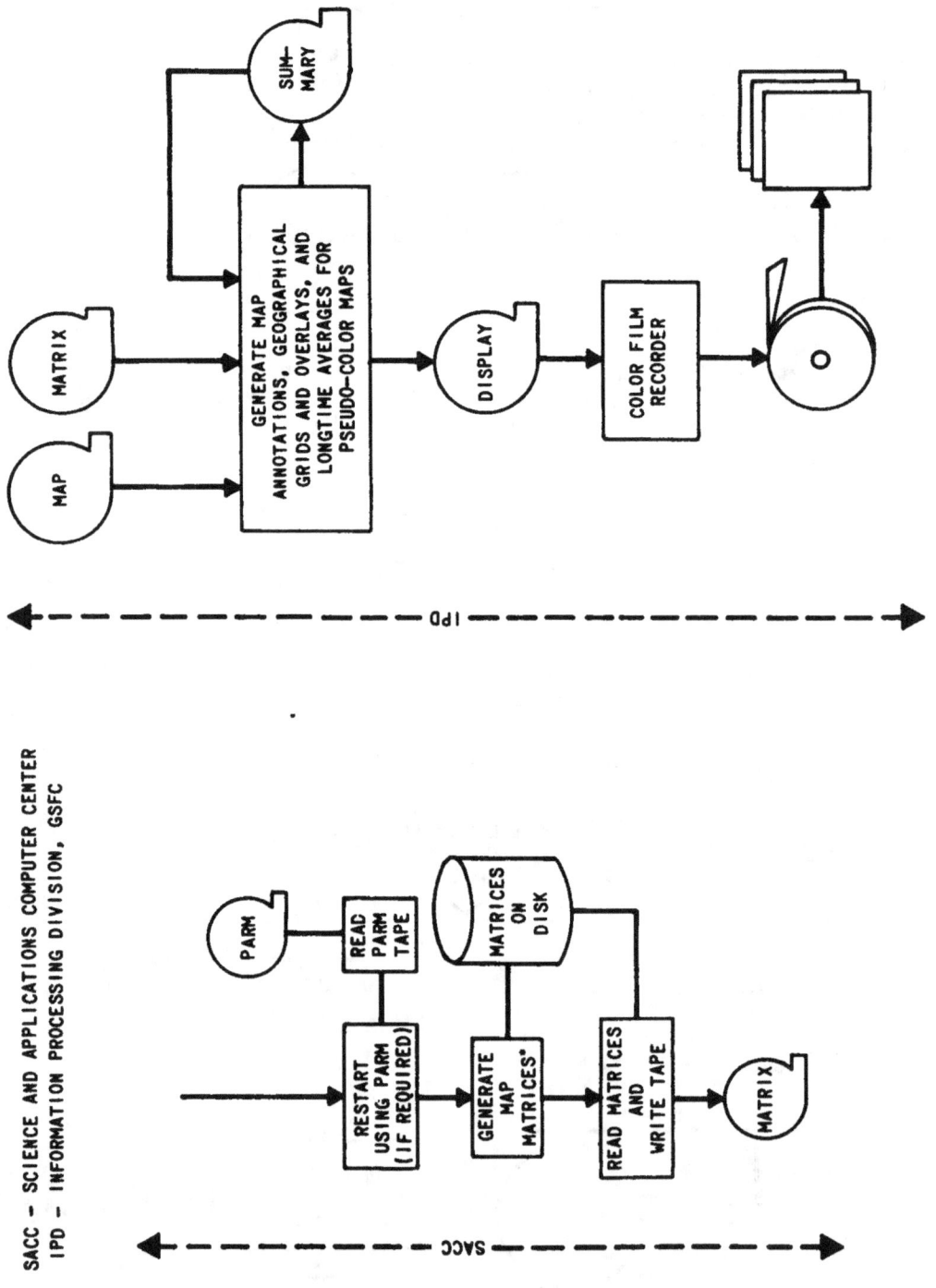

Figure 8-13. SMMR Output Product Flow Chart

Contain north and south polar map projections (SS and 30 tapes) and Mercator projections (LO tapes) of derived parameters in scientific word structure.

The form and content of each of these tapes is specified in a tape specification document for each tape type. Users receiving one of these tapes will receive the appropriate document. See Section 1.5 of this document for details.

8.6.3 Display Products

There are 16 different SMMR map display products. All displays are color annotated maps prepared initially as 105 mm color negatives.

All displays contain either polar stereographic or Mercator map projections. Some displays contain two northern hemisphere polar maps to 50°N and two southern hemisphere maps to 50°S. Figure 8-14 is an example of this format. Some displays contain the same four polar maps but the limits of the maps are 30°N and 30°S, as shown in Figure 8-15. On these map sets the left set of two maps contains the data for the first half of the display period and the right set of two maps contains the data the last half of the display period. Polar displays, containing the parameters averaged over one month, only contain one northern and one southern hemisphere map, as shown in Figure 8-16 (50° map limits) and Figure 8-17 (30° map limits).

Some displays contain two Mercator maps as shown in Figure 8-18. The top map always contains only descending node, or nighttime, data and the bottom map always contains only ascending, or daytime, data. Latitude limits for these maps are 64°N to 64°S. Beneath the maps on all displays is a 32-step color scale. The colors are used in each map to identify parameter values. The number beneath each color chip provide the parameter values for that chip. (A color chip may have from one to six parameter values, depending on the number and complexity of the parameter being displayed.)

Beneath the color scale numbers are the titles of the parameters mapped in the display. Table 8-4 lists the parameter titles for each SMMR color display and their corresponding film specification number. As Table 8-4 indicates, there are from one to four parameters in each map, depending on the display. When more than one parameter is on a map, the area occupied by each is mutually exclusive. Several of the parameters are averaged and mapped at a short time interval and also at one month time periods.

Interpreting the fourth digit from the left in each film specification number in Table 8-4 gives the frequency of output of that display. The frequencies for SMMR are:

$$XXX3XX = \text{produced every three days,}$$

$$XXX4XX = \text{produced every six days, and}$$

$$XXX7XX = \text{produced every month.}$$

On each display the letter (A, B, etc.) before each parameter title is referenced to the same letter at the beginning of one of the lines of color scale annotation.

There are three sets of map resolutions used in the SMMR display products (150 km, 50 km, and 25 km). These resolutions correspond to the nominal IFOV's of the channels which are the principal contributor to the physical parameter presented. The resolution for each parameter appears in paretheses after the display title near the bottom.

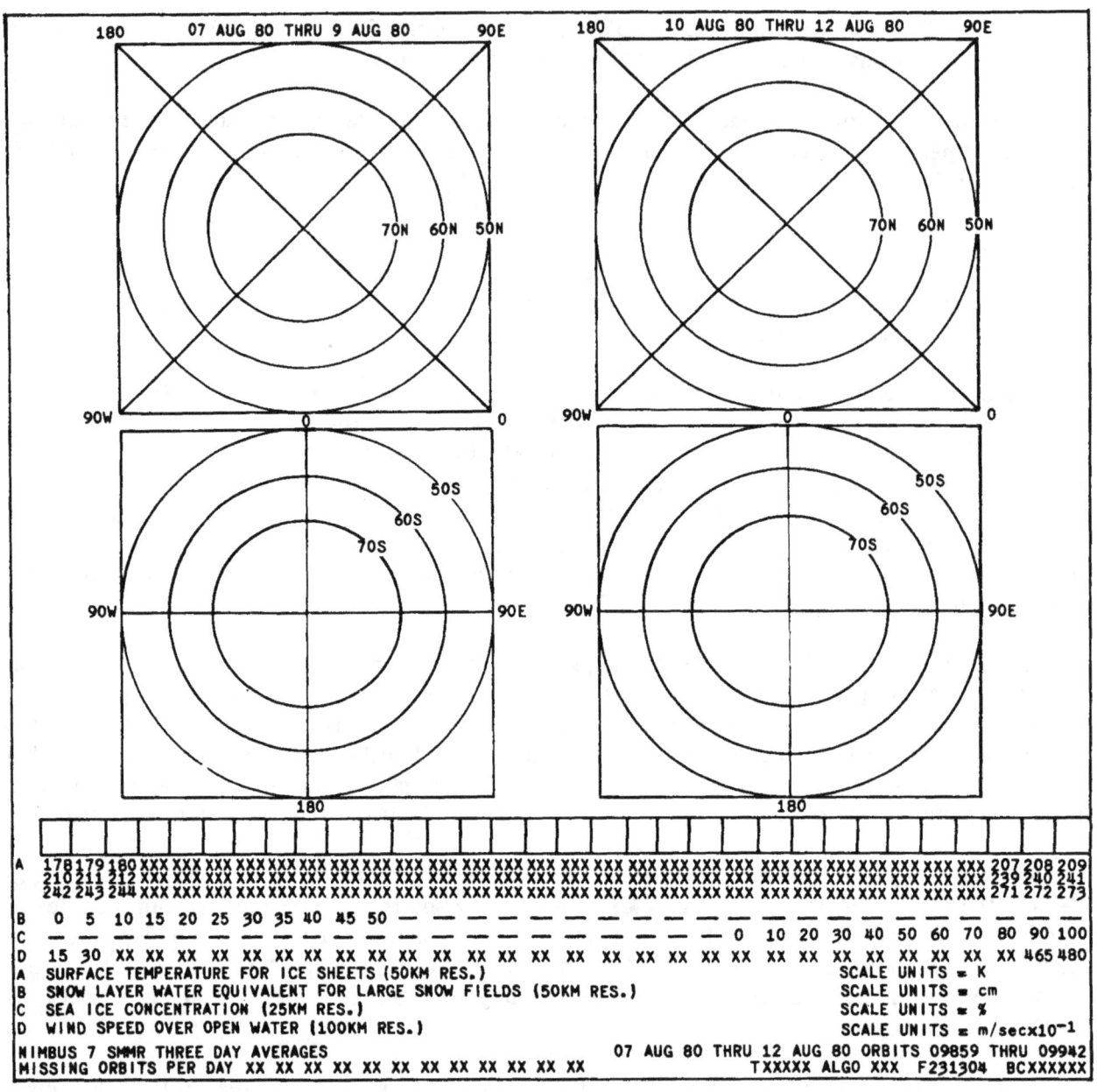

Figure 8-14. Format for SMMR Short-term Polar Displays to 50°N and 50°S

Title and reference information are at the bottom of each display. On the last line on the three and six day displays there is a "missing orbits per day" code followed by a series of two-digit numbers. The first two-digit number is the number of orbits of data missing (for any of several reasons) from the first day of data used to construct the display. The second two-digit number is for the second day, etc.

The items on the right half of the last line are: the physical tape number the data is stored on (TXXXXX), the algorithm reference number used in processing the data (ALGO XXX), the film specification number (F 231XXX), the project data format code (BA, BB, or BC), and the film frame number (XXXXXX).

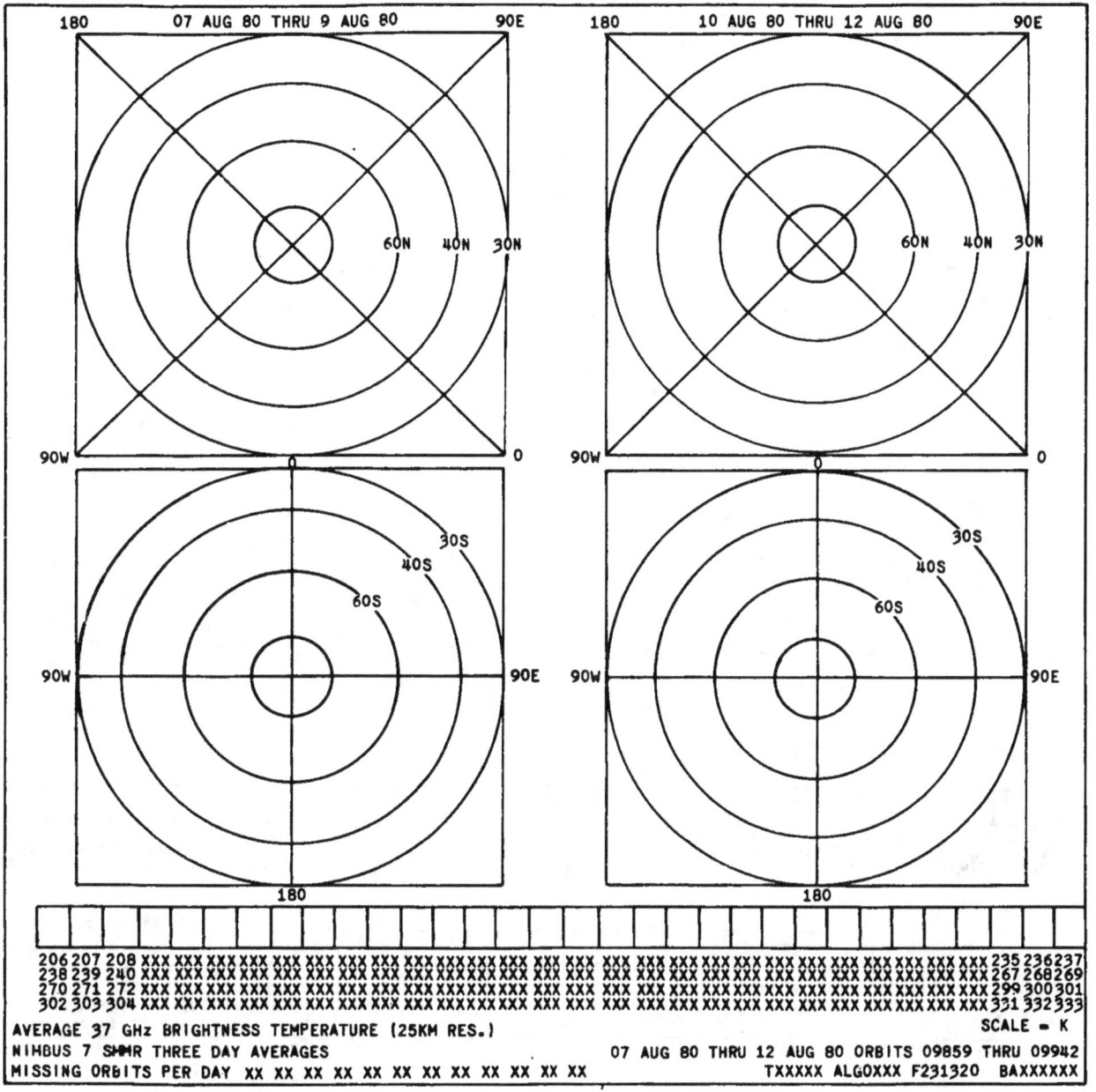

Figure 8-15. Format for SMMR Short-term Polar Displays to 30°N and 30°S

8.6.4 Data Availability

The SMMR experimental data consisting of the magnetic tapes described in Section 8.6.2 and the color film displays illustrated and described in Section 8.6.3 are archived at NSSDC. Initial instrument checkout, algorithm adjustments, and general program debugging is expected to take from six to twelve months, depending on the geophysical parameter. Users requesting SMMR data from the NSSDC should read Section 1.5 of this document for general tape and film ordering information.

Future plans of NOAA-EDIS include archiving data from the first dedicated oceanographic satellite, Seasat-A. A common archive of oceanographic data, including SMMR and CZCS products, is

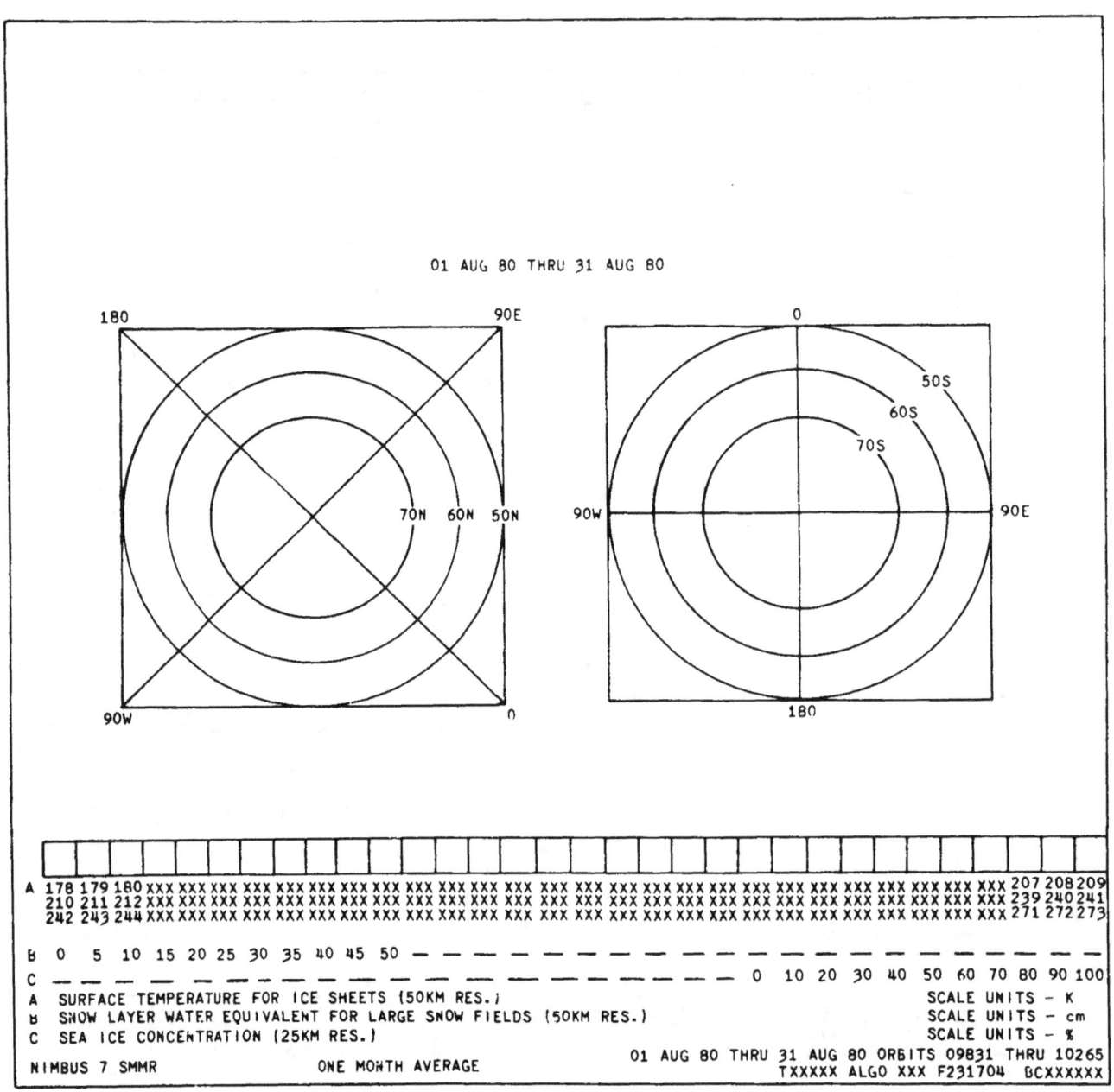

Figure 8-16. Format for SMMR One-month Polar Displays to 50°N and 50°S

under consideration. Beginning on September 1, 1978, and continuing through 1979, the World Meteorological Organization will participate in the First GARP Global Experiment (FGGE). Selected ERB and SMMR data acquired during this period are relevant to the Experiment and are desired for inclusion in the archive. A special effort would be required to get these data to a FGGE data collection center within 45 days of acquisition. If the data are not available for the full period, particular attention is being given to the special observing periods of January 15 to February 13, 1979 and May 10 to June 2, 1979.

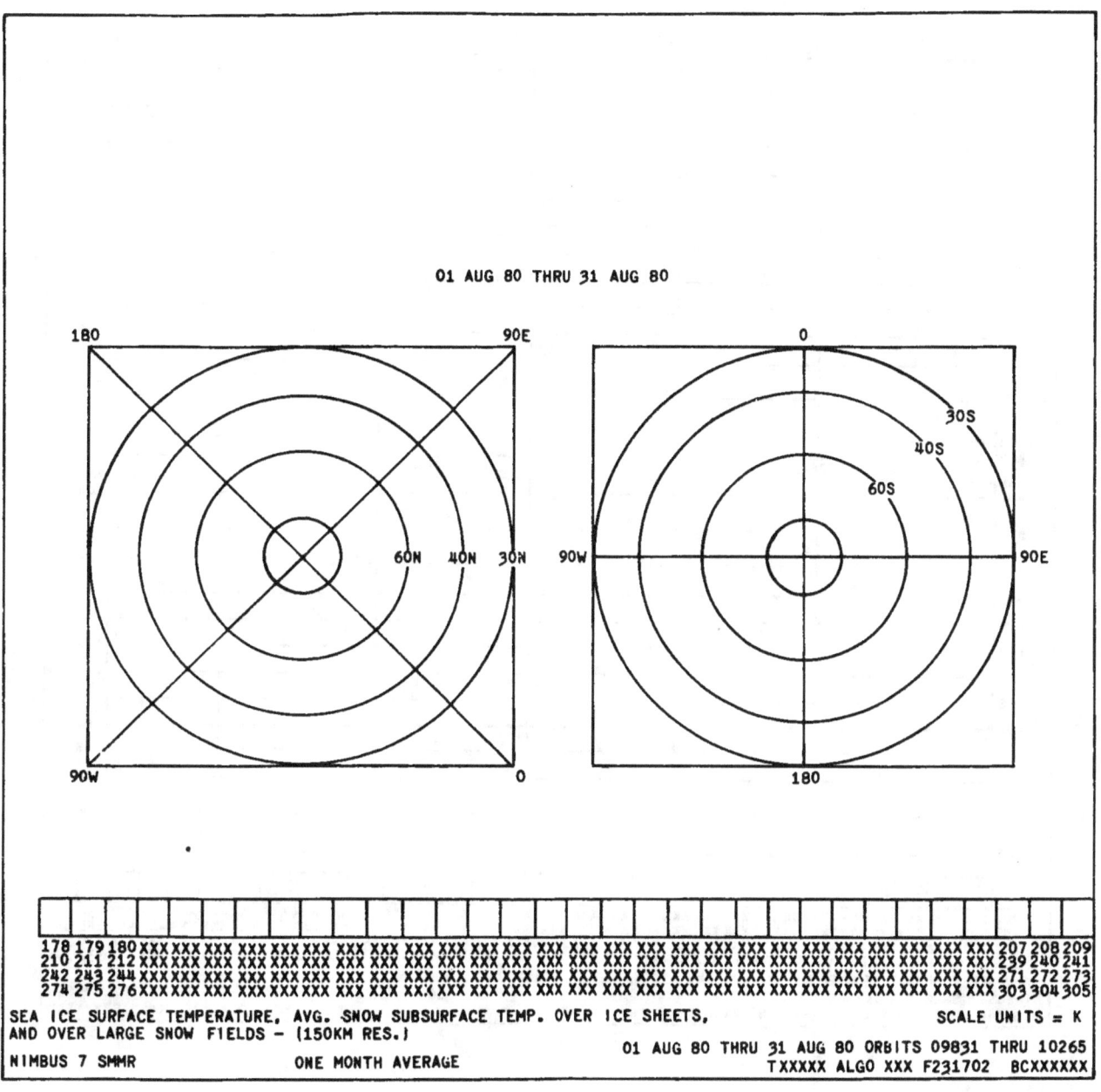

Figure 8-17. Format for SMMR One-month Polar Displays to 30°N and 30°S

8.7 Planned Net Experiment Investigations and Data Applications

There are twenty separately identifiable investigations planned for SMMR NET; eighteen are validation investigations while two are applications. A summary of planned SMMR pre and post launch experiment investigations is given below:

8.7.1 Validation Investigations

8.7.1.1 SMMR Underflights for Ocean/Atmosphere Parameters and New Sea Ice

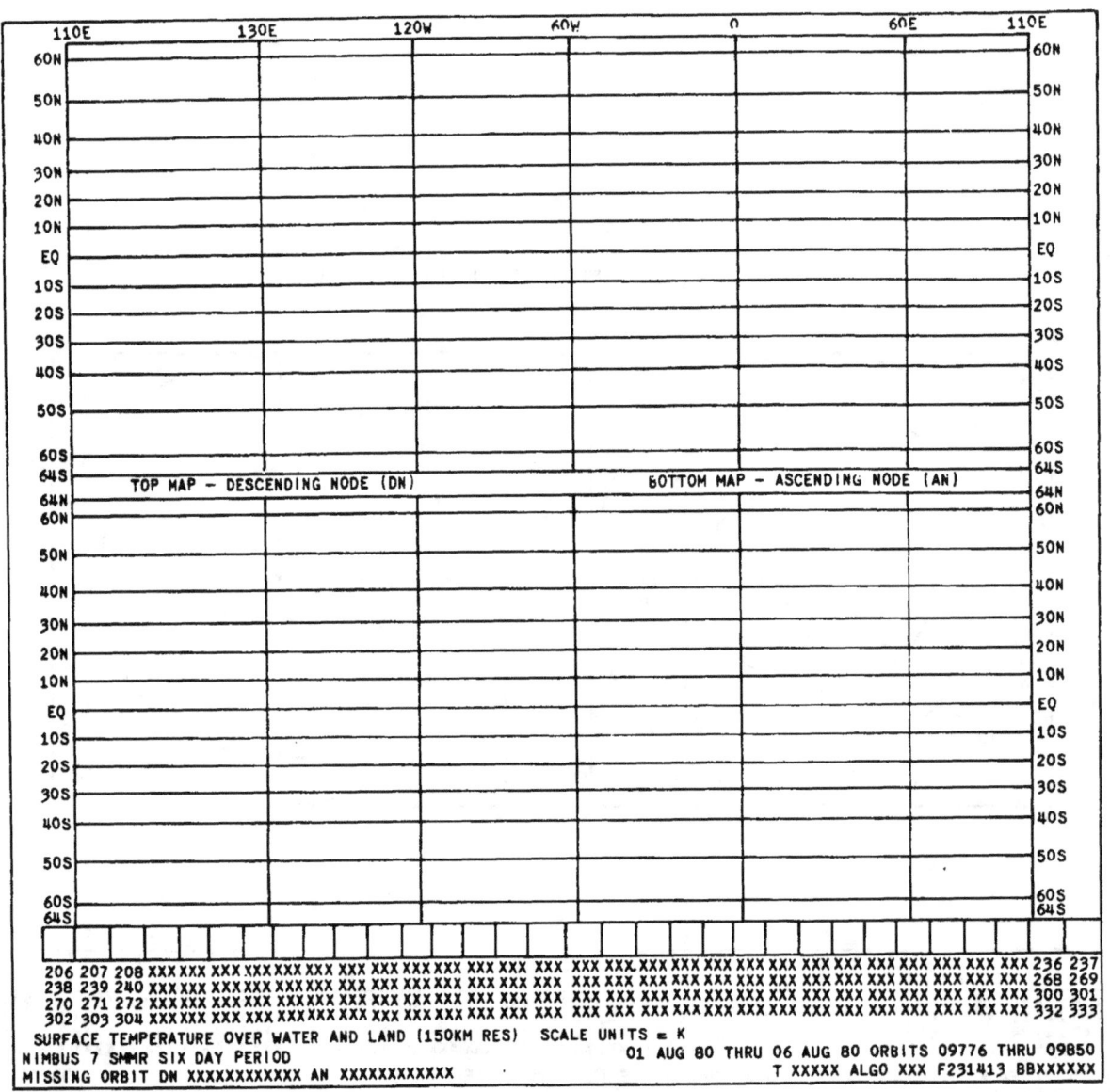

Figure 8-18. Format for SMMR Mercator Displays

This investigation plans on 16 CV-990 underflights, principally in the northern latitudes but with some flights in the Honolulu and Tahiti regions. The objectives of the investigation are to validate retrieval algorithms, calibrate the SMMR and obtain data to determine retrievable continental ice sheet parameters. Dr. Per Gloersen is the principal investigator.

8.7.1.2 Ocean/Atmospheric Parameter Validation Using Underflight and Coastal Radar Data

This investigation plans on using the data obtained from the underflights in the previous investigation. Specific objectives of the investigation are:

Table 8-4
SMMR Film Products

Film Spec Number	Film Product Title
	MAPS
231301	SEA ICE – MULTIYEAR ICE FRACTION
	SNOW FIELDS – AVERAGE SNOW SURFACE TEMPERATURE
	ICE SHEETS – TEMPERATURE λ = 1.7 CM
231302	SEA ICE SURFACE TEMPERATURE,
	AVERAGE SNOW SUBSURFACE TEMP. OVER ICE SHEETS AND LARGE SNOW FIELDS
231702	(same as 231302)
231303	THIN FIRST YEAR ICE FRACTION
	ICE SHEETS – TEMPERATURE FOR λ = 2.8 CM
	DRY SNOW
231304	SURFACE TEMPERATURE FOR ICE SHEETS
	SNOW LAYER WATER EQUIVALENT FOR LARGE SNOW FIELDS
	SEA ICE CONCENTRATION
	WIND SPEED OVER OPEN WATER
231704	(same as 231304 but no "WIND SPEED OVER OPEN WATER")
231410	TOTAL ATMOSPHERIC WATER VAPOR OVER OCEANS
	SOIL MOISTURE
231710	(same as 231410)
231411	RAINFALL RATE OVER OCEANS
	RAIN/NO RAIN OVER LAND
231711	(same as 231411)
231412	TOTAL ATMOSPHERIC LIQUID WATER OVER OCEANS
	OPEN WATER OVER LAND
231712	(same as 231412)
231413	SURFACE TEMPERATURE OVER WATER AND LAND
231713	(same as 231413)
231714	SEA SURFACE WIND SPEED
231320	AVERAGE 37 GHz BRIGHTNESS TEMPERATURE

- Cross-calibration of SMMR and airborne SMMR
- Sea surface wind validation
- Sea surface temperature validation
- Atmospheric water vapor and non-precipitation liquid water validation
- Rainfall rate validation

Dr. Chang is the principal investigator.

8.7.1.3 Validation of Sea Surface Temperature Algorithms Using AXBT Data

This investigation compares sea surface temperatures obtained for Airborne Expendable Bathythemographs (AXBT), with the SMMR simulator and with the spacecraft SMMR instrument, in areas both free and suspect of radio frequency interference (RFI). Objectives are to provide a second-stage validation of the sea surface temperature and to resolve the RFI issue. Dr. J. Mueller is the principal investigator.

8.7.1.4 Validation of the Near Surface Wind Algorithm by Comparison with NOAA Buoy and Ship Data and Correlation with the Ross/Cardone Surface Model

This investigation compares data obtained from NOAA environmental buoys, research vessels and aircraft with SMMR spacecraft data. Another objective is to compare SMMR-derived Near Surface Wind (NSW) with model predictions. To achieve these, the NOAA P-3C aircraft flights will be synchronized with the CV-990 expedition in the Gulf of Alaska. The comparison of NSW's from SMMR and the augmented NSW truth data set will be carried out as a NOAA/SAIL in-house effort. The Ross/Cardone ocean surface model will be used to obtain the NSW's. Mr. Duncan Ross is the principal investigator.

8.7.1.5 Atmospheric Frontal Zone Studies

This investigation provides additional validation of Atmospheric Liquid Water (L), Atmospheric Water Vapor (W) and Rain Rate (R) retrieval algorithms. It is also hoped to identify new retrievable parameters through case studies. Dr. David Staelin, as sole investigator, will analyze appropriate data sets.

8.7.1.6 Validation of Sea Ice Parameter Retrieval Algorithms

This investigation seeks to demonstrate and/or improve accuracy for sea ice and related parameters. This includes the demonstration that the time rate-of-change of C (C) can be correlated with storm patterns. SMMR simulator and associated airborne support instrumentation data will be utilized to produce a sea ice parameter data set. During the investigation, the constants and, if necessary, the functional form of the algorithm will be revised. Dr. William Campbell is the principal investigator.

8.7.1.7 SMMR Underflights for Old Sea Ice, Snow Fields, and Sea Surface Temperature

This investigation involves 12 CV-990 underflights in the phase I of the Spring Experimental Program. Five flights are devoted to Sea Surface Temperature (SST), two flights are devoted to extensive instrumented snow fields in Scandinavia, two flights are devoted to snow courses in North America, and the remainder have miscellaneous missions. The objective of these flights is to obtain data to achieve the following:

- Extend the SST surface truth data base to increase confidence in the accuracy of the SST determinations

- Complete or obtain new data bases for snow fields and other parameters

- Check the calibration of the SMMR for any temporal variation

Dr. Per Gloersen is the principal investigator.

8.7.1.8 Soil Index Studies Based on Antecedent Rainfall

The purpose of this investigation is to correlate the 10 SMMR microwave brightness temperatures with antecedent rainfall amounts as recorded in selected test areas. Suitable data will be acquired at selected test sites for a 12 month period. Dr. Tom Schmugge is the principal investigator.

8.7.1.9 Analysis of Snow Flight Data

This investigation cross correlates airborne SMMR simulator, spacecraft SMMR and surface measurement data. As part of the investigation, it provides post-launch algorithm constants and, if necessary, revises the functional form of the pre-launch snow property algorithm. Snow depth, density, temperatures and other data will be acquired from selected sites in Scandinavia, Canada, Northern USA, Switzerland, and Austria near the CV-990 underflights. These surface measurements will be correlated with radiometric data from the airborne SMMR simulator and the Nimbus 7 SMMR. Dr. Alfred Chang is the principal investigator.

8.7.1.10 SMMR Underflight Development of a Statistical Basis for the Sea Surface Temperature and Sea Ice Algorithms

This investigation is phase 2 of the Spring Experiment Program (see 8.7.1.7). The objectives are to obtain additional SST (under different observational conditions), obtain additional SMMR simulator radiances over selected sea ice and snow field test sites, and finally to provide contingency SST underflights. To achieve this, four flights are planned for SST observations in the Northern Pacific Experiment (NORPAX), in the vicinity of Honolulu and Tahiti. Further, two flights are planned to extend the sea ice and snow field data sets. Dr. Per Gloersen heads the team conducting the investigation.

8.7.1.11 Initial SMMR Ocean Algorithm Comparison with NOAA Surface Data

The objective of this investigation is to determine the error statistics for sea surface temperatures and near-surface wind speeds by comparing these SMMR statistics with a limited available data set obtained from routine NOAA surface data. Dr. Thomas Wilheit is the principal investigator.

8.7.1.12 Snowpack Properties — Correlation of Surface, Aircraft and Spacecraft Data

The objective is to follow the time variation of the snow field test sites studies with aircraft, spacecraft and surface data. In addition, predictions of snow parameters of other large snow-covered areas (e.g., – Siberian) and comparisons will be made with published surface data. Surface truth data sets from foreign and Northern U.S.A. test sites will be extended before and after the scheduled CV-990 underflights for comparison with the SMMR data. Dr. Alfred Chang is the principal investigator.

8.7.1.13 Identification of Additional Observables for Snow Fields

The objective of this study is to extend the multispectral analysis used on the NEMS/SCAMS data to the ten-channel data obtained by SMMR. These studies will use a more comprehensive SMMR data set in order to establish the significance of the observables: mean radio brightness temperature (MTB) and brightness temperature gradient (GTB). Dr. David Staelin is the principal investigator.

8.7.1.14 Snow Accumulation Rates

This study extends the analysis of ESMR-5 radiometric signatures of Greenland and Antarctica to the SMMR data set. In addition, there will be a search for additional snow parameters retrievable from SMMR radiances. The SMMR data will be interpreted on the basis of a volume scattering model which compares snow grain size with snow accumulation rate. Dr. H. Jay Zwally is the principal investigator.

8.7.1.15 SMMR Data as Orthogonal Functions – A Case Study

This study is intended to compress a SMMR data set for a particular geographic location in terms of geographic orthogonal functions and in a fourier time sieves. A SMMR data subset will be analyzed for the purpose. Dr. James Mueller is the principal investigator.

8.7.1.16 Cryosphere Studies in Greenland

The objectives of this investigation are to support ongoing studies of sea ice dynamics and meteorology in the vicinity of Greenland and to maintain an interface with the European Space Agency (ESA) scientists. The study will be accomplished by correlating SMMR data with information from unmanned meteorological stations on the east coast of Greenland and ice buoys in North Greenland. Also, SMMR data obtained over the Greenland ice sheet will be studied comparatively with field measurements and data from airborne active microwave and radar sensors. Dr. Preban Gudmandsen is the principal investigator.

8.7.1.17 Maritime Users of SMMR

It is intended, in this study, to generate a long-term data bank on oceans and shallow seas and to provide SMMR data for Antarctic ice studies to be done by the Scott Polar Research Institute (Cambridge, England). This is accomplished by the following approaches:

- A number of (alternate) algorithms will be generated and correlated with surface data recorded by instrumented surface vessels, oil rigs and data buoys

- The Northern Hemisphere Computer Model, which the U.K. Meteorological Office uses to provide ships with ocean current and sea surface temperatures contour maps will be checked.

- Rain rates, ocean surface temperature, and wind speed retrievals obtained from SMMR data will be compared with similar surface measurement data obtained directly.

- Data will be studied both on temporal and spatial scales to assess the feasibility of foul weather warning systems.

Mr. Peter Windsor is the principal investigator.

8.7.1.18 Snow Field Properties

This study is to determine snow microwave signatures in the range of 1 to 100 GHz and to correlate these signatures with snow parameters. Using prelaunch studies of Swiss and Austrian snow fields, comparison will be made between CV-990 SMMR simulator data and surface measurements of the snow properties. In the postlaunch period, spacecraft SMMR data, obtained over larger snow fields, will be analyzed. Dr. Klaus Kunzi is the principal investigator.

8.7.2 Applications Investigations

8.7.2.1 Orthogonalization of One Year of SMMR Data

The objective of this investigation is to compress the entire SMMR data set in a time series of orthogonal functions. The SMMR data may be compressed one or two orders of magnitude by this technique and be more readily incorporated into climate/dynamics models. Dr. James Mueller heads this investigation.

8.7.2.2 Comprehensive Correlation of SMMR SST's with NOAA Data Sets

The objective is to determine the systematic differences, if any, that exist between SST's obtained from NOAA satellites and ship reports and to determine the cause of such differences. This effort is similar but much more comprehensive than that reported in 8.7.1.11. No principal investigator has been named.

8.8 References and Bibliography

1. W. Nordberg, J. Conaway, and P. Thaddeus, "Microwave Observation of Sea State from Aircraft," Quart. J. Roy. Meteorol. Soc., vol 95, pp. 408–413, 1969.

2. J. P. Hollinger, "Passive Microwave Measurements of the Surface," J. Geophys. Res., vol. 75, pp. 5200–5213, 1970.

3. W. Nordberg, J. Conaway, D. B. Ross, and T. Wilheit, "Measurements of Microwave Emission from a Foam-covered, Wind-driven Sea," J. Atmos. Sci., vol. 28, pp. 429–435, 1971.

4. J. P. Hollinger, "Passive Microwave Measurements of Sea Surface Roughness," IEEE Trans. Geosci. Electron, vol. GE-9, p. 169, 171.

5. W. J. Webster, T. T. Wilheit, D. B. Ross, and P. Gloersen, "Spectral Characteristics of the Microwave Emission from a Wind-driven Foam-covered Sea," J. Geophys. Res., vol. 81, pp. 3095–3099, 1976.

6. T. T. Wilheit, M. G. Fowler, G. Stamback, and P. Gloersen, "Microwave Radiometric Determination of Atmospheric Parameters during the Bering Sea Experiment," Proc. Symp. on the Joint U.S.A./U.S.S.R. Bering Sea Experiment, pp. 15-42 (Gidrometeoizdat, Leningrad, 1975).

7. P. Gloersen, T. T. Wilheit, T. C. Chang, and W. Nordberg, "Microwave Maps of the Polar Ice of the Earth," Bull Am. Meteorol. Soc., vol. 55, pp. 1442-1448, 1974.

8. W. J. Campbell, P. Gloersen, W. Nordberg, and T. T. Wilheit, "Dynamics and Morphology of Beaufort Sea Ice Determined from Satellites, Aircraft, and Drifting Stations," COSPAR Proc. Symp. on Approaces to Earth Survey Problems Through Use of Space Technology, P. Bock ed, pp. 311-329 (Akademie-Verlag, Berlin, 1974).

9. P. Gloersen, and V. V. Salomonson, "Satellites-New Global Observing Techniques for Ice and Snow," J. Glaciology, vol. 15, pp. 373, 1975.

10. W. J. Campbell, W. F. Weeks, R. O. Ramseier, and P. Gloersen, "Geophysical Studies of Floating Ice by Remote Sensing," J. Glaciology, vol. 115, pp. 305-327, 1975.

11. W. J. Campbell, R. O. Ramseier, W. F. Weeks, and P. Gloersen, "An Integrated Approach to the Remote Sensing of Floating Ice," Proc. Third Canadian Symp. on Remote Sensing, pp. 39-72, Sept. 1975.

12. W. J. Campbell, P. Gloersen, W. J. Webster, T. T. Wilheit, and R. O. Ramseier, "Beaufort Sea Ice Zones as Delineated by Microwave Imagery," J. Geophys. Res., vol. 81, pp. 1103-1110, 1976.

13. T. Schmugge, P. Gloersen, T. Wilheit, and F. Geiger, "Remote Sensing of Soil Moisture with Microwave Radiometers," J. Geophys. Res., vol. 79, pp. 317-323, 1976.

14. T. C. Chang, P. Gloersen, T. Schmugge, T. T. Wilheit, and H. J. Zwally, "Microwave Emission from Snow and Glacier Ice," J. Glaciology, vol. 16, pp. 23-39, 1976.

15. T. Wilheit, "The Electrically Scanning Microwave Radiometer (ESMR) Experiment," The Nimbus 5 User's Guide, pp. 59-105, U. S. Govt. Printing Office 1972-735-963/259, 1972.

16. T. Wilheit, "The Electrically Scanning Microwave Radiometer (ESMR) Experiment," The Nimbus 6 User's Guide, pp. 87-108 Goddard Space Flight Center, Greenbelt, MD. 20771, 1975.

17. D. H. Staelin, F. T. Barath, J. C. Blinn III, and E. J. Johnston, "The Nimbus 5 Microwave Spectrometer (NEMS) Experiment," The Nimbus 5 User's Guide pp. 141-157, U.S. Govt. Printing Office 1972-735-963/259, 1972.

18. D. H. Staelin, A. H. Barrett, P. W. Rosenkranz, F. T. Barath, E. J. Johnson, J. W. Waters, A. Wouters, and W. B. Lenoir, "The Scanning Microwave Spectrometer (SCAMS) Experiment," The Nimbus 6 User's Guide, pp. 59-86, Goddard Space Flight Center, Greenbelt, MD. 20771, 1975.

19. N. E. Gaut, "Studies of Atmospheric Water Vapor by Means of Passive Microwave Techniques," Tech. Report 467, M.I.T. Res. Lab Electronics, Cambridge, MA 02139, 1968.

20. N. E. Gaut, and E. C. Reifenstein, III, "Interaction of Microwave Energy with the Atmosphere," Paper No. 70-197 presented at the AIAA Earth Resources Observations and Information Systems Conference, Annapolis, MD., 1970.

21. T. T. Wilheit, J. S. Theon, W. E. Shenk, L. J. Allison, and E. B. Rodgers, "Meteorological Interpretations of the Images from the Nimbus-5 Electrically Scanned Microwave Radiometer," J. Appl. Meteorology, vol. 15, pp. 166-172, 1976.

22. T. T. Wilheit, M. S. V. Rao, T. C. Chang, E. B. Rodgers, J. S. Theon, "A Satellite Technique for Quantitatively Mapping Rainfall Rates over the Ocean," NASA Goddard Space Flight Center Report X-911-75-72, Greenbelt, MD. 20771, 1975.

23. J. A. Lane, and J. A. Saxton, "Dielectric Dispersion in Pure Polar Liquids at Very High Radio Frequencies," Proc. Roy. Soc. London, vol. A213, pp. 400-408, 1952.

24. D. B. Ross, and V. J. Cardone, "Observations of Oceanic White Caps and their Relation to Remote Measurements of Surface Wind Speed," J. Geophys. Res., vol. 79, pp. 444-452, 1974.

25. P. Gloersen, "Surface and Atmosphere Parameter Maps from Earth-Orbiting Microwave Radiometers," Proc. of the XII IEEE Int. Conf. on Communications, Philadelphia, PA., June 1976.

26. P. Gloersen, and F. T. Barath, "A Scanning Multichannel Microwave Radiometer for Nimbus-G and Seasat-A, IEEE Journal of Oceanic Engineering, vol. OE-2, No. 2, April 1977.

27. W. Huston, et al., "Earth Observatory Satellite (EOS) Definition Phase Report," Goddard Space Flight Center Report No. X-401-72-332, August 1971.

SECTION 9

THE TEMPERATURE HUMIDITY INFRARED RADIOMETER (THIR) SUBSYSTEM

by

G. Thomas Cherrix
National Aeronautics and Space Administration
Goddard Space Flight Center
Greenbelt, Maryland 20771

9.1 Introduction

Except for being digitized on board the spacecraft, the Nimbus 7 THIR is of the same design and operation as the THIR flown on Nimbus 4, 5, and 6. The two-channel scanning radiometer is designed to measure earth radiation both day and night from two spectral bands. A 10.5 μm to 12.5 μm (11.5 μm) window channel provides an image of the cloud cover, and temperatures of the cloud tops, land, and ocean surfaces. A 6.5 μm to 7.0 μm (6.7 μm) channel provides information on the moisture and cirrus cloud content of the upper troposphere and stratosphere, and the location of jet streams and frontal systems. The ground resolution at the subpoint is 6.7 km for the 11.5 μm channel and 20 km for the 6.7 μm channel.

9.2 Instrument Description

9.2.1 THIR Operation

The THIR consists of an optical scanner (shown in Figure 9-1) and an electronic module (not shown). The optical scanner provides the necessary scan motion to produce cross-course scanning. It contains the radiometer optics, detectors, preamplifiers, detector bias supply, scan drive, and scan synchronization pulse generator (pip) amplifiers. The electronics module provides the necessary amplification and data processing of detected radiometric signal to achieve the proper levels and formatting compatible with the spacecraft's data processing system (DAPS). The electronics module also contains the necessary switching to respond to the spacecraft commands to THIR, and the appropriate housekeeping telemetry circuits. Table 9-1 gives the specifications of the THIR.

The scanner design uses an elliptically shaped plane scan mirror and primary optics, which are common to both channels (Figure 9-2). The scan mirror, set at an angle of 45 degrees to the scan axis, rotates at 48 rpm and scans in a plane perpendicular to the direction of the satellite motion. The scan mirror rotation is such that, when combined with the velocity vector of the satellite, a right-hand spiral results. Therefore, the field of view scans across the earth from east to west in the daytime when traveling northward, and from west to east at night when traveling southward.

The radiation collected by the primary optics is separated into the two infrared bands by a dichroic filter which spectrally divides the energy into two channels. The 6.7 μm data are reflected off the dichroic mirror, through field stops and relay lens, onto the immersed detector-bolometer flake. The 11.5 μm data pass through the dichroic (transmission portion of the dichroic), an Itran-2 relay lens (which also serves as a long-wavelength blocking filter), a folding mirror, and are focused onto

Figure 9-1. The Tempature Humidity Infrared Radiometer

Table 9-1
THIR Subsystem Specifications

Design Parameter	Channel 1	Channel 2
Wavelength Band of Operation (Half Power Points (microns))	6.5 to 7.0	10.5 to 12.5
Field-of-View (mrad)	20	7
Ground Resolution (Subsatellite Point at 955 Km) (Km)	20	6.7
Collecting Aperture (cm^2)	110	110
Detector (Immersed Bolometer) Size (mm) Time Constant (msec)	0.67 x 0.67 2.7	0.22 x 0.22 1.8
Scan Rate (rps)	0.8	0.8
Dwell Time (msec)	4.2	1.4
Information Bandwidth (Hz)	115	345
Dynamic Range (Target Temperature) (°K)	0 to 270	0 to 330
Performance Characteristics	**Channel 1**	**Channel 2**
Noise Equivalent Irradiance (NEI) (watts/cm^2)	4.35×10^{-10}	3.0×10^{-10}
Noise Equivalent Temperature Differential (NETD) at Indicated Scene Temperature	5.0°K @ 185°K 0.26°K @ 300°K	1.5°K @ 185°K 0.28°K @ 300°K
S/N Ratio at Indicated Scene Temperature	3.8:1 @ 185°K 110:1 @ 270°K	19:1 @ 185°K 375:1 @ 330°K
Physical Characteristics	**Scanner**	**Electronics Module**
Weight (lbs)	14.0	6.0
Size (in.)	7.5 x 7.1 x 15.7 (Excluding sunshield)	7.0 x 6.8 x 6.0
Power Requirements	**Scanner**	**Electronics**
-24.5 vdc (watts)	1.8	5.8
100-Hz Two-Phase Square Wave 5.25 V (watts/phase)	0.1	
Operating Temperature Range	0° to 45°C	

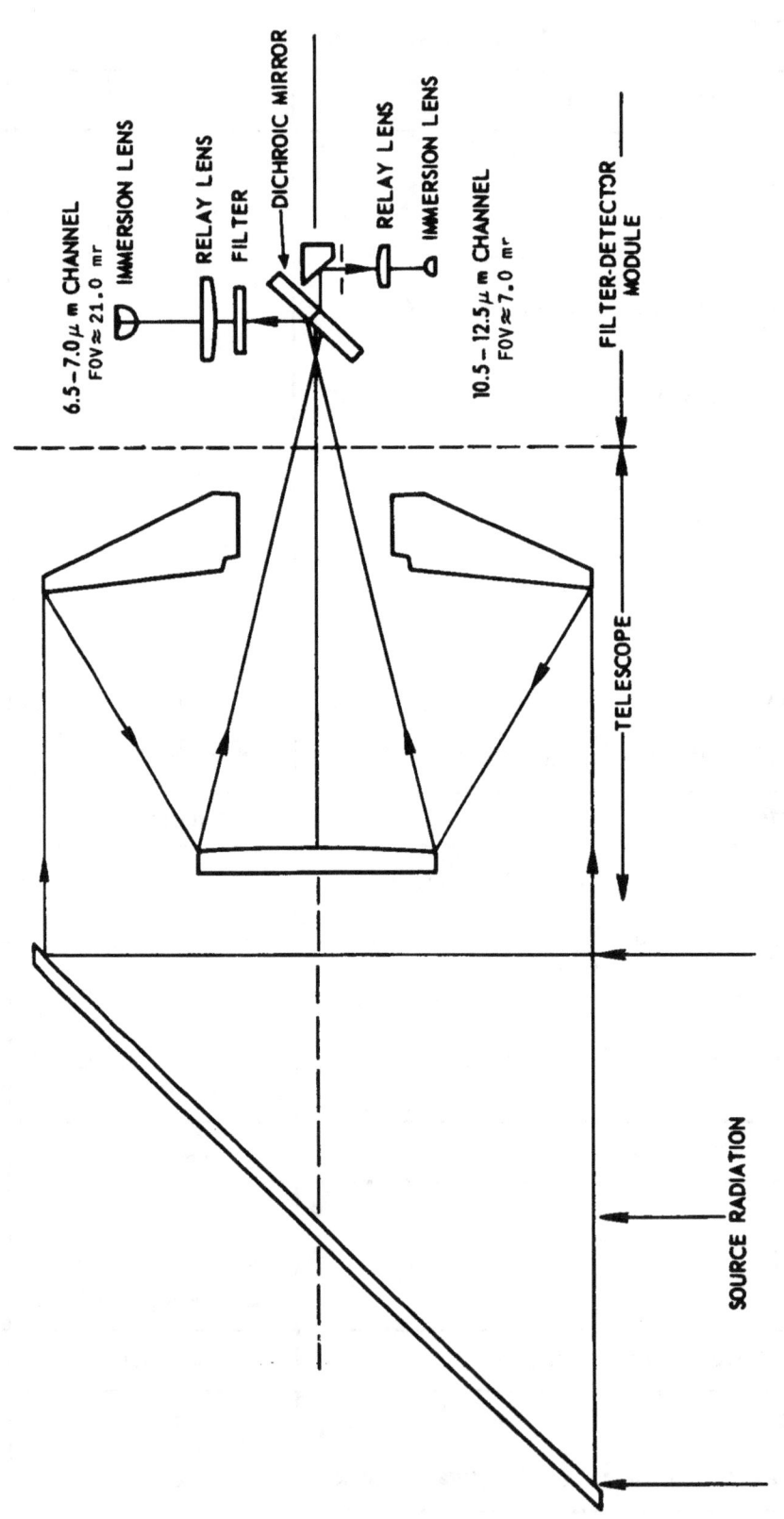

Figure 9-2. THIR Optical Schematic

a germanium-immersed detector-bolometer flake. The field stops at the image plane of each channel define the field of view; 20 mr (1.15 degrees) for the 6.7 µm channel and 7 mr (0.40 degrees) for the 11.5 µm channel. The signals from the detectors are capacitor coupled to the preamplifiers, amplified, and forwarded to the electronics module.

In the electronic module, the signals are further amplified and corrected for detector time constant to provide overall frequency response as required by the subsystem optical resolution. Even though the first stages of amplification are capacitor coupled, the low frequency cutoff (0.5 Hz) of the data bandwidth is so low that a dc restore circuit is necessary to provide a zero signal reference. This occurs during the portion of the scan when the optics are receiving zero radiation (space). The dc restores circuitry also provides additional gain to raise the signal to the desired output level, and filtering to establish proper frequency characteristics. The signals are processed out of the electronics module through buffer amplifiers and into the Digital Information Processor's (DIP) A/D converter.

The tabulated values of the relative spectral response for each channel are shown in Table 9-2, while Figure 9-3 illustrates these data graphically.

9.2.2   Scan Sequence

The radiometer scan mirror continuously rotates the field of view of the detector through 360 degrees in a plane normal to the spacecraft velocity vector. In sequence, the detector views the in-flight blackbody calibration target (which is part of the radiometer housing), outer space, earth, outer space, and returns again to view the radiometer housing. Figure 9-4 illustrates the radiometer timing sequence relative to the angular position of the scan mirror for each scan cycle. The radiometer Z-axis is oriented 5 degrees from the spacecraft zenith. This is done to ensure that the radiometer dc restoration (prior to earth scan) and space check-of-calibration (after earth scan) events will occur when the radiometer is viewing space. This way the correct radiometric data output voltage reference level and the space check-of-calibration of the radiometer will be achieved without ambiguity.

At a scan mirror angle of 5 degrees (referenced to the spacecraft zenith) the radiometer FOV is just starting to leave the scanner housing. At 48 degrees from spacecraft zenith scan mirror position pip No. 1 is generated and the radiometer sync word and calibration sequence is started. At 100 degrees (during the calibration sequence) the radiometer FOV starts to see space fully. At 103.5 degrees the calibration sequence ends and the radiometer dc restores sequence starts. At 110.7 degrees the dc restore sequence ends. The sequence of timing events, starting with the sync word and ending with the radiometer dc restoration, is initiated by scan mirror pip No. 1 and is timed by electronic logic circuits.

At 120.9 degrees (for a nominal 955 km altitude) the earth scan period begins. At 239.1 degrees the earth-scan period ends; the space check-of-calibration period begins. At 250 degrees, the radiometer FOV just starts to see the scanner housing and the space check-of-calibration period ends.

At 302 degrees scan mirror position pip No. 2 is generated and the gain in the 6.7 µm channel is reduced by a factor of 3. (This reduction permits this channel to have a 0°K to 270°K dynamic range and still be capable of being calibrated with a scanner housing reference surface temperature as high as 323°K). At 345 degrees the radiometer FOV is completely filled by the scanner housing and the second (scan housing) check-of-calibration period begins. At

Table 9-2
Relative Spectral Response for the 6.7 μm and 11.5 μm Channels

6.7 μm Channel		11.5 μm Channel			
Wavelength (μm)	Relative Response	Wavelength (μm)	Relative Response	Wavelength (μm)	Relative Response
6.20	0.0000	9.9	0.0248	12.4	0.6546
6.25	0.0071	10.0	0.0295	12.5	0.5303
6.30	0.0141	10.1	0.0769	12.6	0.4257
6.35	0.1013	10.2	0.1996	12.7	0.2591
6.40	0.1884	10.3	0.4333	12.8	0.1071
6.45	0.5103	10.4	0.5871	12.9	0.0407
6.50	0.8322	10.5	0.7550	13.0	0.0147
6.55	0.9135	10.6	0.8355	13.1	0.0000
6.60	0.9948	10.7	0.8927		
6.65	0.9373	10.8	0.8580		
6.70	0.8799	10.9	0.8844		
6.75	0.9393	11.0	0.9224		
6.80	0.9987	11.1	0.9890		
6.85	0.9993	11.2	1.0000		
6.90	1.0000	11.3	0.9928		
6.95	0.9597	11.4	0.9575		
7.00	0.9195	11.5	0.9166		
7.05	0.7165	11.6	0.8888		
7.10	0.5135	11.7	0.9379		
7.15	0.2848	11.8	0.9426		
7.20	0.0562	11.9	0.8985		
7.25	0.0312	12.0	0.8657		
7.30	0.0061	12.1	0.8748		
7.35	0.0031	12.2	0.8288		
7.40	0.0000	12.3	0.7758		

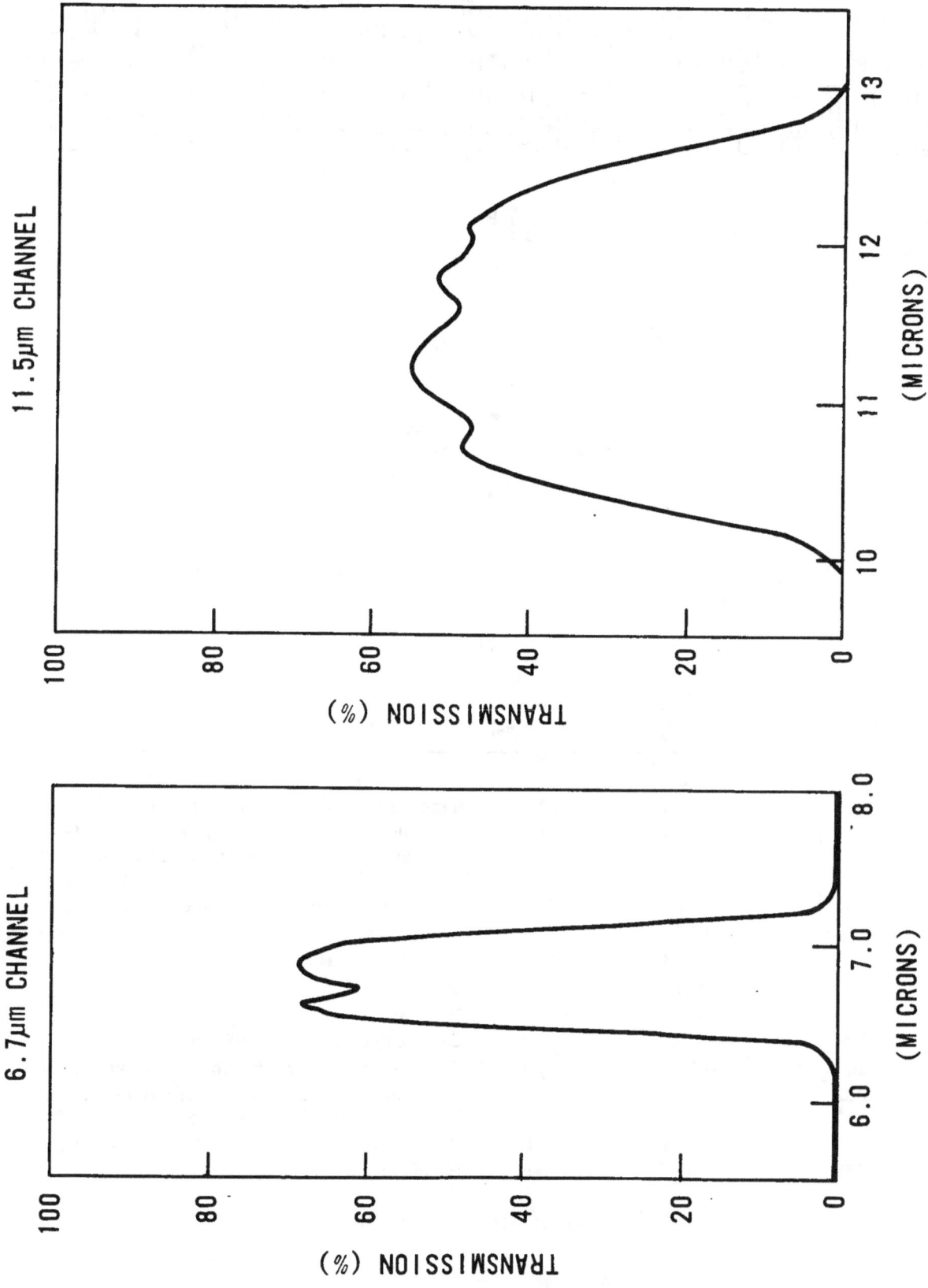

Figure 9-3. Relative Spectral Response (Transmission) of the 6.7 μm and 11.5 μm Channels

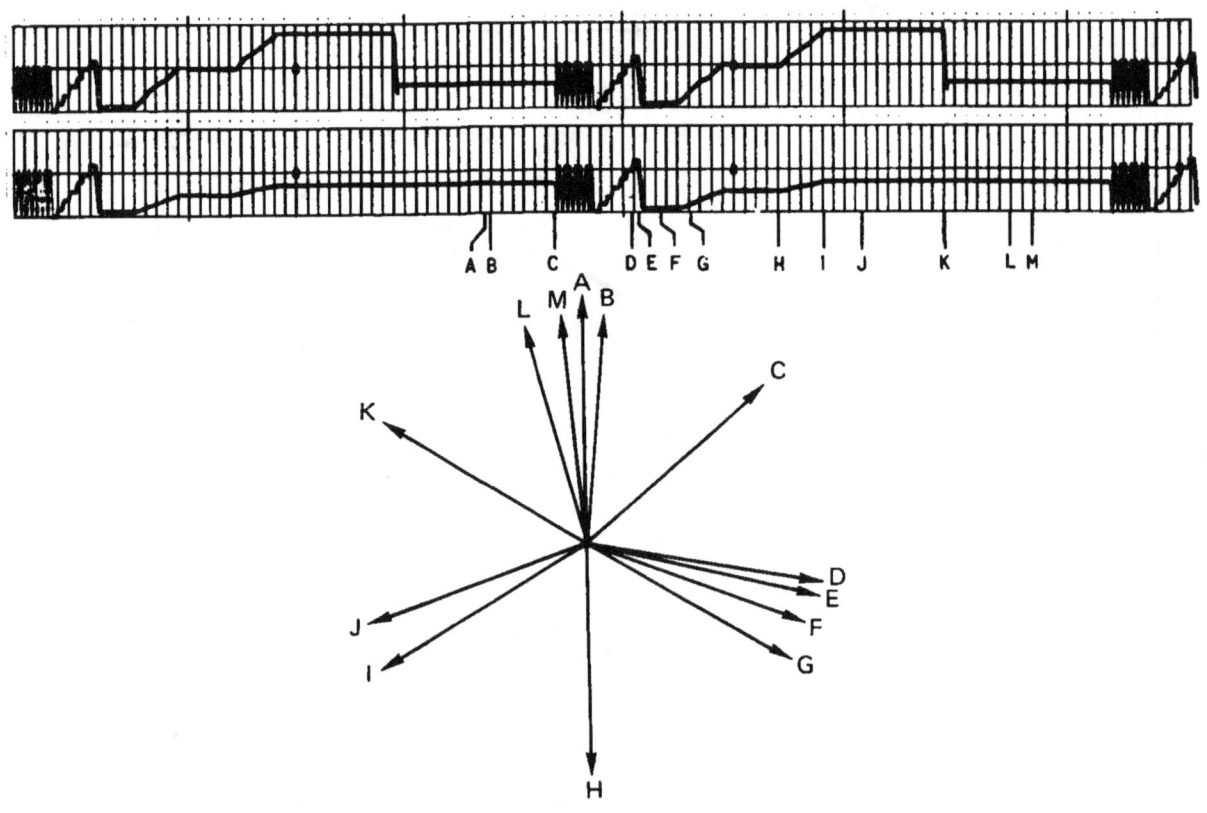

Reference Letter	Angle (degrees)	Time (ms)	Digital Sample (relative)	Event
A	0	0	0	Spacecraft zenith
B	5	17.4	15	Radiometer IFOV just starting to leave housing
C	48	166.7	142	Scan mirror position pip No. 1 occurs and radiometer sync word calibration signal sequence is started. 6.7 µm channel gain returns to normal
D	100	347.5	295	Radiometer IFOV just starting to see all of space
E	103.5	359.4	305	Calibrate signal sequence ends and restore period starts
F	110.7	384.4	327	Restore period ends
G	120.9	419.8	357	Earth scan period begins (955 km orbit)
H	180	625.0	531	Spacecraft nadir
I	239.1	830.2	706	Earth scan period ends (955 km orbit)
J	250	868.9	738	Radiometer IFOV just starting to see housing
K	302	1048.5	891	Scan mirror position pip No. 2 occurs and 6.7 µm gain is attenuated by a factor of 3
L	345	1197.9	1018	Radiometer IFOV completely filled by housing
M	355	1232.6	1048	Radiometer Z-axis

Figure 9-4. THIR Scan Angle Information

355 degrees the scan mirror is parallel with the radiometer Z-axis and the gain of the 6.7 μm channel is returned to normal.

9.2.3 Scan Geometry

For the 11.5 μm channel the scan rate of 48 rpm, combined with the satellite motion, produces nearly contiguous scan line coverage along the subpoint track. As the scan angle from nadir increases there is increasing overlap between consecutive scan lines, reaching 350 percent overlap at the horizon. There is an even greater increase in ground coverage along the scan line (perpendicular to the line of motion of the satellite) as the angle from nadir increases.

Figure 9-5 shows the relationship between the scan sample and ground resolution for the 11.5 μm channel. Figure 9-5a shows this pictorially while Figure 9-5b is a drawing of the relationship. In Figure 9-5a the numbers under each element are nadir angle (in degrees), resolution (in km) along the scan line, and resolution (in km) parallel to the satellite line of motion. At nadir the IFOV of 7 mr (0.40 degrees) for the 11.5 μm channel provides a ground resolution of 6.7 km (3.7 nm). At a 50 degree nadir angle the ground resolution element is approximately 24 km long (east to west) by 15 km wide (north to south).

For the 6.7 μm channel the IFOV of 20 mr (1.15 degrees) at nadir provides a ground resolution of 20 km (10.8 nm). At a 50 degree nadir angle the ground resolution element is approximately 75 km long by 36 km wide.

9.2.4 THIR Data Flow

A simplified block diagram of the THIR/spacecraft data system is given in Figure 9-6. The analog video signals from both THIR channels are input to an analog-to-digital converter which is part of the digital information processor (DIP). The housekeeping data is input to the versatile information processor (VIP) which multiplexes it with data from other sensors and inputs it into DIP. The DIP passes the composite data stream to one of three Goddard Standard Tape Recorders (GSTR) or to the 25 kbs channel of the dual S-Band transponder (for real-time transmission only). When commanded to play back, the GSTR passes the data at a 32-to-1 increased rate (800 kbs) to one of two S-Band transmitters on the high rate channel of the dual S-Band Transponder which broadcasts it to earth.

Figure 9-7 shows the ground data flow in simplified form. The ground station receives the S-Band data stream, demultiplexes it and routes it into the digital data processing system (DDPS) where the THIR data is sent to the data capture processing computer (DCPC) of the Information Processing Division (IPD). A digital tape of THIR data is output and delivered to the THIR processing computer where the video signal (both channels) is blocked into single scan lines; the earth view samples are stripped out and written on the stripped THIR tape (STT). The data is calibrated and located and written on the calibrated located data tape (CLDT). The scan lines are then separated by channels and daytime or nighttime and merged into world montages on high density tape (HDT). The HDT is input to the High Resolution Film Recorder (HRFR) where the four daily world montage film products are generated.

The CLDT's and world montage film products are sent to the National Space Science Data Center (NSSDC) for archival and distribution to the user community.

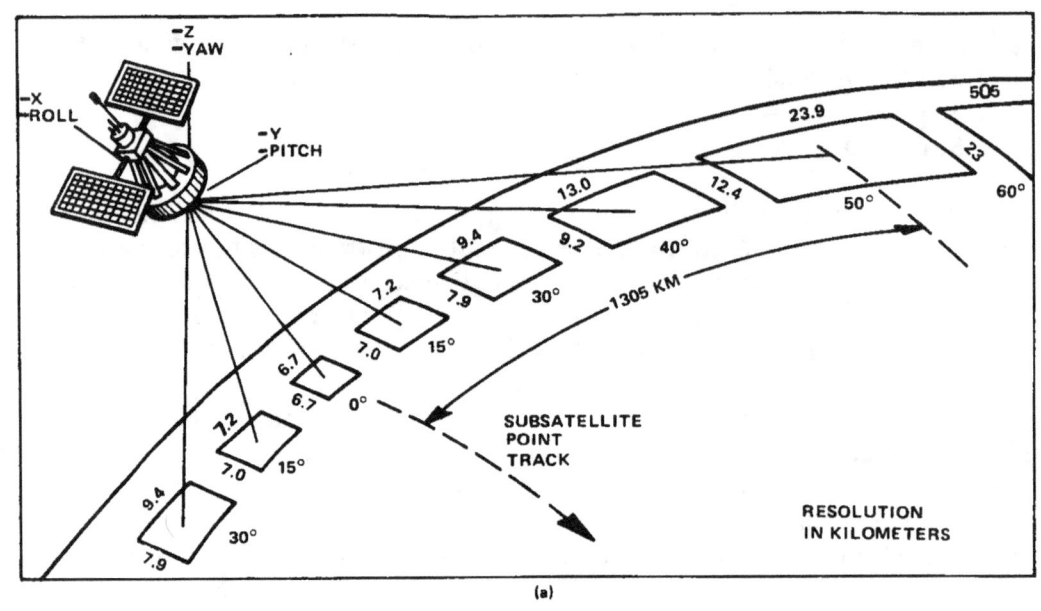

Figure 9-5 Relationship Between Ground Resolution and a Scan Sample with a 0.40° x 0.40° FOV for the THIR 11.5 μm Channel at 955 Km

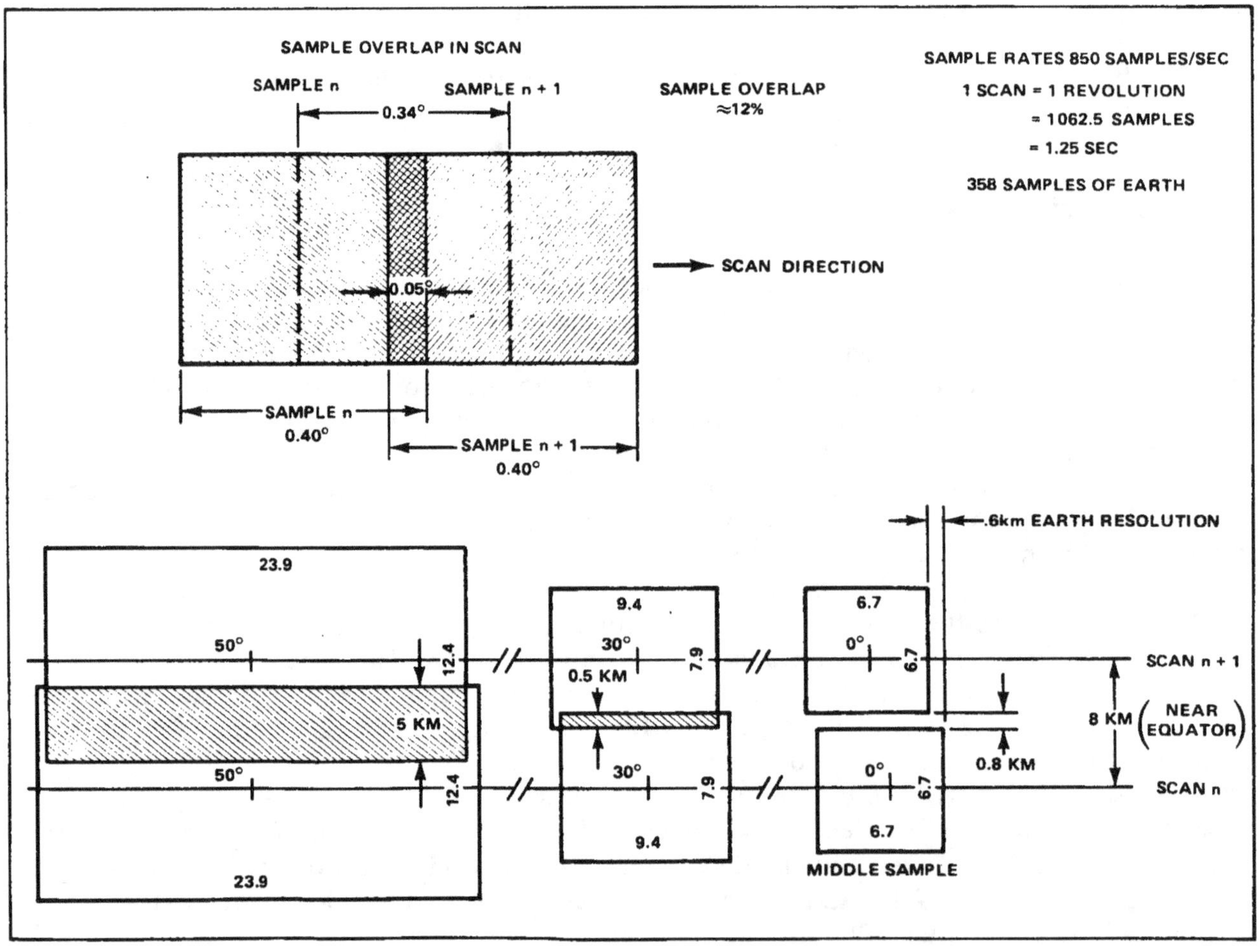

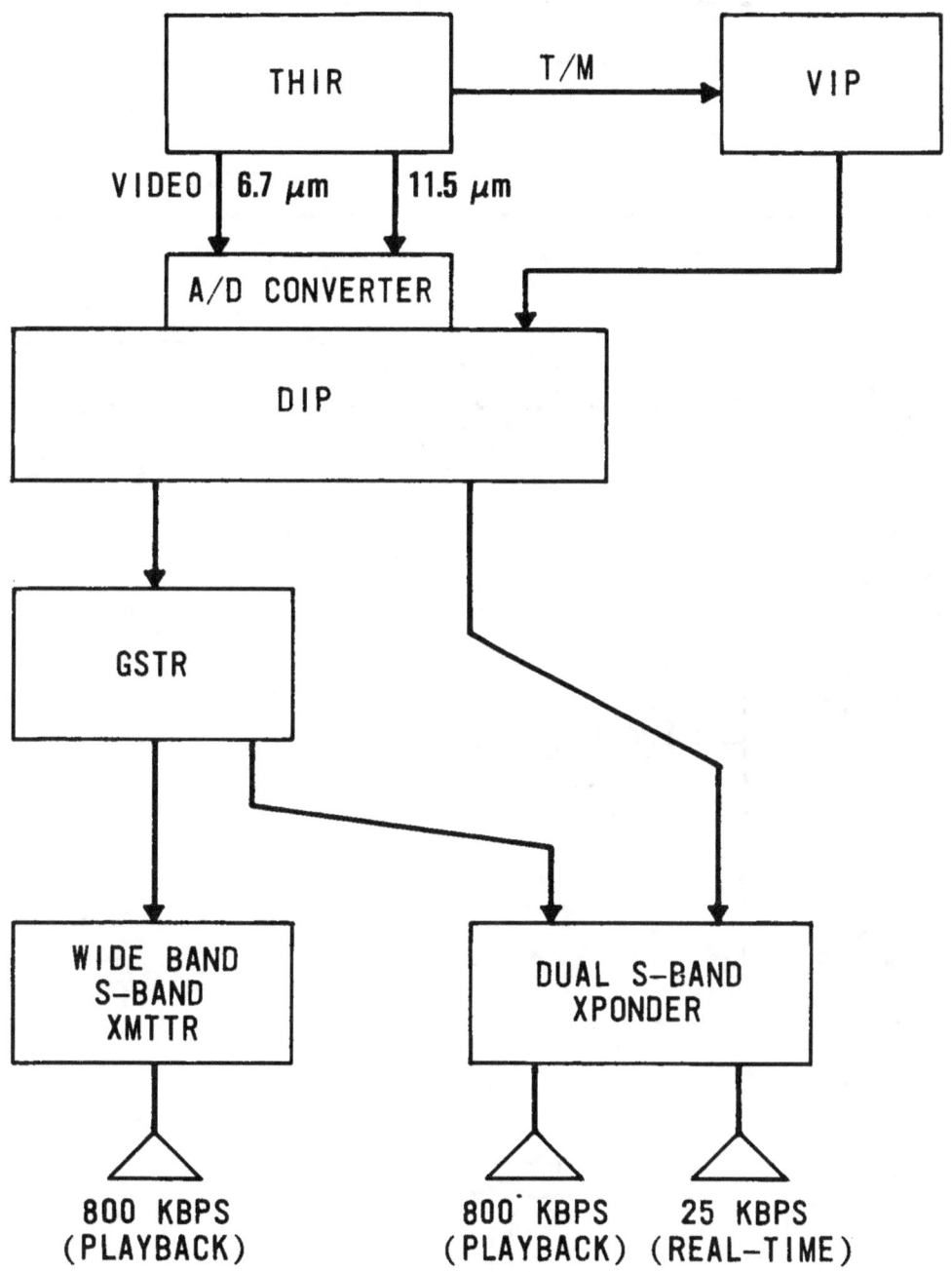

Figure 9-6. Spacecraft/THIR Data Flow

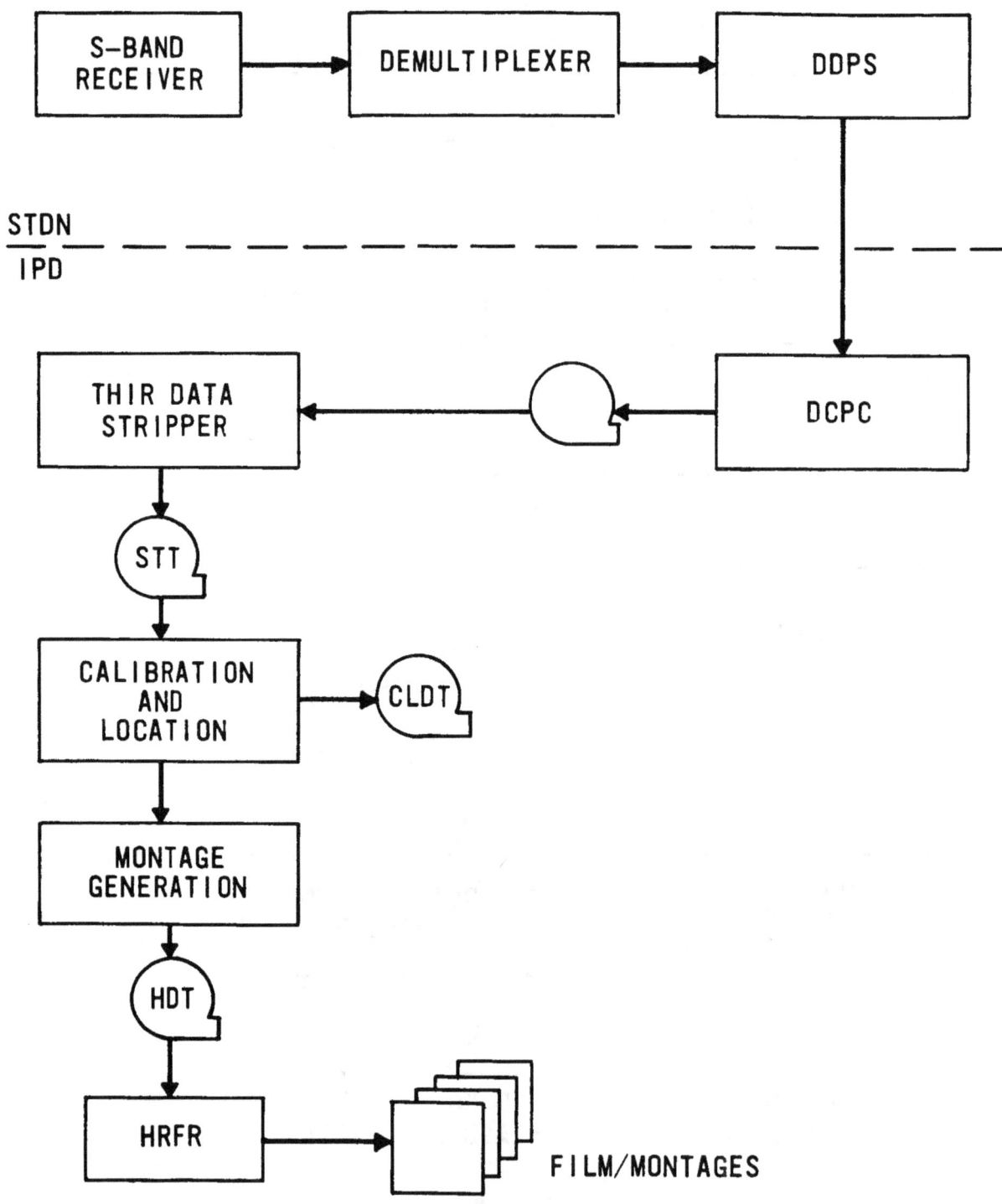

Figure 9-7. Ground/THIR Data Flow

9.3 Calibration

9.3.1 Laboratory Calibration

The main parameters for calibration of all electromagnetic radiation detection devices are essentially the same. Three fundamental quantities must be defined: the effective spectral response, ϕ_λ; the effective radiance, $\overline{N}$; and the equivalent blackbody temperature, T_B. Here ϕ_λ is a composite function involving all of the factors which contribute to the spectral response of the instrument such as filter transmission, mirror reflectances, and the spectral responsivity of the detector.

The effective radiance, $\overline{N}$, is defined as

$$\overline{N} = \int_0^\infty N_\lambda \phi_\lambda d\lambda \tag{1}$$

where N_λ represents the generally non-Planckian radiation from the earth and its atmosphere.

Because of its narrow field of view, the THIR essentially measures beam radiation or radiances toward the satellite along the optical axis. In the pre-flight laboratory calibration, the FOV of the radiometer was filled by a blackbody target whose temperature could be varied and accurately measured over a range of 150°K to 340°K. From the temperature of the blackbody target, T_B, the spectral radiance of the target is determined by the Planck function B_λ. The integration of this function over the effective spectral response, ϕ_λ, yields that portion of the radiance of the target to which the radiometer responds, the "effective radiance," N, given by

$$\overline{N} = \int_0^\infty B_\lambda(T_\lambda) \phi_\lambda d\lambda \tag{2}$$

9.3.2 Equivalent Blackbody Temperature

The effective radiance to which the orbiting radiometer responds may be expressed by

$$\overline{N} = \int_0^\infty N_\lambda \phi_\lambda d\lambda \tag{3}$$

where N_λ is the spectral radiance in the direction of the satellite from the earth and its atmosphere. It is convenient to express the measurement from the orbit in terms of an equivalent temperature of a blackbody filling the field of view which would cause the same response from the radiometer. From Equations 2 and 3 it is seen that this "equivalent blackbody temperature" corresponds to the target temperature, T_B, of the blackbody used in the laboratory calibration. Therefore, the radiometer measurements can be expressed either as values of effective radiance, $\overline{N}$, or as equivalent blackbody temperatures, T_B. The $\overline{N}$ versus T_B function from Equation 2 is given in Table 9-3 for both channels.

9.4 Data Formats and Availability

9.4.1 World Montage

The individual swaths of THIR data are electronically stored until a day (and, separately a night) of data is assembled. Then the data is exposed on 241 mm (9.5 inch) film as a world montage as illustrated in Figure 9-8.

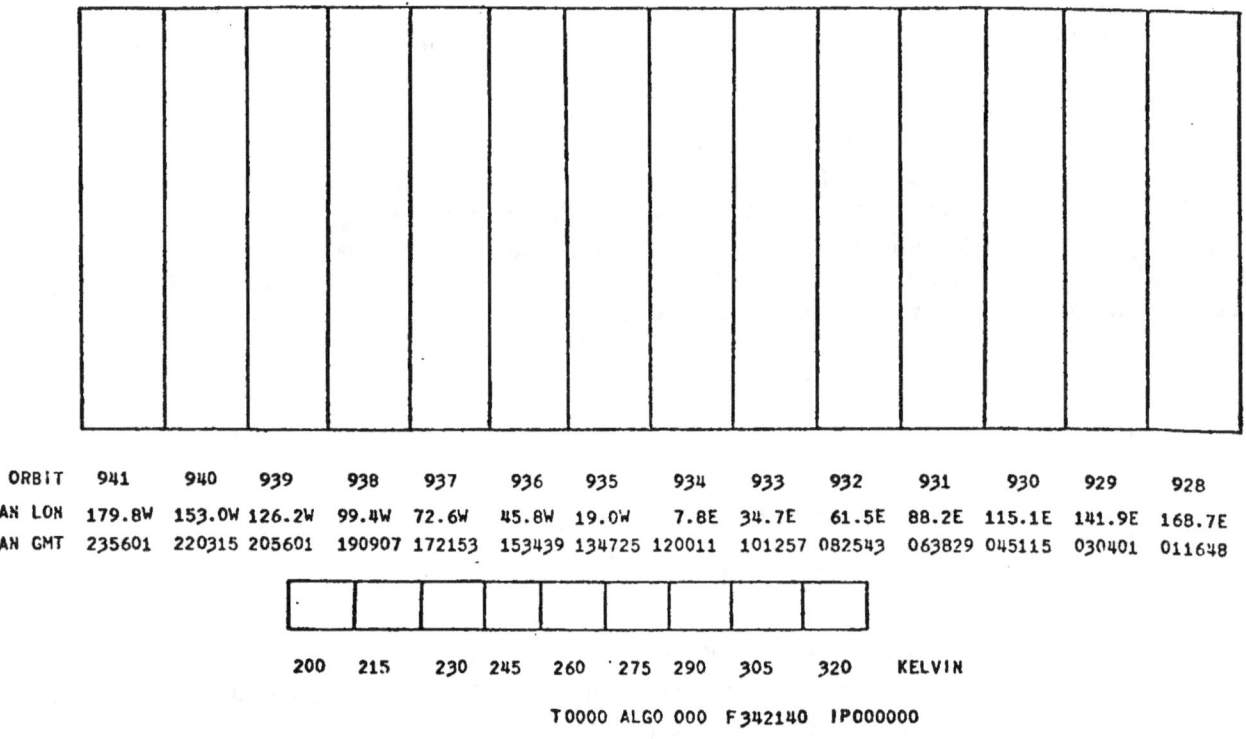

Figure 9-8. THIR Montage Film Display Format

Each display contains appropriate title information identifying the satellite, the channel of data displayed (11.5 μm or 6.7 μm), whether the montage contains daytime or nighttime data, and the date the data was recorded. The 13 or 14 orbits of daytime or nighttime data are in the center of each display. Beneath each daytime orbital swath is its data orbit number plus an ascending node longitude and Greenwich Mean Time (GMT). Beneath each nighttime orbital swath is its data orbit number plus a descending anode longitude and GMT. (See Figure 9-8.)

Beneath the data orbit reference information is a nine-step gray scale. The gray scale is calibrated with the imagery and allows a user to estimate cloud, ocean and land temperatures within each data swath. To the right of center at the bottom of each display is information used mainly for cataloging the data. This information is the physical tape number the data is stored on (TXXXXX), the algorithm reference number used in processing the data (ALGO XXX), the film specification number (F342140), the project data format code (IP), and the film frame number (XXXXXX).

There are four daily THIR montage displays. Each is displayed using the format described in this section. The four displays are an 11.5 μm or 6.7 μm nighttime display, and an 11.5 μm or 6.7 μm daytime display.

9.4.2 Tape Data

Whenever observations are required at the full capability of the THIR, the data available on magnetic tape should be utilized. The CLDT is generated at the same time as the data base used for producing the world montage film products. The CLDT replaces the THIR-NMRTs produced for the Nimbus 4, 5 and 6 THIRs. However, the CLDT does not resemble the NMRT in format. The CLDT has the following general characteristics:

- 9 track 1600 BPI

- Data expressed in 8, 16 and 32-bit word sizes

- Bi-spectral (both 6.7 μm and 11.5 μm data in the same records)

- Measurements expressed in radiance values (conversion table to temperature (K) included in header record)

- One tape covers half a GMT day

- One file covers one data orbit

- GMT is given for the nadir sample of each scan

- Latitude and longitude values are given for each sixth data point

THIR CLDT's are generated for all THIR data collected and deemed to be of sastifactory quality.

The 11.5 μm channel THIR data are also formatted on tape to supply cloud cover statistics for the ERB and the SBUV/TOMS experiments. These tape formats were designed specifically for these experiments, but users might find other applications. The THIR clouds-ERB tapes contain mean radiances and rms deviations of clouds at each of four altitude levels for each ERB subtarget area. The THIR clouds-TOMS tapes contain the same information for each SBUV and/or TOMS IFOV. (Conversion from radiances to temperatures is available in the CLDT documentation.)

9.4.3 Data Availability

The THIR data are available from the NSSDC as computer-produced daily world montage film products and as computer compatible digital tapes. Users requesting THIR data should read Section 1.5 of this document for general tape and film ordering information.

The world montage displays (described in Section 9.4.1) are available as 241 mm by 241 mm positive or negative black and white transparencies or as 241 mm black and white positive prints.

When requesting THIR montages, a user should specify the display medium (film or print), the channel desired (6.7 μm or 11.5 μm), the diurnal requirement (daytime, nighttime, or both), and the user's start and stop dates for data.

The digital tapes (described in Section 9.4.2) are available on a 9-track 1600 bpi tape format. Furnished with each tape type is a tape specification document describing the record and file content and word format of each tape type. Table 1-5 in Section 1.5 of this Guide provides tape specification numbers for the THIR tapes. Request these documents by tape specification number and tape name (acronym).

Table 9-3
Effective Radiance ($\overline{N}$) versus Equivalent Blackbody Temperature (T_B)

Blackbody Temperature (°K)	Effective Radiance (w/m² ster)	
	6.7 μm Channel	11.5 μm Channel
150	0.0039	0.2827
160	0.0094	0.4758
170	0.0204	0.7536
180	0.0407	1.135
190	0.0755	1.639
200	0.1317	2.281
210	0.2180	3.079
220	0.3446	4.046
230	0.5236	5.194
240	0.7685	6.532
250	1.094	8.070
260	1.516	9.813
270	2.050	11.71
280	2.714	13.93
290	3.524	16.31
300	4.498	18.90
310	5.652	21.70
320	7.002	24.71
330	8.563	27.92
340	10.35	31.35
350	12.38	34.96

APPENDIX A
ABBREVIATIONS AND ACRONYMS

AC	Attitude Control
ACE	Actuator Control Electronics
ACS	Attitude Control System
A/D	Analog to Digital
ADPE	Automatic Data Processing Equipment
A/E	Absorptivity to Emmissivity
AF	Audio Frequency
AFC	Automatic Frequency Control
AGC	Automatic Gain Control
AGE	Aerospace Ground Equipment
AGREE	Advisory Group on Reliability Electronic Equipment
ALGO	Algorithm
ANC	Automatic Nutation Control System
AOS	Acquisition of Satellite
APL	Applied Physics Laboratory
ATC	Active Thermal Control
ATDM	Asychronous Time Division Multiplexing
ATFE	Advanced Thermal Control Flight Experiment
ATHC	Active Thermal Control
AVHRR	Advanced Very High Resolution Radiometer
AXBT	Airborne Expandable Bathythemographs
BANAT	Beta–Aerosol Number Density Archive Tape
BAT	Beta Archive Tape (SAM II)
BCD	Binary Coded Decimal

BCU	Bench Checkout Unit
ber	Bit Error Rate
BOT	Beginning of Tape
bpi	Bits per second
BSA	Bit Synchronization Acquisition
BTC	Binary Time Code
BW	Bandwidth
CAI	Computer Assisted Instruction
CAT	Catalog Archive Tape, Cross-section Archive Tape
CBTT	Calibrated Brightness Temperature Tapes
CBW	Constant Bandwidth
CDA	Command and Data Acquisition
CDHS	Command and Data Handling System
CDP	Central Data Processor
CDT	Calibrated Data Tape
CDU	Computer Display Unit
C.G.	Center of Gravity
Chan	Channel
CH_4	Methane
CHRT	Channel Reference Tone
CIC	Command Interface Control
CITE	Control Item Test Equipment
CJ	Clock Jitter
CL	Clock
CLB	Control Logic Box

CLC	Channel Level Control
CLDT	Calibrated Location Data Tape
CLOUDS-E	Cloud Cover Tape For ERB
CLOUDS-T	Cloud Cover Tape for TOMS
cm	centimeter
CM	Command Memory
CMD	Command
CMS	Composite Multiplex Signal
CNR	Carrier to Noise Ratio
CO	Carbon Monoxide
CO_2	Carbon Dioxide
CO_2N	Narrow Carbon Dioxide Channel
CO_2W	Wide Carbon Dioxide Channel
Coeff	Coefficient
CR	Control Room
CRB	Change Review Board
CRCST	Calibrated Radiance, Chlorophyll, Sediment and Temperature Tape
CRT	Cathode Ray Tube
CSS	Coarse Sun Sensor
CTC	Comprehensive Terminal C-Band
CTEC	Communication Test Equipment Console
CTS	Comprehensive Terminal S-Band
CW	Carrier Wave
C/WM^{-2}	Counts/Watt Meter^{-2}
CZCS	Coastal Zone Color Scanner

D/A	Digital to Analog
DAC	Digital to Analog Converters
DACU	Data Acquisition and Control Unit
DAPS	Data Processing System
dB	Decibel
DCA	Detector Capsule Assembly
DD	Digital Data
DDHS	Digital Data Handling System
DDP	Digital Data Processor
DEC	Declination
DEMOD	Demodulator
DEMUX	Demultiplex
DEOF	Double End of File
DET	Detector
DFVLR	Radiative Transfer Computer Programs at Joint Research Center (European CZCS related)
DHE	Data Handling Equipment
DIP	Digital Information Processor
DIR	Direction
DISC	Discriminator
DLO	Double Local Oscillator
DMA	Direct Memory Access
DOC	Digital Operational Controller
DOMES	Deep Ocean Mining Experiment Stations
DP	Digital Processor
DPS	Digital Processor System
DPT	Digital Picture Terminal

DR&A	Data Reduction and Analysis
DRAG	Drag Coefficients – Dimensionless
DRSS	Data Relay Satellite System
DSAS	Digital Solar Aspect Sensor
DSS	Digital Sun sensor
DSU	Data Switching Unit
DTC	Digital Time Code
DTL	Diode-Transistor Logic
DTS	Data Transmission System
EBR	Electron Beam Recorder
ECH	Earth Coverage Horn
ED	Experiment Data
EIRP	Effective Isotropic Radiated Power
ELM	Electronic Module
EMC	Electromagnetic Compatibility
EME	Environmental Measurements Experiment
EMI	Electromagnetic Interference
Enab	Enable
ENFLX	Energy Flux – (Watt/m^2)
EOF	End of File
EOT	End of Tape
EQ	Equator
ERB	Earth Radiation Budget
ERL	Environmental Research Laboratories
ERP	Effective Radiated Power
ESA	European Space Agency

ESMR	Electrically Scanning Microwave Radiometer
ESSA	Environmental Science Services Administration
ETR	Eastern Test Range
EURASEP	European Association of Scientists for Environmental Protection
EUROMET	European Sub-NET (On CZCS)
EVM	Earth Viewing Module
EXT	Extinction
FC	Frequency Carrier
FCHP	Feedback-Controlled Heat Pipe
FD	Frequency Diversity
FDM	Frequency Division Multiplex
FDX	Full Duplex
FET	Field Effect Transistor
FEU	Frame Housing Electronics Unit
FGGE	First GARP Global Experiment
FHA	Frame Housing Assembly
FMECA	Failure Mode Effects and Critical Analysis
FNWC	Fleet Numerical Weather Control
FOB	Field Optical Bench
FOV	Field of View
FRT	Flight Readiness Test
FSK	Frequency Shift Keying
FSS	Fine Sun Sensor
FT	Frequency Translation
GAC	Ground Attitude Control

GARP	Global Atmospheric Research Program
GEN	Generator
FWF	Fleet Weather Facility
GHz	Gigahertz
GIE	Ground Instrument Equipment
GISS	Goddard Institute for Space Studies
GMT	Greenwich Mean Time
GND	Ground
GOES	Geostationary Operational Environmental Satellite
GPE	Ground Processing Equipment
GR&RR	Goddard Range and Range Rate
GSE	Ground Support Equipment
GSFC	Goddard Space Flight Center
GSFT	Geographic Season Filter
GST	Ground System Test
GSTR	Goddard Standard Tape Recorder
GTB	Brightness Temperature Gradient
HEATR	Heating Rate — Degree K/sec
HDRSS	High Data Rate Storage System
HDX	Half Duplex
HEMI	Hemisphere
HF	High Frequency (3 to 30 MHz)
HIRS	High Resolution Infrared Radiation Sounder
HNO_3	Nitric Acid
HTRS	Heaters

Hz	Hertz (cycles per second)
H_2O	Water vapor
ICD	Interface Control Document
ID	Identification
IDB	Input Data Block
IEU	Interface Electronic Unit
IF	Intermediate Frequency
I/F	Interface
IFC	Inflight Calibration Source
IFOV	Instantaneous Field of View
IGS	Inertial Guidance System
ILT	Image Location Tape
IMU	Inertial Measurement Unit
Inhib	Inhibit
INT	Interferometer
I/O	Input/Output
IP	Ionospheric Propagation
IPAT	Inverted Profile Archive Tape
IPC	International Pysheliometric Comparison
IPCIV	International Pysheliometric Intercomparison IV
IPD	Information Processing Division
IPF	Image Processing Facility
IPS	International Pysheliometric Scale
ips	inches per second
IR	Infrared

IRA	Inertial Reference Assembly
IRU	Inertial Reference Unit
JASIN	Joint Air-Sea Interaction Experiment
JRC	Joint Research Center (of the European Community)
K	Kelvin
kbs	kilobits per second
kg	kilogram
KHz	Kilohertz
km	kilometer
KPPS	Thousand Pulses Per Second
LaRC	Langley Research Center
Landsat	Land Satellite (formerly ERTS)
LAT	Latitude
LC	Launch Complex
LCC	Launch Control Center
LCS	Large Core Storage
LDT	LIMS Data Tape
LEC	Load Event Counter
LEDS	Light Emitting Diodes
LIDAR	Light Detection and Ranging
LIMS	Limb Infrared Monitor of the Stratosphere
LMFC	Load Major Frame Counter
LMFCD	Load MFC (Major Frame Counter) Divider
LO	Land Ocean
LOC	Launch Operations Center

LOG	Logarithm
LOP	Launch Operations Plan
LOS	Line of Sight, Loss of Signal
LRIR	Limb Radiance Inversion Radiometer
LSB	Least Significant Bit
LTE	Local Thermodynamic Equilibrium
L/V	Launch Vehicle
LVDT	Linear Variable Differential Transformer
L.W.	Long Wave
LWSCB	Longwave Scanning Channel Blackbody
MA	Multiple Access
MAT	Map Archival Tape
MATRIX	Mapped Data Matrix
mb	Millibar
MERC	Mercator
MetOCC	Meteorological Operations Control Center
MFC	Major Frame Counter
MHz	Megahertz
MI	Modulation Index
mm	Millimeter
MOD	Modulator
mr	Milliradians
ms	Millisecond
MSB	Most Significant Bit
MTB	Mean Radio Brightness Temperature

MUX	Multiplex
MV	Millivolt
NASA	National Aeronautics and Space Administration
NCAR	National Center for Atmospheric Research
NDPF	NASA Data Processing Facility
NEI	Noise Equivalent Irradiance
NEMS	Nimbus E Microwave Spectrometer
NESS	National Environmental Satellite Service
NET	Nimbus Experiment Team(s), Noise Equivalent Temperature
NETD	Noise-Equivalent Temperature Difference
NFOV	Narrow Field of View
NIP	Normal Incidence Pysheliometer
n.m.	Nautical Mile
nm	nanometer
NMC	National Meteorological Center
NMRT	Nimbus Meteorological Radiation Tape
NO	Nitric Oxide
NO_2	Nitrogen Dioxide
NOAA	National Oceanic and Atmospheric Administration
NOPS	Nimbus Observation Processing System
NORPAX	Northern Pacific Experiment
NOS	National Ocean Survey
NP	North Pole
NPR	Noise Power Ratio
NSSDC	National Space Science Data Center

NSW	Near Surface Wind
N_2O	Nitrous Oxide
O_3	Ozone
OC	Orbit Control
OCC	Operations Control Center
OCS	Ocean Color Scanner (U-2 A/C version) (CZCS)
OMP	Optical Mechanical Package
ORTHP	Orthogonal Pressure Function — mb
OSC	Oscillator
OSR	Optical Solar Reflector
PA	Power Amplifier
PACRAD	Primary Absolute Cavity Radiometer
PAM	Phase Amplitude Modulation
PAM-II	Preliminary Aerosol Monitor
PARM	Parameter Tape
PC	Printed Circuit
PCA	Polar Cap Absorption
PCL	Program Control Logic
PCM	Pulse Code Modulation
PCM-DHE	PCM Telemetry Data Handling Equipment
PCU	Power Control Unit
PDFC	Project Data Format Code
PFM	Pulse Frequency Modulation
PIP	Position Pulse
PLL	Phase Lock Loop

PM	Phase Modulated
PMEL	Pacific Marine Environmental Laboratory (NOAA)
PMMR	Passive Multichannel Microwave Radiometer
PMR	Pressure Modulated Radiometer
PMT	Photomultiplier Tube
PN	Pseudo Noise
PNT	Pointing
POP	Project Operational Plan
PPMV	Parts per Million by Volume
pps	Pulses per Second
PRES	Pressure
PROG	Program
PRP	Photo Reference Pickup
PRS	Parabolic Reflector Subsystem
PRT	Platinium Resistance Thermometer
PRU	Power Regulation Unit
PS	Power Supply
PSD	Power Spectral Density
psi	Pounds Per Square Inch
PSK	Phase-Shift Keying
PSP	Program Support Plan
pt	Point
R	Roll
rads	Radians
RADTMO	Radiometric Transfer Computer Program at JRC

RAT	Radiance Archive Tape
RCMP	Regression Coefficients for Mean Precipitation — Dimensionless
RCMT	Regression Coefficients for Mean Temperature — Degree K/meter
RCPOP	Regression Coefficients for Probability of Precipitation — Percent/meter
RCV	Receive
RCV-ECH	Receive-Earth Coverage Horn
RCVR	Receiver
RDAT	Raw Data Archive Tape
REG	Register
RF	Radio Frequency
RFI	Radio Frequency Interference
RFSW	RF Switch
RGA	Rate Gyro Assembly
R-H	Relative Humidity — Percent
rms	Root Mean Square
RMU	Remote Multiplex Unit
ROT	Receive Only Terminal
rpm	Revolutions Per Minute
RSM	Reference Sensor Model (ERB-related)
RUT-S	Raw Units Tape — SBUV
RUT-T	Raw Units Tape — TOMS
RT	Real Time
RVDT	Rotary Variable Differential Transformer
SACC	Science and Applications Computer Center

SAGE	Stratospheric Aerosol and Gas Experiment
SAM II	Stratospheric and Aerosol Measurement II
SAMS	Stratospheric and Mesospheric Sounder
SAMSO	Space and Missile Systems Organizaton
SAS	Solar Aspect Sensor
SBUV/TOMS	Solar Backscattered Ultraviolet/Total Ozone Mapping Spectrometer
S/C	Spacecraft
SC	Signal Conditioner, Spacecraft Clock
SCAMP	Small Command Antenna Medium Power
SCAMS	Scanning Microwave Spectrometer
SCAR	Scientific Committee on Antarctic Research
SCAT	Summary Cross-Section Archival Tape
SCE	Spacecraft Command Encoder
SCO	Subcarrier Oscillator
SCP	Solid Cryogen Package
SCR	Surface Composition Radiometer
SCU	Signal Conditioning Unit
SEC	Seconds
SECC	Shortwave Earth Flux Channel Ratio
SEMS	Space Environment Monitor System
SEU	Sensor Electronics Unit
SHF	Super High Frequency
SMAT	Stacked Map Archival Tape
SMMR	Scanning Multichannel Microwave Radiometer
SMR	Switching Mode Regulator

SNR	Signal-To-Noise Ratio
SPDT	Single Pole Double Throw
SR	Sunrise
SSB	Single Sideband
SST	Sea Surface Temperature
STC	System Test Console
STDN	Space Tracking and Data Network
Sr	Steradian
SYNC	Synchronous
SYNCOM	Synchronous Communication
SYS	System
TACH	Tachometer
TAT	Antenna Temperature Tape
TGS	Transportable Ground Station, Trigline Sulphate
TH	Thermal
THIR	Temperature-Humidity Infrared Radiometer
TIROS	Television and Infrared Observation Satellite
TLM	Telemetry
TM	Telemetry, Temperature Monitor, and Time Mode
TMP	Atmospheric Temperature – degree K
UFO	User Formatted Output (Tape)
UHF	Ultra High Frequency
UV	Ultraviolet
V/A	Vacuum to Air Ratio
VAR	Variable

VCO	Voltage Controlled Oscillator
VCXO	Voltage Controlled Crystal Oscillator
VHF	Very High Frequency
VHR	Very High Resolution
VHRR	Very High Resolution Radiometer
VHRRE	Very High Resolution Radiometer Experiment
VIP	Versatile Information Processor
VLF	Very Low Frequency
V/T	Vacuum Thermal Test
VTVM	Vacuum Tube Voltmeter
W	Watts, West
WB	Wideband
WBVCO	Wideband Voltage-Controlled Oscillator
WD	Wideband Data
WFOV	Wide Field of View
WM^{-2}	Watts per square meter
WRR	World Radiometric Reference
XMTR	Transmitter
ZIP	Zonal Information Processor
°	Degrees
Δ	Delta, Change
μm	Micrometers
μs	Microseconds
%	Percent
λ	Wavelength

APPENDIX B

NIMBUS EXPERIMENT TEAM (NET) MEMBERS

B.1 CZCS NET

Dr. Warren Hovis (elected team leader)
NOAA/NESS Code S-32
Room 135, FOB-4
Washington, D. C. 20233

Dr. John Apel
Pacific Marine Environmental Laboratory
3711 15th Avenue NW
Seattle, Washington 98105

Mr. Dennis Clark
NOAA/NESS Environmental Science Group
Room 810-D, Code 533
World Weather Building
Washington, D. C. 20233

Dr. Sayed El-Sayed
Texas A & M University
Department of Oceanography
College Station, Texas 77843

Dr. Howard Gordon
Department of Physics
University of Miami
Coral Gables, Florida 33124

Mr. Frank P. Anderson
CSIR
National Research Institute for Oceanography
P. O. Box 320
Stellenbosch 7600
South Africa

Bruno Sturm
Att: Delegation of the Commission of
    European Communities
Suite 707
2100 M Street, N. W.
Washington, D. C. 20037

Dr. Robert Wringley
AMES Research Center
National Aeronautics and Space Administration
Moffett Field, California 94035

Dr. Charles Yentsch
Director of Research
Bigelow Laboratory of Ocean Sciences
McKown Point
West Boothbay Harbor, Maine 04574

B.2 ERB NET

Dr. Herbert Jacobowitz (elected team leader)
NOAA/NESS (S313)
Room 701, World Weather Building
Washington, D. C. 20233

Dr. K. L. Coulson
Department of Land, Air and Water Resources
University of California
Davis, California 95616

Mr. John Hickey
The Eppley Laboratory Inc.
12 Sheffield Avenue
Newport, Rhode Island 02840

Dr. Frederick House
Department of Physics and Atmospheric Science
Drexel University
32nd and Chestnut Streets
Philadelphia, Pennsylvania 19104

Dr. Andrew P. Ingersoll
Division of Geological and Planetary Sciences, 170-25
California Institute of Technology
Pasadena, California 91125

Mr. Louis Smith
Mail Stop 324
NASA/Langley Research Center
Hampton, Virginia 23665

Dr. Larry Stowe
National and Atmospheric Administration
NESS, S32, FOB-4
Washington, D. C. 20233

Dr. Thomas Vonder Haar
Department of Atmospheric Sciences
Colorado State University
Fort Collins, Colorado 80523

B.3 LIMS NET

Dr. James M. Russell, III (elected co-team leader)
Mail Stop 401A
NASA/Langley Research Center
Hampton, Virginia 23665

Dr. John C. Gille (elected co-team leader)
National Center for Atmospheric Research
P. O. Box 3000
Boulder, Colorado 80303

Dr. Roland Drayson
Research Activities Bldg.
University of Michigan
Ann Arbor, Michigan 48105

Dr. Herbert Fischer
Meteorologisches Institute
Theresienstr.,37
D-8000 Munchen 2
Federal Republic of Germany

Dr. Andre Girard
Onera-92320
Chatillon, France

Dr. Conway Leovy
Department of Atmospheric Sciences and Geophysics
College of Arts and Sciences
University of Washington
Seattle, Washington 98195

Dr. Walter G. Planet
Physics Branch, S321-B
Satellite Experiment Laboratory
NOAA/NESS
Suitland, Maryland 20233

Dr. Ellis Remsberg
Lidar Applications Section
Mail Stop 401A
NASA/ Langley Research Center
Hampton, Virginia 23665

Dr. John Harries
National Physical Laboratory
Middlesex Teddington TW11 OLW
United Kingdom

Dr. Fredrick House
Department of Physics and Atmospheric Sciences
Drexel University
32nd and Chestnut Street
Philadelphia, Pennsylvania 19104

B.4 SAM II NET

Dr. M. P. McCormick (elected team leader)
Mail Stop 475
NASA/Langley Research Center
Hampton, Virginia 23665

Dr. Gerald W. Grams
School of Geophysical Sciences
Georgia Tech
Atlanta, Georgia 30332

Dr. Benjamin M. Herman
Institute of Atmospheric Physics
University of Arizona
Tucson, Arizona 85721

Dr. Theodore J. Pepin
Department of Physics and Astronomy
University of Wyoming
Laramie, Wyoming 82071

Dr. Philip B. Russell
Research Physicist
Atmospheric Sciences
Stanford Research Institute
Menlo Park, California 94025

B.5 SAMS NET

Dr. John Houghton (team leader)
Department of Atmospheric Physics
University of Oxford
Clarendon Laboratory
Parks Road
Oxford, OX1 3PU United Kingdom

Dr. M. Ackerman
Institut D'Aeronomie Spatial de Belgique
3 Avenue Circularie
B-1180 Bruxelles.
Belgium

Dr. J. E. Harries
National Physical Laboratory
Teddington
Middlesex TW11 OLW
United Kingdom

Dr. K. H. Stewart
Meteorological Office
London Road
Bracknell, Berks.
United Kingdom RG12 2SZ

Mr. H. Yates
NESS/NOAA, U. S. Department of Commerce
Federal Office Building 4
Suitland, Maryland 20233

B.6 SBUV/TOMS NET

Dr. D. Heath (team leader)
Code 912
Goddard Space Flight Center
Greenbelt, Maryland 20771

Dr. Arthur Belmont
Research Division HQM 251
Control Data Corporation
P. O. Box 1249
Minneapolis, Minnesota 55440

Dr. Derek Cunnold
Research Associate, 54-1517
Department of Meteorology
Massachusetts Institute of Technology
Cambridge, Massachusetts 02139

Dr. Alex Green
ICASS
221 Space Sciences Research Building
University of Florida
Gainsville, Florida 32611

Dr. William Imhof
Space Science Laboratory
Department 52-12, Building 205
Lockheed Palo Alto Research Laboratory
3251 Hanover Street
Palo Alto, California 94394

Mr. A. Krueger
Code 912
Goddard Space Flight Center
Greenbelt, Maryland 20771

Dr. Carlton Mateer
Atmospheric Environment Service
4905 Dufferin Street
Downsview, Ontario M3H 5T4
Canada

Mr. Alvin J. Miller
World Weather Building
NOAA/NESS
Washington, D.C. 20233

B.7 SMMR NET

Dr. P. Gloersen (elected team leader)
Code 913
Goddard Space Flight Cenetr
Greenbelt, Maryland 20771

Mr. Frank Barath
Mail Station 183-701
Jet Propulsion Laboratory
4800 Oak Grove Drive
Pasadena, California 91103

Dr. William Campbell
Ice Dynamics Project, USGS
113 Thomas Hall
University of Puget Sound
Tacoma, Washington 98416

Dr. Preben Gudmandsen
Electromagnetics Institute
348 Technical University of Denmark
KD-2800 Lyngby
Denmark

Dr. K. F. Kunzi
Institute of Applied Physics
Sidlerstrasse 5
3012 Berne
Switzerland

Dr. Rene O. Ramseier
Deputy Manager–Applications
SSPO, Room 1195
520 Preston Street, K1A OY7
Ottawa, Ontario
Canada

Mr. Duncan Ross
NOAA/SAIL
15 Rickenbacher Causeway
Miami, Florida 33149

Dr. David Staelin
Research Laboratory of Electronics
Massachusetts Institute of Technology
77 Massachusetts Avenue
Cambridge, Massachusetts 02139

Dr. T. Wilheit
Code 953
Goddard Spcae Flight Center
Greenbelt, Maryland 20771

Dr. E. P. L. Windsor
British Aircraft Corporation Limited
Guided Weapons Division, Electronics and Space Systems Group
Filton House
Bristol, United Kingdom

www.ingramcontent.com/pod-product-compliance
Lightning Source LLC
Chambersburg PA
CBHW081719170526
45167CB00009B/3637